MULTIVECTORS
and
CLIFFORD ALGEBRA
in
ELECTRODYNAMICS

MULTIVECTORS and CLIFFORD ALGEBRA in ELECTRODYNAMICS

Bernard Jancewicz
University of Wroclaw

Published by

World Scientific Publishing Co. Pte. Ltd.
P O Box 128, Farrer Road, Singapore 9128

USA office: World Scientific Publishing Co., Inc.
687 Hartwell Street, Teaneck, NJ 07666, USA

UK office: World Scientific Publishing Co. Pte. Ltd.
73 Lynton Mead, Totteridge, London N20 8DH, England

Library of Congress Cataloging-in-Publication Data

Jancewicz, Bernard.
Multivectors & Clifford Algebra in Electrodynamics/author,
Bernard Jancewicz.
p., cm.
Bibliography: p.
ISBN 9971502909

1. Electrodynamics – Mathematics. 2. Differential forms. 3. Clifford algebras. I. Title. II. Title: Multivectors & Clifford Algebra in Electrodynamics.
QC631.J36 1989
537.6 – dc19 88-37999

Printed in Singapore by JBW Printers & Binders Pte. Ltd.

PREFACE

In the last two decades a new way of presenting electrodynamics has been proposed based on a broad use of differential forms, see e.g. Refs. 1,2,3. Such a formulation represents a deep synthesis of formulae and simplifies many deductions.

However, other mathematical concepts exist which are useful in physics in general, and in electromagnetism in particular, namely multivectors. They have an additional merit in comparison with differential forms in that they appeal more readily to geometric intuition. Just as a vector, in the process of abstraction, arises from a straight line segment with an orientation, a bivector originates from a plane segment with an orientation, and a trivector — from a solid body with an orientation. Multivector is a common name for these concepts. Their connection with straight lines, planes and bodies gives them the additional advantage of being easily depicted in illustrations. When dealing with differential forms, however, authors rarely propose pictures to the reader; an excellent exception is the book "Gravitation"[1] in which exterior forms are presented as specific "slicers".

Multivectors were first considered by Hermann Grassmann in the second half of the last century under the name "Ausdehnung" (extension). Using them, Grassmann created an algebraic structure which is known nowadays as exterior algebra. From this, and due to the works of Ellie Cartan at the turn of the century, the theory of differential forms was developed which is of particular use for multidimensional integrals.

The difference between k-vectors and k-forms in modern mathematics is the same as that between the vectors and elements of the dual space. If a scalar product is given in the underlying vector space, a one-to-one correspondence exists between k-forms and k-vectors which allows reciprocal replacements. In classical electrodynamics, the three-dimensional Euclidean space or the four-dimensional space-time (Minkowski space) are sufficient for developing the theory, so we only make use of multivectors without resorting to differential forms. Such an attitude is assumed in presenting the subject even for integrals.

Furthermore, the introduction of so-called nonhomogeneous multivectors, that is, combinations of multivectors of different ranks, is possible. The set of such quantities with a special product forms Cliffors algebra which is named after William Kingdon Clifford who considered such a structure at the end of the nineteenth century. Clifford algebra allows for further simplifications in electromagnetic theory including the possibility of writing down all the Maxwell equations in a single formula as shown for the vacuum by Marcel Riesz.[4]

This book presents classical electrodynamics in such a formulation. It is based mostly on David Hestenes' works, see e.g. Refs. 5,6,7 and puts great emphasis on the interpretational advantages of the formalism. For instance, the action of the electromagnetic field on electric charges can be better understood when the magnetic field is treated as a bivector rather than as an axial vector quantity (Sec. 1.1). The similarity between electric field E and magnetic induction B, on the one hand, and between electric induction D and magnetic field H on the other (the so-called Lorentz-Abraham analogy) becomes more complete since the boundary conditions at an interface between two media confirm it (Sec. 1.4). The notion of force surfaces for the magnetic field is introduced as a counterpart to the force lines known for the electrostatic field — in this way, the intimate link between the magnetic field and the electric current as its source becomes more apparent (Secs. 1.5–1.7 and 2.5). The symmetries of the action of a rectilinear electric current and a magnetic moment is easier to comprehend when the latter is treated as a bivector rather than as an axial vector quantity (Sec. 3.2). Moreover, the plane of oscillations of an electromagnetic wave with linear polarization can be chosen without any element of arbitrariness (Sec. 4.4 and Fig. 93) in contrast to the descriptions hitherto presented.

The possibility of writing all Maxwell equations in a single formula is demonstrated in Sec. 1.3 for an isotropic medium with given electric permittivity and magnetic permeability. This equation is then used for finding the most general solutions, in which the dependence on the space variable **r** takes place through the argument $\mathbf{n} \cdot \mathbf{r}$ where **n** is a constant unit vector — we call them plane fields (Chap. 4). Among them are the harmonic plane waves — in Sec. 4.4, we introduce two generalizations of the circularly polarized waves. An alternative to the Poincaré sphere is proposed to describe all possible elliptic polarizations (Sec. 4.5 and Fig. 98). Chapter 4 is highly influenced by the works of Kuni Imaeda.[21] The single-formula Maxwell equation is solved in Sec. 5.2 for a plane-stratified medium when the plane wave propagates perpendicularly to the layers. The synthetic Maxwell equation has Clifford-algebra-valued fundamental solutions which can be used for finding electromagnetic fields produced by electric and magnetic dipoles with arbitrary time dependence without resorting to the Hertz potentials (Sec. 5.3).

The book is intended to be self-contained, therefore it begins with an introductory Chapter 0 devoted to multivectors and Clifford algebra for the three-dimensional space. Mathematical notions are also presented in Secs. 6.2 and 6.3 with the modifications necessary for the higher dimension and for the pseudo-Euclidean metric of the Minkowski space. (In a four-dimensional space, non-simple bivectors exist, and in the non-Euclidean metric, the invariant bona fide norm cannot be ascribed to the multivectors.) So persons not interested in relativistics formulation are not forced to learn the more sophisticated formalism in Chapter 0. Several remarks are directed to readers familiar with differential geometry about correspondence between the notions used in this book and the notions employed in the context of differential forms. These remarks are oversimplified — I apologize to readers for this.

This book is addressed to a broad audience of physicists and engineers including undergraduate students. Because of the very elementary level of presentation, it can be treated as an introductory course to electrodynamics although a new mathematical language is used in the area. The knowledge of electromagnetism is assumed in the amount usually presented during the general physics courses. The SI system of units is employed for expressing the physical quantities and formulae. Each chapter is followed by a list of problems which sometimes contain elements of reasoning which are essential but easy to reproduce.

I wish to express my gratitude to many persons who were helpful to me when this book was in the making. First of all, I am indebted to Garret Sobczyk for turning my attention to the works of Hestenes, for demonstrating the many advatnages of multivectors and for numerous discussions on the subject. I am grateful to Wojciech Kopczyński for many invaluable remarks which helped to improve the presentation, to Czesław Jasiukiewicz for his help in preparing the computer drawings (Figs. 72 and III.4, III.5), to Krystyna Polaczek for making the photo of magnets (Fig. 86), to Piotr Garbaczewski for a critical reading of the manuscript, to Una Maclean-Hańćkowiak for her assistance in the preparation of the English text and — last, but not least — to my wife Maria for her patience while this book was being written.

Wrocław, March 1987 Bernard Jancewicz

TABLE OF CONTENTS

Chapter 1 ELECTROMAGNETIC FIELD

Chapter 2 ELECTROMAGNETIC POTENTIALS

Chapter 3 CHARGES IN THE ELECTROMAGNETIC FIELD

Chapter 4 PLANE ELECTROMAGNETIC FIELDS

Chapter 5 VARIOUS KINDS OF ELECTROMAGNETIC WAVES

Chapter 6 SPECIAL RELATIVITY

Chapter 7 RELATIVITY AND ELECTRODYNAMICS

MULTIVECTORS
and
CLIFFORD ALGEBRA
in
ELECTRODYNAMICS

Chapter 0

MATHEMATICAL PRELIMINARIES

§1. Bivectors

The principal concept in the mathematical formalism used in this book is the bivector. In analogy to it, a trivector and a quadrivector — that is, other multivectors — will be treated later. Before presenting the definition of bivector, some auxiliary notions should be introduced.

A plane closed curve which does not intersect with itself will be called a *loop*. A loop with distinguished orientation or sense (we denote this by an arrow on it) will be called a *directed loop*. Of course, only two distinct orientations of a given loop are possible — we call them *opposite*. The orientation of a loop may also be described by an ordered triple of its points, see Fig. 1, in which the two opposite orientations are pointed out in two described manners (the points of the ordered triple are denoted as A_1, A_2, A_3).

The reader probably knows intuitively whether two directed loops lying one inside the other on the same plane have the same or opposite orientations. We present here a strict definition for this. Two directed loops L and L', one of which lies inside the other, have *compatible orientations* if the points of the ordered triple $\{A_1, A_2, A_3\}$ on L can be connected with the points of some ordered triple $\{A'_1, A'_2, A'_3\}$ on L' by segments $A_iA'_i$ not intersecting with themselves nor with the curves L and L' (see Fig. 2a). If such a construction

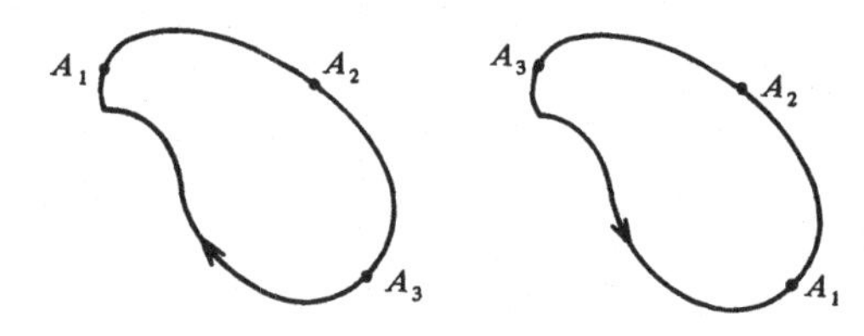

Fig. 1.

is impossible, the two directed loops L, L' have *opposite orientations* (see Fig. 2b).

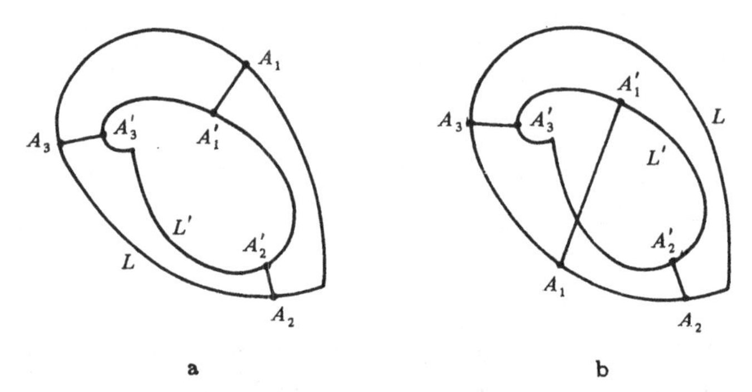

Fig. 2.

For the more general case, when one of the loops is not contained in the other, the definition is that: Two directed loops L and L' have *compatible orientations* if their orientations are compatible with that of a third directed loop L'' surrounding both L and L' (see Fig. 3). If such a loop L'' does not exist, the loops L and L' have *opposite orientations.*[1)]

The set of all directed loops with compatible orientations, lying on a given plane is called the *orientation of the plane.* Only two distinct orientations are possible for a plane — we call them opposite. A plane with the distinguished orientation is called a *directed plane.* Indicating a directed loop on a plane makes it the directed plane.

When the plane is rotated around an arbitrary point on it, any other point traces a circle. The sense of the rotation gives it the orientation, therefore the

---

[1)]Remark for readers with a more mathematical background: Compatibility of orientations of the directed loops is an equivalence relation in a set $\mathcal{L}$ of all directed loops on a given plane. This relation divides the set $\mathcal{L}$ into two abstraction classes — these are two possible orientations of the plane.

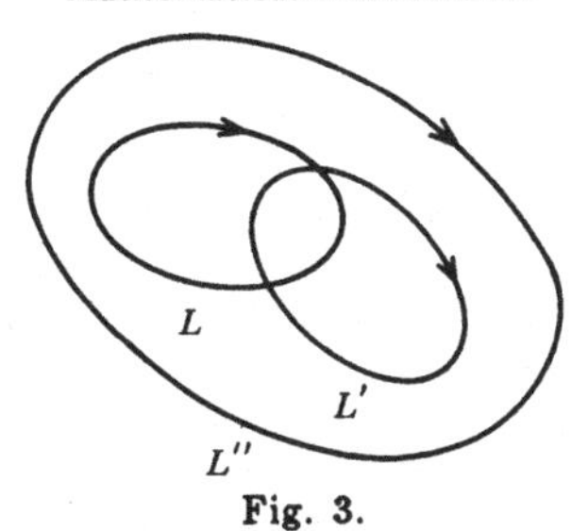

Fig. 3.

circle is a directed loop. All the directed circles traced during one rotation have, of course, compatible orientations. In this manner, indicating the sense of the rotation of the plane makes it the directed plane.

Two parallel directed planes have *compatible orientations* if during a parallel translation, which overlaps the planes, the directed loops determining the orientation of one plane transform into the directed loops determining the orientation of the other. The family of all parallel directed planes filling the space, which have compatible orientations, is called the *two-dimensional direction.*

When the parallel planes are rotated around a common axis orthogonal to all of them, all the planes obtain compatible orientations. In this way, we notice that each rotation in the three-dimensional space determines some two-dimensional direction.

A *volutor*[1] is a geometrical object with two relevant features: (i) *two-dimensional direction* and (ii) *magnitude*, interpreted as an area. One volutor may have many geometrical images, *geometrical image* being a segment of some plane from the family of parallel planes forming the two-dimensional direction of the volutor. The plane segment has a definite area equal to the magnitude of the volutor. Therefore, the geometrical image of a volutor can be taken as a flat figure[2] with an arrow placed on its boundary (Fig. 4a).

The orientation of the volutor will also be symbolized by a curved arrow placed somewhere in the interior of the figure; the arrow can be arbitrarily shifted and rotated within the figure (see Fig. 4b).

A good physical model of a volutor is a plane electric circuit. Its magnitude is the area encompassed by the circuit and its orientation is given by the sense of the current.

---

[1]Volutor, from the Latin *volutare* = to roll, formed in analogy to vector, tensor, or spinor.

[2]A pedantic reader may ask whether the figure must be connected and simply-connected, that is, roughly speaking, consisting of one piece and devoid of holes. The answer to this question is positive, since we choose figures which have loops as their boundaries.

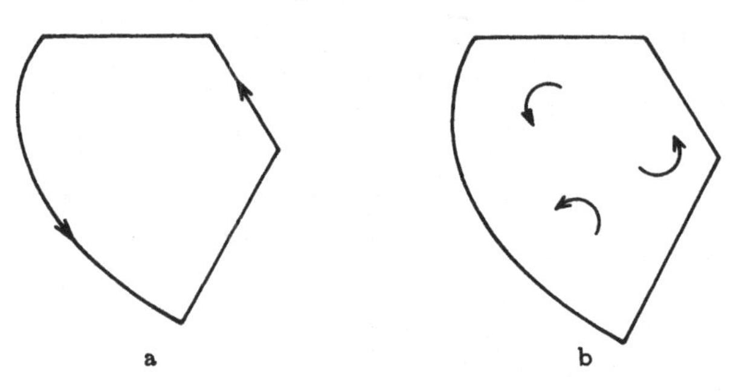

Fig. 4.

We shall denote a volutor by a boldface letter with a caret above it, e.g. $\hat{\mathbf{B}}$. The magnitude of $\hat{\mathbf{B}}$ will be denoted as $|\hat{\mathbf{B}}|$. Two volutors $\hat{\mathbf{B}}$ and $\hat{\mathbf{B}}'$ with the same magnitude and opposite directions will be called *opposite*, and written $\hat{\mathbf{B}} = -\hat{\mathbf{B}}'$. Zero is a special kind of volutor, because it only has magnitude which, of course, is zero, and no direction.

An example of a volutorial physical and/or geometrical quantity is a plane angle, since it is always connected with the plane in which it is measured. Its volutorial orientation may be taken from its physical rotation in one sense or another. Therefore, we shall write $\hat{\boldsymbol{\varphi}}$ instead of φ whenever we treat angle as a volutor.

A volutor can be obtained from the two vectors **a** and **b** in the following manner. First, we choose which vector is to be first; let it be **a**. Then we put the initial point of the second vector **b** at the tip point of the first vector **a**, and draw two parallel lines passing through the two other ends of the vectors, which produces a parallelogram as in Fig. 5. The vectors **a** and **b** lying on the boundary of the parallelogram determine its orientation. Thus, the parallelogram is a geometric image of the volutor $\hat{\mathbf{B}}$ to be obtained. $\hat{\mathbf{B}}$ is called the *outer product of the vectors* **a** and **b**, and is denoted by a wedge sign between the factors:

$$\hat{\mathbf{B}} = \mathbf{a} \wedge \mathbf{b}\,. \tag{1}$$

Of course, its magnitude is given by the formula

$$|\hat{\mathbf{B}}| = |\mathbf{a} \wedge \mathbf{b}| = |\mathbf{a}||\mathbf{b}|\sin\alpha \tag{2}$$

where α is the angle between **a** and **b**, and $|\mathbf{a}|$ is the magnitude of **a**.

This possibility of representing $\hat{\mathbf{B}}$ as the outer product of two vectors led to the use of the word *bivector* for volutors. In modern mathematics, a bivector is a notion which turns out to be a finite sum of the outer products (1), that

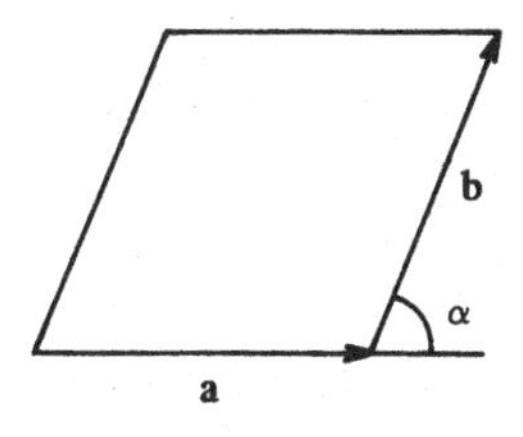

Fig. 5.

is, a finite sum of volutors (the addition of volutors is to be defined later), the term *simple bivector* is used as a synonym for volutor.[1)]

No doubt the reader agrees that every volutor can be represented in the form (1), and that this representation is non-unique, since one can construct many parallelograms parallel to themselves with the same area and orientation but not necessarily with parallel sides. The representation (1) is called the *factorization of a volutor in the outer product.*

We can see from the construction described that the outer product is *anticommutative*:

$$\mathbf{a} \wedge \mathbf{b} = -\mathbf{b} \wedge \mathbf{a} \tag{3}$$

(see Fig. 6). Moreover, for the parallel vectors $\mathbf{a}$ and $\mathbf{b}$ (we write this as $\mathbf{a} \| \mathbf{b}$), one obtains $\mathbf{a} \wedge \mathbf{b} = 0$.

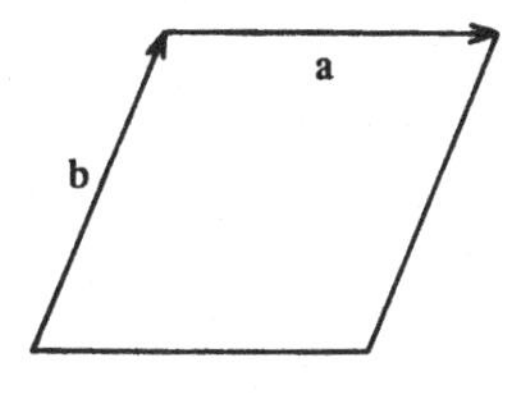

Fig. 6.

Now, it is time to define the addition of volutors. We found our construction on the demand that the outer product of vectors is *distributive* under the addition

$$\mathbf{a} \wedge (\mathbf{b} + \mathbf{c}) = \mathbf{a} \wedge \mathbf{b} + \mathbf{a} \wedge \mathbf{c} \ . \tag{4}$$

This property enables us to obtain a volutor as the result of the addition of two volutors: We merely take the right-hand side of (4) as defined by the

[1)]Non-simple bivectors exist in spaces of four and more dimensions. We are going to discuss this question in Chap. 6 in the context of four-dimensional space-time.

left-hand side. Of course, we can use Eq. (4) only for volutors $\widehat{\mathbf{B}}$ and $\widehat{\mathbf{C}}$, represented as outer products with a common factor. This, however, can be achieved for any volutors $\widehat{\mathbf{B}}$ and $\widehat{\mathbf{C}}$; we need only make use of the fact that in three-dimensional space, two nonparallel planes always intersect along some straight line. We choose a vector **a** lying on this line, represent both volutors in the form $\widehat{\mathbf{B}} = \mathbf{a} \wedge \mathbf{b}$, $\widehat{\mathbf{C}} = \mathbf{a} \wedge \mathbf{c}$, juxtapose the obtained parallelograms in such a way that the common side has opposite orientations if treated as a boundary of $\widehat{\mathbf{B}}$ or $\widehat{\mathbf{C}}$, and then join the two other sides parallel to the common one (that is, to vector **a**) by two lines parallel to $\mathbf{b} + \mathbf{c}$ (see Fig. 7). In this way, we obtain the sum $\widehat{\mathbf{B}} + \widehat{\mathbf{C}} = \mathbf{a} \wedge (\mathbf{b} + \mathbf{c})$. One can say that the addition of two volutors is performed by bringing the summands to a common factor.

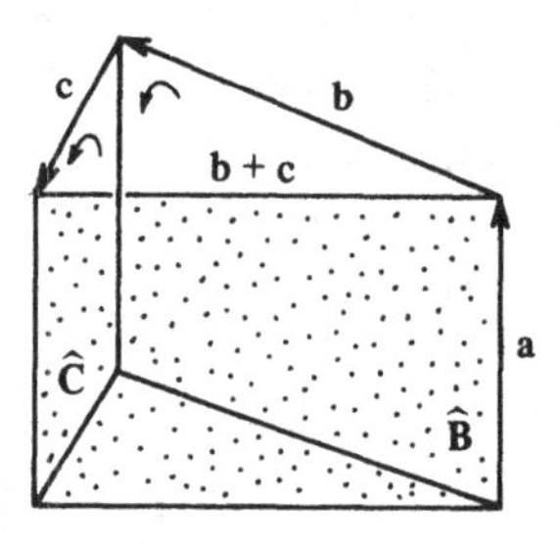

Fig. 7.

The addition of parallel volutors $\widehat{\mathbf{B}}$ and $\widehat{\mathbf{C}}$ is illustrated in Fig. 8 for the same orientation and in Fig. 9 for opposite orientations. The common factor is easy to find in such a case — it may be any vector parallel to the common plane of $\widehat{\mathbf{B}}$ and $\widehat{\mathbf{C}}$.

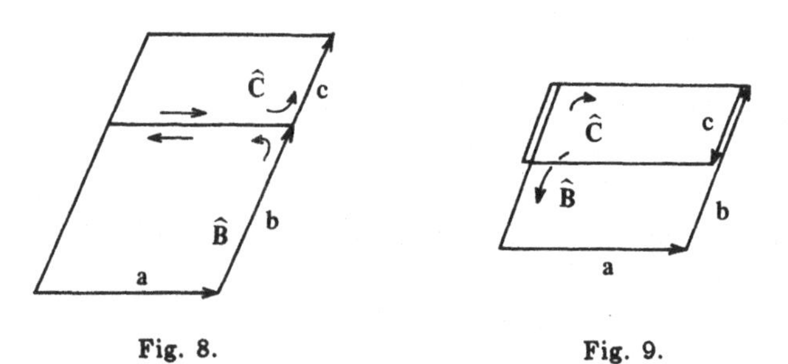

Fig. 8. Fig. 9.

Now, we see that, in three-dimensional space, the set of bivectors is identical to the set of volutors — or, in other words, each sum of volutors can be represented as the outer product of two vectors.

Another operation can be defined for volutors, namely, multiplication by real scalars. Now, the volutor $\lambda\widehat{\mathbf{B}}$ for a real λ has the magnitude $|\lambda||\widehat{\mathbf{B}}|$, the family of parallel planes the same as $\widehat{\mathbf{B}}$, and the orientation the same as $\widehat{\mathbf{B}}$ for positive λ and opposite to $\widehat{\mathbf{B}}$ for negative λ. The outer product of vectors is *homogeneous* under multiplication by scalars:

$$(\lambda\mathbf{a}) \wedge \mathbf{b} = \lambda(\mathbf{a} \wedge \mathbf{b}) = \mathbf{a} \wedge (\lambda\mathbf{b}) \ . \tag{5}$$

Now we look more closely at the nonuniqueness of the factorization of a volutor. Let $\mathbf{a}'$ and $\mathbf{b}'$ be other vectors giving, in the outer product $\widehat{\mathbf{B}} = \mathbf{a}' \wedge \mathbf{b}'$, the same volutor as in (1). The vectors $\mathbf{a}'$ and $\mathbf{b}'$ have to lie in the plane determined by $\mathbf{a}$ and $\mathbf{b}$, and hence can be expressed as the linear combinations

$$\begin{aligned} \mathbf{a}' &= \alpha_{11}\mathbf{a} + \alpha_{12}\mathbf{b} \ , \\ \mathbf{b}' &= \alpha_{21}\mathbf{a} + \alpha_{22}\mathbf{b} \ . \end{aligned} \tag{6}$$

Making use of properties (4) and (5), we arrive at

$$\mathbf{a}' \wedge \mathbf{b}' = \alpha_{11}\alpha_{21}\mathbf{a} \wedge \mathbf{a} + \alpha_{12}\alpha_{21}\mathbf{b} \wedge \mathbf{a} + \alpha_{11}\alpha_{22}\mathbf{a} \wedge \mathbf{b} + \alpha_{12}\alpha_{22}\mathbf{b} \wedge \mathbf{b} \ .$$

The products $\mathbf{a} \wedge \mathbf{a}$ and $\mathbf{b} \wedge \mathbf{b}$ are zeroes, so we omit them; and, due to (3), we may write

$$\mathbf{a}' \wedge \mathbf{b}' = (\alpha_{11}\alpha_{22} - \alpha_{12}\alpha_{21})\mathbf{a} \wedge \mathbf{b} \ .$$

If the product $\mathbf{a}' \wedge \mathbf{b}'$ is to be equal to $\mathbf{a} \wedge \mathbf{b}$, the expression in the brackets must be one. This means that the determinant of transformation (6) is equal to one:

$$\alpha_{11}\alpha_{22} - \alpha_{12}\alpha_{21} = 1 \ . \tag{7}$$

This is the only restriction for factorization (1).

It is worthwhile to ponder whether or not the sum of two volutors as defined by (4) depends on the choice of vectors $\mathbf{a}, \mathbf{b}$ or $\mathbf{c}$, which represent the summands. Now, another possible common factor $\mathbf{a}'$ must satisfy $\mathbf{a}' = \mu\mathbf{a}$ for some real μ, whereas $\mathbf{b}'$ and $\mathbf{c}'$ may be $\mathbf{b}' = \alpha_1\mathbf{a} + \alpha_2\mathbf{b}$, $\mathbf{c}' = \beta_1\mathbf{a} + \beta_2\mathbf{c}$ for real α_i, β_i. Then equalities $\mu\alpha_2 = 1$, $\mu\beta_2 = 1$ follow from the demands $\mathbf{a}' \wedge \mathbf{b}' = \mathbf{a} \wedge \mathbf{b}$, $\mathbf{a}' \wedge \mathbf{c}' = \mathbf{a} \wedge \mathbf{c}$ and, using them, we may write the left-hand side of (4) for new vectors as follows

$$\begin{aligned} \mathbf{a}' \wedge (\mathbf{b}' + \mathbf{c}') &= \mu\mathbf{a} \wedge (\alpha_1\mathbf{a} + \alpha_2\mathbf{b} + \beta_1\mathbf{a} + \beta_2\mathbf{c}) \\ &= \mathbf{a} \wedge (\mu\alpha_2\mathbf{b} + \mu\beta_2\mathbf{c}) = \mathbf{a} \wedge (\mathbf{b} + \mathbf{c}) \ . \end{aligned}$$

Therefore, the left-hand side of (4) remains unchanged.

Since bivectors (in 3-dimensional space, as we already mentioned, they can be identified with volutors) can be added and multiplied by real scalars and these operations satisfy the usual properties (addition is commutative and associative, and multiplication by a scalar is distributive under addition — we leave their verification to the reader), we may state that the set of bivectors forms a linear space. A question of its dimension arises. In order to answer it, we note that when the vectors $\mathbf{e}_1, \mathbf{e}_2, \mathbf{e}_3$ form the orthonormal basis in the basic vector space, and the vectors $\mathbf{a}, \mathbf{b}$ have the representation

$$\mathbf{a} = a_1\mathbf{e}_1 + a_2\mathbf{e}_2 + a_3\mathbf{e}_3\,, \quad \mathbf{b} = b_1\mathbf{e}_1 + b_2\mathbf{e}_2 + b_3\mathbf{e}_3\,,$$

we may write their outer product as

$$\mathbf{a}\wedge\mathbf{b} = (a_1b_2 - a_2b_1)\mathbf{e}_1\wedge\mathbf{e}_2 + (a_3b_1 - a_1b_3)\mathbf{e}_3\wedge\mathbf{e}_1 + (a_2b_3 - a_3b_2)\mathbf{e}_2\wedge\mathbf{e}_3\,. \quad (8)$$

Thus, we have shown that an arbitrary bivector $\widehat{\mathbf{B}} = \mathbf{a}\wedge\mathbf{b}$ can be represented as a linear combination of the three volutors $\mathbf{e}_1 \wedge \mathbf{e}_2$, $\mathbf{e}_3 \wedge \mathbf{e}_1$, $\mathbf{e}_2 \wedge \mathbf{e}_3$, and they are linearly independent (their planes are orthogonal to each other). In this way, we have the answer to our question: The linear space of bivectors is three-dimensional. We take the named triple of volutors $\mathbf{e}_i \wedge \mathbf{e}_j$ as the basis in it.

After introducing the scalar coefficients

$$B_{ij} = a_ib_j - a_jb_i \qquad (9)$$

for $i, j \in \{1,2,3\}$, we can rewrite (8) as

$$\widehat{\mathbf{B}} = B_{12}\mathbf{e}_1 \wedge \mathbf{e}_2 + B_{31}\mathbf{e}_3 \wedge \mathbf{e}_1 + B_{23}\mathbf{e}_2 \wedge \mathbf{e}_3\,.$$

The coefficients B_{ij} are called *coordinates* of the bivector $\widehat{\mathbf{B}}$. The indices in the products $\mathbf{e}_i \wedge \mathbf{e}_j$ can be used in any order; therefore, we can write

$$\widehat{\mathbf{B}} = B_{21}\mathbf{e}_2 \wedge \mathbf{e}_1 + B_{13}\mathbf{e}_1 \wedge \mathbf{e}_3 + B_{32}\mathbf{e}_3 \wedge \mathbf{e}_2$$

as well, and Eq. (9) also applies in this case. Hence we obtain

$$B_{ij} = -B_{ji}\,. \qquad (10)$$

We may also add the coefficients B_{ii} with the same indices, but Eqs. (9) and (10) imply $B_{ii} = 0$. Thus, due to (10), only three numbers are independent among the nine possible B_{ij}, $i, j \in \{1,2,3\}$.

Using the summation sign, we may write the bivector $\hat{\mathbf{B}}$ in two ways

$$\hat{\mathbf{B}} = \sum_{\substack{i,j=1 \\ i>j}}^{3} B_{ij}\mathbf{e}_i \wedge \mathbf{e}_j = \frac{1}{2}\sum_{i,j=1}^{3} B_{ij}\mathbf{e}_i \wedge \mathbf{e}_j \; . \tag{11}$$

The coefficient $\frac{1}{2}$ emerged in the second sum because the equal terms $B_{12}\mathbf{e}_1 \wedge \mathbf{e}_2$ and $B_{21}\mathbf{e}_2 \wedge \mathbf{e}_1$ appear in it in pairs.

Equation (9) also gives the coordinates of $\mathbf{a} \wedge \mathbf{b}$ in our bivector basis:

$$(\mathbf{a} \wedge \mathbf{b})_{ij} = a_i b_j - a_j b_i \; . \tag{12}$$

Now it is visible that for transformations of the vector basis $\mathbf{e}_1, \mathbf{e}_2, \mathbf{e}_3$, the bivector coordinates transform as products of the vector coordinates. Readers who are familiar with tensors will recognize that the bivector coordinates form a tensor of the second rank, and Eq. (10) states that it is an antisymmetric tensor. Thus we see that a second rank antisymmetric tensor can be related to each bivector. And vice versa: a bivector may be related to each second rank antisymmetric tensor B_{ij}, by means of Eq. (11). Due to the linear independence of the bivectors $\mathbf{e}_1 \wedge \mathbf{e}_2$, $\mathbf{e}_2 \wedge \mathbf{e}_3$ and $\mathbf{e}_3 \wedge \mathbf{e}_1$, this correspondence between bivectors and antisymmetric tensors of second rank is one-to-one.

Three-dimensional space has the peculiar property that for any plane all straight lines perpendicular to it are parallel to each other. This fact can be used for attaching a vector $\mathbf{B}$ to each volutor $\hat{\mathbf{B}}$, with an additional convention for choosing the orientation of the vector, called the *right-handed screw rule.* Now the prescription for mapping $\hat{\mathbf{B}}$ into $\mathbf{B}$ is as follows: Vector $\mathbf{B}$ has the magnitude of volutor $\hat{\mathbf{B}}$; its direction is perpendicular to the (2-dimensional) direction of $\hat{\mathbf{B}}$; and its orientation is connected with the orientation of $\hat{\mathbf{B}}$ by the right-handed screw rule, as shown in Fig. 10. Of course, the mapping of $\hat{\mathbf{B}}$ into $\mathbf{B}$ is one-to-one; therefore we can introduce the reverse mapping $\mathbf{B} \to \hat{\mathbf{B}}$ which is called a *Hodge map* and is denoted by a star: $*\mathbf{B} = \hat{\mathbf{B}}$. We also say that $\hat{\mathbf{B}}$ is *dual to* $\mathbf{B}$.

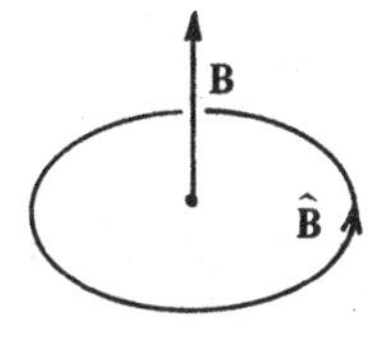

Fig. 10.

Notice that in the Hodge map the (right-handed) vector product is transformed into the outer product of vectors:

$$*(\mathbf{a} \times \mathbf{b}) = \mathbf{a} \wedge \mathbf{b} \,. \tag{13}$$

Therefore it should not seem strange to us that the three independent coordinates (12) of the outer product are so similar to the three components of the vector product. One can show (see Problem 2) that the Hodge map is linear.

In the right-handed[1] orthonormal basis $\mathbf{e}_1, \mathbf{e}_2, \mathbf{e}_3$ for vectors, the following equations

$$*\mathbf{e}_1 = \mathbf{e}_2 \wedge \mathbf{e}_3 \,, \quad *\mathbf{e}_2 = \mathbf{e}_3 \wedge \mathbf{e}_1 \,, \quad *\mathbf{e}_3 = \mathbf{e}_1 \wedge \mathbf{e}_2 \tag{14}$$

are satisfied. They can be written in a single equation

$$*\mathbf{e}_i = \frac{1}{2} \sum_{j,k=1}^{3} \varepsilon_{ijk} \mathbf{e}_j \wedge \mathbf{e}_k$$

using the *totally antisymmetric symbol:*

$$\varepsilon_{ijk} = \begin{cases} 1 & \text{if } i, j, k \text{ is an even permutation of } 1,2,3\,, \\ -1 & \text{if } i, j, k \text{ is an odd permutation of } 1,2,3\,, \\ 0 & \text{if any two indices have equal value}\,. \end{cases}$$

For an arbitrary vector $\mathbf{B} = \sum_{i=1}^{3} B_i \mathbf{e}_i$, due to the linearity of the Hodge map, we obtain from (14)

$$*\mathbf{B} = *(B_1 \mathbf{e}_1 + B_2 \mathbf{e}_2 + B_3 \mathbf{e}_3) = B_1 \mathbf{e}_2 \wedge \mathbf{e}_3 + B_2 \mathbf{e}_3 \wedge \mathbf{e}_1 + B_3 \mathbf{e}_1 \wedge \mathbf{e}_2 \,.$$

As we know, $*\mathbf{B} = \widehat{\mathbf{B}}$, so after comparing this with (11), we obtain the following relation between the components of the vector $\mathbf{B}$ and of the bivector $\widehat{\mathbf{B}}$:

$$B_1 = B_{23} = -B_{32} \,, \quad B_2 = B_{31} = -B_{13} \,, \quad B_3 = B_{12} = -B_{21} \,. \tag{15}$$

Adopting the convention of summation over repeating indices without the sum sign, we write this as

$$B_j = \frac{1}{2} \varepsilon_{jkl} B_{kl} \quad \text{or} \quad B_{kl} = \varepsilon_{klj} B_j \,.$$

[1] We shall, in Sec. 3, turn again to the orientation of the basis in 3-dimensional space. For now, we use the meaning expressed by the rule: Take the right-handed screw, place it parallel to $\mathbf{e}_3$ and turn from $\mathbf{e}_1$ to $\mathbf{e}_2$; if this rotational motion is accompanied by the translational motion in a positive direction of $\mathbf{e}_3$, the basis is called *right-handed.*

Let us now give some thought to what is going on with three vectors connected by the vector product

$$\mathbf{B} = \mathbf{a} \times \mathbf{b} \tag{16}$$

when some of them are reversed. We see that it is impossible to reverse all three of them and still retain relation (16) among the reversed vectors. Therefore, we have to designate the vectors which are to be reversed and those which are not. The vectors which remain unchanged under such operations are called *pseudovectors* or *axial vectors*. Other vectors, that is, ones reversing their signs, are called *polar vectors*. So if we assume that $\mathbf{a}$ and $\mathbf{b}$ are polar vectors in (16), then $\mathbf{B}$ must be an axial vector.

Of course, the volutor $\widehat{\mathbf{B}} = \mathbf{a} \wedge \mathbf{b}$ remains fixed under the reversion of the factors $\mathbf{a}$ and $\mathbf{b}$. Comparing the behaviour of the vectors and volutors under reversions, we must note that only axial vectors should be mapped through the Hodge map onto bivectors.

§2. Products of Bivectors

Let us define the following operation

$$\mathbf{a} \cdot (\mathbf{b} \wedge \mathbf{c}) = (\mathbf{a} \cdot \mathbf{b})\mathbf{c} - (\mathbf{a} \cdot \mathbf{c})\mathbf{b} = \mathbf{w} \tag{17}$$

(where $\mathbf{a} \cdot \mathbf{b}$ and $\mathbf{a} \cdot \mathbf{c}$ are ordinary scalar products of vectors) and call it the *inner product of a vector,* $\mathbf{a}$*, and a volutor* $\mathbf{b} \wedge \mathbf{c}$. This expression does not depend on the choice of vectors $\mathbf{b}$ and $\mathbf{c}$, which represent the volutor in the outer product (see Problem 3). Several facts are significant:

1. Product (17) is distributive with respect to the addition of both factors (see Problem 3).
2. The operation gives a vector lying on the plane determined by $\mathbf{b}$ and $\mathbf{c}$.
3. The result $\mathbf{w}$ of the product is perpendicular to $\mathbf{a}$, since

$$\mathbf{w} \cdot \mathbf{a} = (\mathbf{a} \cdot \mathbf{b})(\mathbf{c} \cdot \mathbf{a}) - (\mathbf{a} \cdot \mathbf{c})(\mathbf{b} \cdot \mathbf{a}) = 0 \ .$$

4. When $\mathbf{a}$ is perpendicular to $\mathbf{b} \wedge \mathbf{c}$ (we write this as $\mathbf{a} \perp \mathbf{b} \wedge \mathbf{c}$), then $\mathbf{a} \cdot (\mathbf{b} \wedge \mathbf{c}) = 0$, which means that only the component $\mathbf{a}_{\parallel}$ of $\mathbf{a}$ parallel to $\mathbf{b} \wedge \mathbf{c}$ makes a nonzero contribution to (17):

$$\mathbf{a} \cdot (\mathbf{b} \wedge \mathbf{c}) = \mathbf{a}_{\parallel} \cdot (\mathbf{b} \wedge \mathbf{c}) \ .$$

5. Take the vector $\mathbf{c}$ perpendicular to $\mathbf{a}$ (this is possible because of the arbitrariness of the factorization of the volutor). Then (17) along with Point 4 yields

$$\mathbf{w} = (\mathbf{a}_{\|} \cdot \mathbf{b})\mathbf{c} \,. \tag{18}$$

Now, choose the vector $\mathbf{b}$ parallel to $\mathbf{a}_{\|}$, with the same orientation (this would imply, if necessary, reversion of the previously chosen vector $\mathbf{c}$). Then $\mathbf{b} \perp \mathbf{c}$ and $\mathbf{a}_{\|} \cdot \mathbf{b} \geq 0$, hence Eq. (18) now says that $\mathbf{w}$ has the direction of $\mathbf{c}$.

6. Taking $\mathbf{b}$ and $\mathbf{c}$ as in Point 5, we have $|\mathbf{b}\wedge\mathbf{c}| = |\mathbf{b}||\mathbf{c}|$, and the magnitude of $\mathbf{w}$ is

$$|\mathbf{w}| = (\mathbf{a}_{\|} \cdot \mathbf{b})|\mathbf{c}| = |\mathbf{a}_{\|}||\mathbf{b}||\mathbf{c}| = |\mathbf{a}_{\|}||\mathbf{b} \wedge \mathbf{c}| \,,$$

that is,

$$|\mathbf{w}| = |\mathbf{a}||\mathbf{b} \wedge \mathbf{c}|\cos\theta \tag{19}$$

where θ is an angle between $\mathbf{a}$ and $\mathbf{b} \wedge \mathbf{c}$. Equation (19) resembles the scalar product of vectors, especially if one writes it in the form

$$|\mathbf{a} \cdot \widehat{\mathbf{B}}| = |\mathbf{a}||\widehat{\mathbf{B}}|\cos\theta \,, \tag{20}$$

where $\widehat{\mathbf{B}} = \mathbf{b} \wedge \mathbf{c}$.

Figure 11 illustrates the defined product. We see from it, among other things, that the volutor $\mathbf{a}_{\|} \wedge \mathbf{w}$ has the same direction as the volutorial factor $\widehat{\mathbf{B}} = \mathbf{b} \wedge \mathbf{c}$. Taking the magnitudes into account as well, we can express this observation in the equality $\mathbf{a}_{\|} \wedge \mathbf{w} = |\mathbf{a}_{\|}|^2\widehat{\mathbf{B}}$. When $\widehat{\mathbf{B}}$ is not factorized into a parallelogram of the outer product of vectors, the relation between the factors and the result is illustrated in Fig. 12.

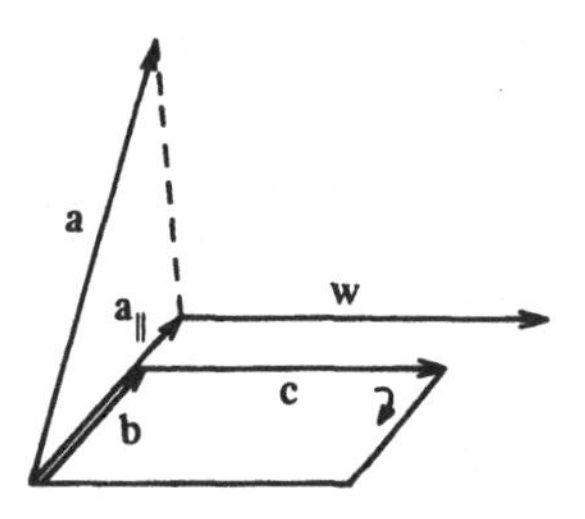

Fig. 11.

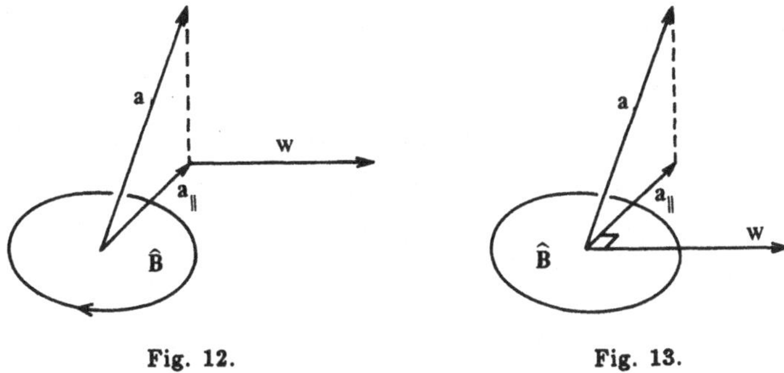

Fig. 12. Fig. 13.

An easy way to memorize the direction of the product $\mathbf{a}\cdot\widehat{\mathbf{B}}$ is the following "recipe": Project $\mathbf{a}$ on the plane of $\widehat{\mathbf{B}}$, then turn the result by the angle $\pi/2$ in the sense given by $\widehat{\mathbf{B}}$ (see Fig. 13).

We shall also use the *inner product of the volutor and vector* in reversed order:

$$(\mathbf{b}\wedge\mathbf{c})\cdot\mathbf{a} = \mathbf{b}(\mathbf{c}\cdot\mathbf{a}) - \mathbf{c}(\mathbf{b}\cdot\mathbf{a})\,. \tag{21}$$

As we see, the result is opposite to that of (17). Here we notice that such an inner product is *anticommutative*:

$$\mathbf{a}\cdot\widehat{\mathbf{B}} = -\widehat{\mathbf{B}}\cdot\mathbf{a}\,. \tag{22}$$

The inner product of a vector $\mathbf{a}$ with an arbitrary bivector $\widehat{\mathbf{B}} = \widehat{\mathbf{B}}_1 + \ldots + \widehat{\mathbf{B}}_n$ expressed as a sum of volutors $\widehat{\mathbf{B}}_1, \ldots, \widehat{\mathbf{B}}_n$ is defined by additivity:

$$\mathbf{a}\cdot\widehat{\mathbf{B}} = \mathbf{a}\cdot\widehat{\mathbf{B}}_1 + \ldots + \mathbf{a}\cdot\widehat{\mathbf{B}}_n\,.$$

The inner product (17) can be expressed in the following way by the components of its factors

$$\mathbf{a}\cdot\widehat{\mathbf{B}} = (a_3B_{31} + a_2B_{21})\mathbf{e}_1 + (a_1B_{12} + a_3B_{32})\mathbf{e}_2 + (a_2B_{23} + a_1B_{13})\mathbf{e}_3$$

(see Problem 4) which for separate coordinates may be written as

$$(\mathbf{a}\cdot\widehat{\mathbf{B}})_k = a_jB_{jk}\,. \tag{23}$$

Analogously, $(\widehat{\mathbf{B}}\cdot\mathbf{a})_k = B_{kj}a_j$. It is worth noting the similarity to the scalar product of vectors expressed by the coordinates, $\mathbf{a}\cdot\mathbf{b} = a_jb_j$.

If the volutor $\mathbf{b}\wedge\mathbf{c}$ is replaced by the (pseudo) vector $\mathbf{b}\times\mathbf{c}$, the identity $\mathbf{a}\cdot(\mathbf{b}\wedge\mathbf{c}) = -\mathbf{a}\times(\mathbf{b}\times\mathbf{c})$ holds (see Problem 5) which can be written more briefly as

$$\mathbf{a}\cdot\widehat{\mathbf{B}} = -\mathbf{a}\times\mathbf{B}\,. \tag{24}$$

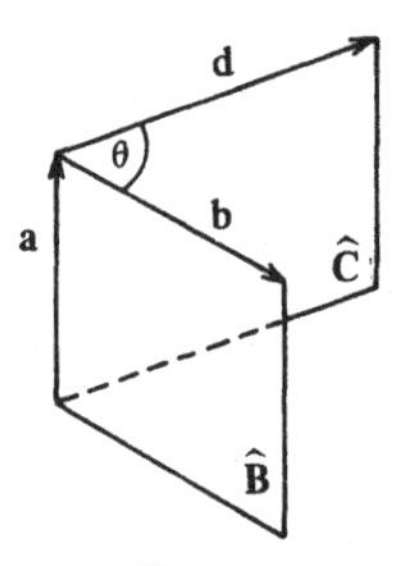

Fig. 14.

Now we define the next operation, namely, the *inner product of two volutors* $\hat{\mathbf{B}} = \mathbf{a} \wedge \mathbf{b}$ and $\hat{\mathbf{C}} = \mathbf{c} \wedge \mathbf{d}$:

$$\hat{\mathbf{B}} \cdot \hat{\mathbf{C}} = (\mathbf{a} \wedge \mathbf{b}) \cdot \hat{\mathbf{C}} = \mathbf{a} \cdot (\mathbf{b} \cdot \hat{\mathbf{C}}) \,. \tag{25a}$$

Due to (17) and (21), it is possible to prove that the equality

$$\hat{\mathbf{B}} \cdot \hat{\mathbf{C}} = \hat{\mathbf{B}} \cdot (\mathbf{c} \wedge \mathbf{d}) = (\hat{\mathbf{B}} \cdot \mathbf{c}) \cdot \mathbf{d} \tag{25b}$$

is also satisfied. The result of this product is a scalar. This operation is commutative and does not depend on the factorization of $\hat{\mathbf{B}}$ and $\hat{\mathbf{C}}$ in the outer products (see Problem 8).

Definition (25) gives for the basis vectors:

$$(\mathbf{e}_i \wedge \mathbf{e}_j) \cdot (\mathbf{e}_k \wedge \mathbf{e}_l) = \delta_{il}\delta_{jk} - \delta_{ik}\delta_{jl} \tag{26}$$

where δ_{ij} is the *Kronecker symbol*

$$\delta_{ij} = \begin{cases} 1 & \text{for } i = j \\ 0 & \text{for } i \neq j \,. \end{cases}$$

Owing to (26), one finds that the coordinates of any bivector may be calculated as follows:

$$B_{ij} = -(\mathbf{e}_i \wedge \mathbf{e}_j) \cdot \hat{\mathbf{B}} \,, \tag{27}$$

(see Problem 9).

Let us find how the product $\hat{\mathbf{B}} \cdot \hat{\mathbf{C}}$ is expressed by the magnitudes of the factors. We know that in three-dimensional space it is possible to find a common edge for any two volutors. So let $\hat{\mathbf{B}} = \mathbf{a} \wedge \mathbf{b}$, $\hat{\mathbf{C}} = \mathbf{a} \wedge \mathbf{d}$ and $\mathbf{b} \perp \mathbf{a} \perp \mathbf{d}$ (see Fig. 14), and let the angle between $\mathbf{b}$ and $\mathbf{d}$ be θ (this is also the angle between the volutors $\hat{\mathbf{B}}$ and $\hat{\mathbf{C}}$).[1] Then (25) yields

$$\hat{\mathbf{B}} \cdot \hat{\mathbf{C}} = (\mathbf{a} \cdot \mathbf{d})(\mathbf{a} \cdot \mathbf{b}) - (\mathbf{b} \cdot \mathbf{d})(\mathbf{a} \cdot \mathbf{a}) \,.$$

[1] An angle between two volutors is defined so that the common edge in their factorization has the same orientation if treated as boundary of them separately. The volutors are now juxtaposed differently than they were for addition.

We use the assumed perpendicularity $\mathbf{a} \perp \mathbf{b}$ and $\mathbf{a} \perp \mathbf{d}$:

$$\widehat{\mathbf{B}} \cdot \widehat{\mathbf{C}} = -|\mathbf{a}|^2(\mathbf{b} \cdot \mathbf{d}) = -|\mathbf{a}|^2|\,\mathbf{b}\,||\,\mathbf{d}\,|\cos\theta = -|\,\mathbf{a}\,||\,\mathbf{b}\,||\,\mathbf{a}\,||\,\mathbf{d}\,|\cos\theta$$

that is,

$$\widehat{\mathbf{B}} \cdot \widehat{\mathbf{C}} = -|\,\widehat{\mathbf{B}}\,||\,\widehat{\mathbf{C}}\,|\cos\theta\ . \tag{28}$$

We see that the signature here is different than that for the scalar product of vectors (for the acute angle $\widehat{\mathbf{B}} \cdot \widehat{\mathbf{C}} < 0$, for the obtuse angle $\widehat{\mathbf{B}} \cdot \widehat{\mathbf{C}} > 0$). In particular, the inner square turns out to be nonpositive:

$$\widehat{\mathbf{B}} \cdot \widehat{\mathbf{B}} = -|\,\widehat{\mathbf{B}}\,|^2\ . \tag{29}$$

The scalar product of pseudovectors can be related to the above inner product of their dual volutors:

$$\widehat{\mathbf{B}} \cdot \widehat{\mathbf{C}} = -\mathbf{B} \cdot \mathbf{C} \tag{30}$$

which follows immediately from (28). If we want to express the inner product of volutors by their coordinates, we obtain $\widehat{\mathbf{B}} \cdot \widehat{\mathbf{C}} = \frac{1}{2} B_{lk} C_{kl}$ due to (11) and (26). The inner product of two bivectors expressed as sums of volutors may be defined by additivity.

We can define one more operation for bivectors. For $\widehat{\mathbf{B}} = \mathbf{b} \wedge \mathbf{a}$ and $\widehat{\mathbf{C}} = \mathbf{c} \wedge \mathbf{d}$, the expression

$$\widehat{\mathbf{B}} \dot{\wedge} \widehat{\mathbf{C}} = (\mathbf{a} \cdot \mathbf{c})\mathbf{b} \wedge \mathbf{d} - (\mathbf{a} \cdot \mathbf{d})\mathbf{b} \wedge \mathbf{c} - (\mathbf{b} \cdot \mathbf{c})\mathbf{a} \wedge \mathbf{d} + (\mathbf{b} \cdot \mathbf{d})\mathbf{a} \wedge \mathbf{c} \tag{31}$$

is called the *mingled product of two volutors.* It does not depend on the factorizations of $\widehat{\mathbf{B}}$ and $\widehat{\mathbf{C}}$ in the outer products (see Problem 10). It is easy to see that this product gives a bivector (a volutor in three-dimensional space) and that it is *anticommutative.*[1)]

The above can be simplified when factorization with parallel factors is used. Let $\mathbf{b} \perp \mathbf{a} \parallel \mathbf{c} \perp \mathbf{d}$, then $\mathbf{a} \cdot \mathbf{d} = \mathbf{b} \cdot \mathbf{c} = 0 = \mathbf{a} \wedge \mathbf{c}$, and in (31)

$$\widehat{\mathbf{B}} \dot{\wedge} \widehat{\mathbf{C}} = (\mathbf{a} \cdot \mathbf{c})\mathbf{b} \wedge \mathbf{d}$$

remains. For vectors $\mathbf{a}$ and $\mathbf{c}$ with the same orientation, we see that $\widehat{\mathbf{B}} \dot{\wedge} \widehat{\mathbf{C}}$ has the direction of the volutor $\mathbf{b} \wedge \mathbf{d}$ (see Fig. 15). The magnitude of $\widehat{\mathbf{B}} \dot{\wedge} \widehat{\mathbf{C}}$ is

$$|\,\widehat{\mathbf{B}} \dot{\wedge} \widehat{\mathbf{C}}\,| = |\,\mathbf{a}\,||\,\mathbf{c}\,||\,\mathbf{b} \wedge \mathbf{d}\,| = |\,\mathbf{a}\,||\,\mathbf{c}\,||\,\mathbf{b}\,||\,\mathbf{d}\,|\sin\theta$$

[1)] These two properties, along with a verifiable Jacobi identity, mean that the set of bivectors with this operation is a Lie algebra.

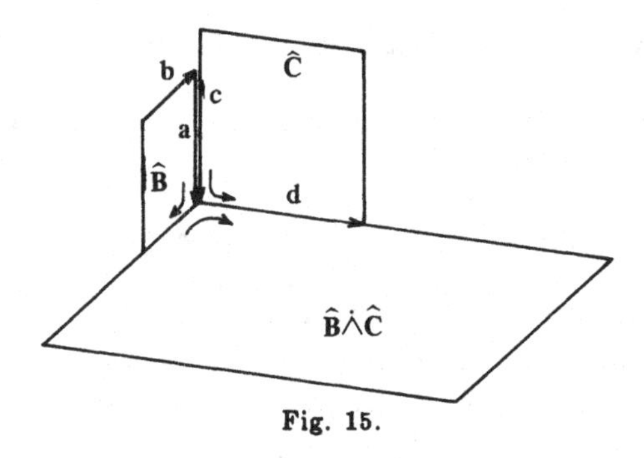

Fig. 15.

where θ is an angle between vectors $\mathbf{b}$ and $\mathbf{d}$ and, simultaneously, between volutors $\widehat{\mathbf{B}}$ and $\widehat{\mathbf{C}}$. Thus

$$|\widehat{\mathbf{B}}\dot{\wedge}\widehat{\mathbf{C}}| = |\widehat{\mathbf{B}}||\widehat{\mathbf{C}}|\sin\theta \ .$$

In particular, the mingled square of any volutor is zero: $\widehat{\mathbf{B}}\dot{\wedge}\widehat{\mathbf{B}} = 0$.

Expression (31) can be transformed by appropriately assembling the terms

$$\widehat{\mathbf{B}}\dot{\wedge}\widehat{\mathbf{C}} = \mathbf{b} \wedge [(\mathbf{a}\cdot\mathbf{c})\mathbf{d} - (\mathbf{a}\cdot\mathbf{d})\mathbf{c}] - \mathbf{a} \wedge [(\mathbf{b}\cdot\mathbf{c})\mathbf{d} - (\mathbf{b}\cdot\mathbf{d})\mathbf{c}]$$

and using (17)

$$\widehat{\mathbf{B}}\dot{\wedge}\widehat{\mathbf{C}} = (\mathbf{b}\wedge\mathbf{a})\dot{\wedge}\widehat{\mathbf{C}} = \mathbf{b}\wedge(\mathbf{a}\cdot\widehat{\mathbf{C}}) - \mathbf{a}\wedge(\mathbf{b}\cdot\widehat{\mathbf{C}}) \ . \tag{32}$$

Similarly, with a different assembly of the terms and using (21), one obtains

$$\widehat{\mathbf{B}}\dot{\wedge}\widehat{\mathbf{C}} = \widehat{\mathbf{B}}\dot{\wedge}(\mathbf{c}\wedge\mathbf{d}) = (\widehat{\mathbf{B}}\cdot\mathbf{c})\wedge\mathbf{d} - (\widehat{\mathbf{B}}\cdot\mathbf{d})\wedge\mathbf{c} \ .$$

It is possible to ascertain that in the Hodge correspondence, the mingled product is related to the vector product of pseudovectors as follows:

$$*(\mathbf{B}\times\mathbf{C}) = -\widehat{\mathbf{B}}\dot{\wedge}\widehat{\mathbf{C}} \ . \tag{33}$$

If one wants to express the mingled product in terms of coordinates of volutors, one obtains

$$(\widehat{\mathbf{B}}\dot{\wedge}\widehat{\mathbf{C}})_{ik} = B_{ij}C_{jk} - B_{kj}C_{ji}$$

(see Problem 10).

In the physical part of this book, we shall often change formulas with axial vectors into corresponding formulas with bivectors. We know from the foregoing considerations that three possibilities exist for the vector product, depending on whether any of the factors are axial vectors, and if so, how many.

(I) If both factors are polar vectors, the result is an axial vector and we use Eq. (13):

$$*(\mathbf{a}\times\mathbf{b}) = \mathbf{a}\wedge\mathbf{b}\ .$$

(II) If one factor — let it be **a** — is a polar vector and the other one — let it be **B** — is an axial vector, the result is a polar vector and we apply Eq. (24):

$$\mathbf{a}\times\mathbf{B} = -\mathbf{a}\cdot\widehat{\mathbf{B}}\ .$$

(III) If both factors are axial vectors, the result is also an axial vector and one should observe equality (33):

$$*(\mathbf{A}\times\mathbf{B}) = -\widehat{\mathbf{A}}\dot{\wedge}\widehat{\mathbf{B}}\ .$$

The mingled product of two bivectors treated as sums of volutors may be defined by additivity.

§3. Trivectors

We call *three-dimensional orientation* a combination of rotational motion and translatory motion, during which the latter cannot be parallel to the plane of rotation. Only two such orientations are possible in three-dimensional space; they are shown in Fig. 16. Two kinds of screws correspond to them: (i) those having a right-hand thread, (ii) those with a left-hand thread.

Fig. 16.

Three-dimensional space with a definite three-dimensional orientation is called *three-dimensional direction.* If this notion is considered in a space of a higher dimension, it is defined as a whole family of parallel three-dimensional spaces with compatible orientations.

A *trivector* is a geometric object with two significant features: (i) *three-dimensional direction* and (ii) *magnitude* interpreted as volume.[1] If we restrict ourselves to the three-dimensional space, only two different directions

[1] Strictly speaking, this is a *simple trivector.* A trivector is a sum of such objects. Non-simple trivectors can occur only in spaces of five dimensions or more.

are possible; they correspond to the two orientations shown in Fig. 16, and we call them opposite directions. As a notation of a trivector we choose a letter with a caret above it: e.g. $\hat{T}$. The magnitude of $\hat{T}$ will be denoted by $|\hat{T}|$. Two trivectors $\hat{T}$ and $\hat{T}'$ with the same magnitude and opposite directions will be called *opposite* and written $\hat{T} = -\hat{T}'$. The zero trivector, of course, has no direction.

The *geometrical image* of a trivector is a solid body[1)] with orientation marked by two arrows inside the body: one straight, the other one curved and encompassing the first one (see Fig. 17). The orientation of a trivector can be marked also by a curved arrow on the boundary of the body — that is, on its surface (see Fig. 18). This is a counterpart of the orientation of a volutor, which is marked on its boundary (Fig. 4a). This curved arrow symbolizes the rotational motion; translational motion is to be represented by an arrow left inside the body. Hence Figs. 17 and 18 depict trivectors with the same orientation, which we call *right-handed.* The arrows on the surface of the body of Fig. 18, viewed from the outside, have counter-clockwise orientation. Thus, for a trivector with *left-handed orientation,* the corresponding curved arrows on the surface of the body will have clockwise orientation. We also see from Fig. 18 that if the solid body representing a trivector is a polyhedron, its faces get naturally a two-dimensional orientation; therefore they can be treated as volutors (compare with Problem 1).

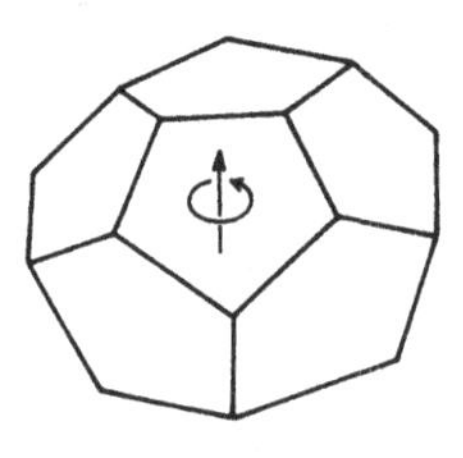

Fig. 17.

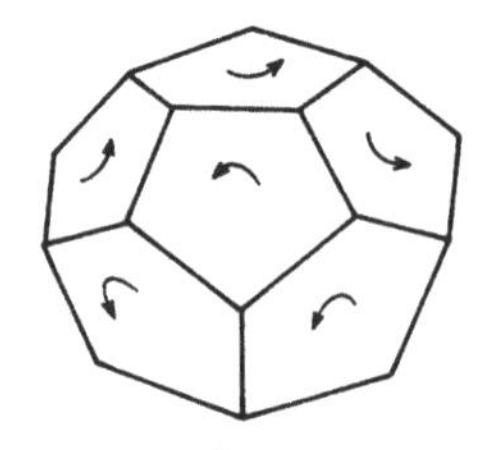

Fig. 18.

The shape of the solid body, of course, is not important for the geometric image of a given trivector.

An example of trivectorial geometrical quantity could be a solid angle if it is given three-dimensional orientation, for instance, by adding a curved arrow on its spherical cap (see Fig. 19).

A trivector can be obtained from a vector $\mathbf{c}$ and a volutor $\hat{\mathbf{B}}$ in the following manner. We put the tip point of vector $\mathbf{c}$ at the boundary of the geometric

[1)]We shall choose connected and simply connected bodies although it does not seem to make any difference.

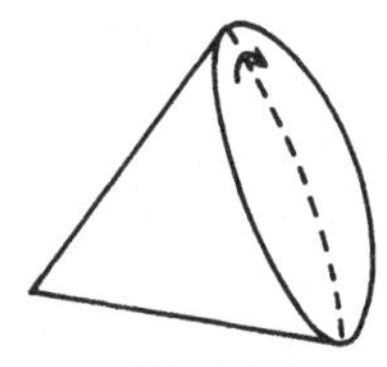

Fig. 19.

image of $\hat{\mathbf{B}}$ and shift it parallelly along this boundary so as to obtain an oblique cylinder with the base $\hat{\mathbf{B}}$ (see Fig. 20). In this way, all the relevant features of a trivector $\hat{T}$ are given: its orientation is the rotational motion taken from $\hat{\mathbf{B}}$ along with the translatory motion taken from $\mathbf{c}$, and its magnitude is the volume of the cylinder:

$$|\hat{T}| = |\hat{\mathbf{B}}||\mathbf{c}|\sin\alpha \tag{34}$$

where α is the angle between $\mathbf{c}$ and $\hat{\mathbf{B}}$. The obtained operation ascribing $\hat{T}$ to the factors $\mathbf{c}$ and $\hat{\mathbf{B}}$ is called the *outer product of vector with volutor* and is denoted by a wedge sign:

$$\hat{T} = \mathbf{c} \wedge \hat{\mathbf{B}} .$$

We also define the *outer product of volutor with vector* (in opposite order) by the formula

$$\hat{\mathbf{B}} \wedge \mathbf{c} = \mathbf{c} \wedge \hat{\mathbf{B}} , \tag{35}$$

which implies that this product is *commutative.*

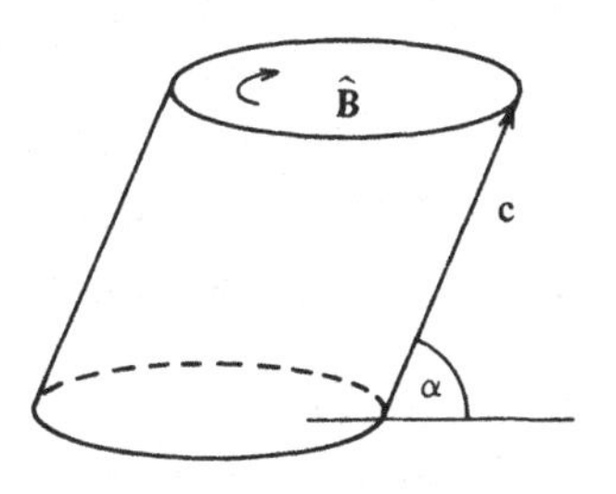

Fig. 20.

A trivector can also be treated as the outer product of three vectors: $\hat{T} = \mathbf{c} \wedge (\mathbf{a} \wedge \mathbf{b})$, and this fact justifies its name. The expression $\mathbf{c} \wedge (\mathbf{a} \wedge \mathbf{b})$ is symmetric under cyclic changes of factors:

$$\mathbf{c} \wedge (\mathbf{a} \wedge \mathbf{b}) = \mathbf{a} \wedge (\mathbf{b} \wedge \mathbf{c}) = \mathbf{b} \wedge (\mathbf{c} \wedge \mathbf{a}) \tag{36}$$

(see Problem 11). If we apply Eq. (35) to the last term, we obtain

$$\mathbf{c} \wedge (\mathbf{a} \wedge \mathbf{b}) = (\mathbf{c} \wedge \mathbf{a}) \wedge \mathbf{b} \ ,$$

which means that the outer product of three vectors is *associative*. Thus we may write (36) without brackets:

$$\widehat{T} = \mathbf{c} \wedge \mathbf{a} \wedge \mathbf{b} = \mathbf{a} \wedge \mathbf{b} \wedge \mathbf{c} = \mathbf{b} \wedge \mathbf{c} \wedge \mathbf{a} \ . \tag{37}$$

Trivectors can be multiplied by scalars under a quite obvious rule: Trivector $\lambda\widehat{T}$ has the magnitude $|\lambda||\widehat{T}|$ and the same orientation as $\widehat{T}$ for $\lambda > 0$, or opposite to $\widehat{T}$ for $\lambda < 0$. It is easy to check that the outer product of vectors with volutors satisfies

$$(\lambda \mathbf{c}) \wedge \widehat{\mathbf{B}} = \lambda(\mathbf{c} \wedge \widehat{\mathbf{B}}) = \mathbf{c} \wedge (\lambda \widehat{\mathbf{B}}) \ . \tag{38}$$

We may also add trivectors. If $\widehat{T}$ and $\widehat{T}'$ have the same orientation, $\widehat{T} + \widehat{T}'$ has the orientation of both summands and the magnitude $|\widehat{T}| + |\widehat{T}'|$. If $\widehat{T}$ and $\widehat{T}'$ have opposite orientations, $\widehat{T} + \widehat{T}'$ has the orientation of the greater summand and the magnitude $||\widehat{T}| - |\widehat{T}'||$. As may be verified, the outer product of vectors with bivectors is *distributive with respect to the addition* of both factors:

$$(\mathbf{c} + \mathbf{d}) \wedge \widehat{\mathbf{B}} = \mathbf{c} \wedge \widehat{\mathbf{B}} + \mathbf{d} \wedge \widehat{\mathbf{B}}, \quad \mathbf{c} \wedge (\widehat{\mathbf{B}} + \widehat{\mathbf{D}}) = \mathbf{c} \wedge \widehat{\mathbf{B}} + \mathbf{c} \wedge \widehat{\mathbf{D}} \ . \tag{39}$$

The factorization (37) of a given trivector is not unique. If we take

$$\begin{aligned} \mathbf{a}' &= \alpha_{11}\mathbf{a} + \alpha_{12}\mathbf{b} + \alpha_{13}\mathbf{c} \\ \mathbf{b}' &= \alpha_{21}\mathbf{a} + \alpha_{22}\mathbf{b} + \alpha_{23}\mathbf{c} \\ \mathbf{c}' &= \alpha_{31}\mathbf{a} + \alpha_{32}\mathbf{b} + \alpha_{33}\mathbf{c} \ , \end{aligned}$$

the outer product of these new vectors can be transformed with the use of properties (38) and (39), which gives

$$\mathbf{a}' \wedge \mathbf{b}' \wedge \mathbf{c}' = (\det \mathrm{A}) \mathbf{a} \wedge \mathbf{b} \wedge \mathbf{c} \tag{40}$$

where A is a matrix with elements α_{ij} and $\det A$ is its determinant. If the product $\mathbf{a}' \wedge \mathbf{b}' \wedge \mathbf{c}'$ is to be equal to $\mathbf{a} \wedge \mathbf{b} \wedge \mathbf{c}$, then $\det A$ must be equal to one. This is the only condition limiting the arbitrariness of representing $\widehat{T}$ in form (37).

Since we may add trivectors and multiply them by scalars, and these operations (as one may check) satisfy the appropriate conditions, the set of trivectors forms a linear space. We have already mentioned that in three-dimensional space, all trivectors have the same direction up to the sign, so they are parallel. We conclude from this that the set of trivectors is a one-dimensional linear space. Hence it is sufficient to choose one basic trivector. Let $\{\mathbf{e}_1, \mathbf{e}_2, \mathbf{e}_3\}$ be the orthonormal basis in the underlying space of vectors. One may form six outer products from them, since there are six permutations of the three elements. All of them are proportional with the coefficient ± 1, which may be included in the equation

$$\mathbf{e}_i \wedge \mathbf{e}_j \wedge \mathbf{e}_k = \varepsilon_{ijk} \mathbf{e}_1 \wedge \mathbf{e}_2 \wedge \mathbf{e}_3 \tag{41}$$

with the use of the totally antisymmetric symbol ε_{ijk}. The indices i, j and k in (41) need not be different. If some of them coincide, both sides of (41) are equal to zero. In this way, equality (41) covers all 27 possible values of i, j, k.

We choose one of the six mentioned trivectors as the basic element in the linear space of trivectors. Let it be

$$\hat{I} = \mathbf{e}_1 \wedge \mathbf{e}_2 \wedge \mathbf{e}_3 \ . \tag{42}$$

If its orientation is right-handed, the basis $\{\mathbf{e}_1, \mathbf{e}_2, \mathbf{e}_3\}$ of the vector space is called *right-handed*; otherwise the basis is called *left-handed*. In this way, we introduce the *orientation of the basis* for vector space (see the Footnote on Pg.77). An example of a right-handed basis would be three neighbouring fingers of the right hand spaced-out as in Fig. 21 and enumerated in succession from the thumb. Similarly spaced-out and enumerated fingers of the left hand form an example of a left-handed basis.

Fig. 21.

All orthonormal bases with the same orientation are related to each other through orthogonal linear transformations with determinants equal to one.

We know from linear algebra that such transformations are rotations. No rotation can change a right hand into a left hand; therefore the two orientations described are essentially different.

It is possible to verify that an orthonormal basis $\{\mathbf{e}_1, \mathbf{e}_2, \mathbf{e}_3\}$ is right-handed if, and only if, the traditional condition $\mathbf{e}_1 \times \mathbf{e}_2 = \mathbf{e}_3$ is satisfied with the right-hand vector product.

All trivectors are proportional to the basic trivector (42):

$$\hat{T} = t\hat{I} . \tag{43}$$

The real number t is called a *coordinate* of the trivector $\hat{T}$. The mapping $t \rightarrow \hat{T}$ for a given basis $\{\mathbf{e}_1, \mathbf{e}_2, \mathbf{e}_3\}$ is one-to-one. It is called a *Hodge map for trivectors* and denoted by a star: $*t = \hat{T}$. We say also that $\hat{T}$ is *dual to t*. It follows from (43) that

$$*t = t\hat{I} = \hat{I}t . \tag{44}$$

It is easily seen from this that the Hodge map is linear.

When we represent three vectors $\mathbf{a}_j = a_{ji}\mathbf{e}_i$ in terms of the basis $\{\mathbf{e}_1, \mathbf{e}_2, \mathbf{e}_3\}$ and introduce the matrix A with elements a_{ji}, then due to (40) we obtain

$$\mathbf{a}_1 \wedge \mathbf{a}_2 \wedge \mathbf{a}_3 = (\det A)\mathbf{e}_1 \wedge \mathbf{e}_2 \wedge \mathbf{e}_3 = (\det A)\hat{I} .$$

In this way, we see that the trivector $\hat{T} = \mathbf{a}_1 \wedge \mathbf{a}_2 \wedge \mathbf{a}_3$ has its coordinate $t = \det A$.

If all factors in the expression $\hat{T} = \mathbf{a} \wedge \mathbf{b} \wedge \mathbf{c}$ are replaced by their opposites: $\mathbf{a}' = -\mathbf{a}, \mathbf{b}' = -\mathbf{b}, \mathbf{c}' = -\mathbf{c}$, the trivector $\hat{T}$ also changes its sign: $\hat{T}' = \mathbf{a}' \wedge \mathbf{b}' \wedge \mathbf{c}' = -\hat{T}$. If, during this process, we keep the basic trivector constant, Eq. (43) is invariant when the scalar t changes its sign: $t' = -t$. Therefore, such a scalar behaves differently from other scalars (for instance, obtained as scalar products of vectors) under the reversal of vectors. For this reason, we distinguish it by the name *pseudoscalar*. Comparing the behaviour of scalars, pseudoscalars and trivectors under the reversion of vectors, we must note that only pseudoscalars should be related to trivectors by the Hodge map.

Let us express the coordinate of the trivector $\mathbf{c} \wedge \hat{\mathbf{B}}$ by the coordinates of $\mathbf{c}$ and $\hat{\mathbf{B}}$:

$$\mathbf{c} \wedge \hat{\mathbf{B}} = (c_1\mathbf{e}_1 + c_2\mathbf{e}_2 + c_3\mathbf{e}_3) \wedge (B_{23}\mathbf{e}_2 \wedge \mathbf{e}_3 + B_{31}\mathbf{e}_3 \wedge \mathbf{e}_1 + B_{12}\mathbf{e}_1 \wedge \mathbf{e}_2) .$$

After performing the operations, we obtain, due to (41) and (42):

$$\mathbf{c} \wedge \hat{\mathbf{B}} = (c_1 B_{23} + c_2 B_{31} + c_3 B_{12})\hat{I} . \tag{45}$$

This means that the coordinate t of the trivector $\mathbf{c} \wedge \widehat{\mathbf{B}} = t\widehat{I}$ has the form $t = \frac{1}{2}\varepsilon_{ijk}c_i B_{jk}$.

Having taken into account the Hodge map for bivectors, expressed — among others — in Eq. (15), we may rewrite (45) as

$$\mathbf{c} \wedge \widehat{\mathbf{B}} = (\mathbf{c} \cdot \mathbf{B})\widehat{I}\,, \tag{46}$$

or

$$\mathbf{c} \wedge (*\mathbf{B}) = *(\mathbf{c} \cdot \mathbf{B})\,. \tag{47}$$

If, moreover, we represent $\widehat{\mathbf{B}} = \mathbf{a} \wedge \mathbf{b}$, we obtain

$$\mathbf{c} \wedge (\mathbf{a} \wedge \mathbf{b}) = *[\mathbf{c} \cdot (\mathbf{a} \times \mathbf{b})]\,. \tag{48}$$

In this way, we ascertain that the outer product of three vectors is a trivector dual to the so-called *mixed product* of the three vectors.

§4. Products of Trivectors

We define the following operation

$$\widehat{\mathbf{f}} = \mathbf{a} \cdot (\mathbf{b} \wedge \mathbf{c} \wedge \mathbf{d}) = (\mathbf{a} \cdot \mathbf{b})\mathbf{c} \wedge \mathbf{d} + (\mathbf{a} \cdot \mathbf{c})\mathbf{d} \wedge \mathbf{b} + (\mathbf{a} \cdot \mathbf{d})\mathbf{b} \wedge \mathbf{c} \tag{49}$$

and call it the *inner product of a vector* $\mathbf{a}$ *and a trivector* $\widehat{T} = \mathbf{b} \wedge \mathbf{c} \wedge \mathbf{d}$. This expression does not depend on the choice of vectors $\mathbf{b}, \mathbf{c}, \mathbf{d}$ representing the trivector $\widehat{T}$ (see Problem 13). Notice the following facts:

1. Product (49) is distributive with respect to the addition of both factors (see Problem 13).
2. The result is a bivector perpendicular to $\mathbf{a}$, since

$$\begin{aligned}
\mathbf{a} \cdot \widehat{\mathbf{f}} &= (\mathbf{a} \cdot \mathbf{b})\mathbf{a} \cdot (\mathbf{c} \wedge \mathbf{d}) + (\mathbf{a} \cdot \mathbf{c})\mathbf{a} \cdot (\mathbf{d} \wedge \mathbf{b}) + (\mathbf{a} \cdot \mathbf{d})\mathbf{a} \cdot (\mathbf{b} \wedge \mathbf{c}) \\
&= (\mathbf{a} \cdot \mathbf{b})(\mathbf{a} \cdot \mathbf{c})\mathbf{d} - (\mathbf{a} \cdot \mathbf{b})(\mathbf{a} \cdot \mathbf{d})\mathbf{c} + (\mathbf{a} \cdot \mathbf{c})(\mathbf{a} \cdot \mathbf{d})\mathbf{b} \\
&\qquad - (\mathbf{a} \cdot \mathbf{c})(\mathbf{a} \cdot \mathbf{b})\mathbf{d} + (\mathbf{a} \cdot \mathbf{d})(\mathbf{a} \cdot \mathbf{b})\mathbf{c} - (\mathbf{a} \cdot \mathbf{d})(\mathbf{a} \cdot \mathbf{c})\mathbf{b} \\
&= 0\,.
\end{aligned}$$

3. Take $\mathbf{b}$ to have the direction of $\mathbf{a}$, and $\mathbf{c}$, $\mathbf{d}$ to be perpendicular to $\mathbf{b}$. Then $\mathbf{a} \perp \mathbf{c}$, $\mathbf{a} \perp \mathbf{d}$ and definition (49) gives

$$\widehat{\mathbf{f}} = \mathbf{a} \cdot \widehat{T} = (\mathbf{a} \cdot \mathbf{b})\widehat{\mathbf{B}} \tag{50}$$

where $\widehat{\mathbf{B}} = \mathbf{c} \wedge \mathbf{d}$. Since $\mathbf{a} \cdot \mathbf{b} \geq 0$, the volutors $\widehat{\mathbf{f}}$ and $\widehat{\mathbf{B}}$ have the same direction. Also note that $\mathbf{a} \wedge \widehat{\mathbf{f}} = |\,\mathbf{a}\,|^2\widehat{T}$.

4. The magnitude of $\mathbf{f}$, due to (50), is

$$|\hat{\mathbf{f}}| = (\mathbf{a}\cdot\mathbf{b})|\hat{\mathbf{B}}| = |\mathbf{a}||\mathbf{b}||\hat{\mathbf{B}}| = |\mathbf{a}||\hat{T}|\,,$$

since for $\hat{T} = \mathbf{b}\wedge\hat{\mathbf{B}}$ with $\mathbf{b}\perp\hat{\mathbf{B}}$ one has $|\hat{T}| = |\mathbf{b}||\hat{\mathbf{B}}|$. This is a counterpart of Eq. (20) and of the scalar product of vectors. Why is the angle between $\mathbf{a}$ and $\hat{T}$ missing from it? It is because all vectors in three-dimensional space are parallel to any trivector, and that angle is zero.

We also introduce the *inner product of a trivector and a vector* (in reverse order):

$$(\mathbf{b}\wedge\mathbf{c}\wedge\mathbf{d})\cdot\mathbf{a} = \mathbf{b}\wedge\mathbf{c}(\mathbf{d}\cdot\mathbf{a}) + \mathbf{d}\wedge\mathbf{b}(\mathbf{c}\cdot\mathbf{a}) + \mathbf{c}\wedge\mathbf{d}(\mathbf{b}\cdot\mathbf{a})\,. \tag{51}$$

This implies that the inner product of vectors with trivectors is *commutative*.

Let $\hat{I} = \mathbf{e}_1\wedge\mathbf{e}_2\wedge\mathbf{e}_3$ for the right-handed orthonormal basis. Then (49) gives

$$\begin{aligned}\mathbf{B}\cdot\hat{I} &= (\mathbf{B}\cdot\mathbf{e}_1)\mathbf{e}_2\wedge\mathbf{e}_3 + (\mathbf{B}\cdot\mathbf{e}_2)\mathbf{e}_3\wedge\mathbf{e}_1 + (\mathbf{B}\cdot\mathbf{e}_3)\mathbf{e}_1\wedge\mathbf{e}_2\\ &= B_1\mathbf{e}_2\wedge\mathbf{e}_3 + B_2\mathbf{e}_3\wedge\mathbf{e}_1 + B_3\mathbf{e}_1\wedge\mathbf{e}_2\,.\end{aligned}$$

We may write this as $\mathbf{B}\cdot\hat{I} = *\mathbf{B}$ and, due to the commutativity of this product, as

$$*\mathbf{B} = \hat{I}\cdot\mathbf{B}\,. \tag{52}$$

This equation expresses the observation that the Hodge map depends on some unit trivector; that is, the transition from axial vectors to their dual bivectors depends on the choice of orientation in three-dimensional space.

Multiply (52) by pseudoscalar t: $t(*\mathbf{B}) = (t\hat{I})\cdot\mathbf{B}$. Due to (44) and the linearity of the Hodge map (52)

$$*(t\mathbf{B}) = (*t)\cdot\mathbf{B}\,.$$

Thus, we see that the inner product of a trivector with a pseudovector is connected via the Hodge map to the ordinary multiplication of the corresponding pseudoscalar with the same pseudovector. If we put polar vector $\mathbf{a}$ in place of pseudovector $\mathbf{B}$ in the last equation, we obtain

$$*(t\,\mathbf{a}) = (*t)\cdot\mathbf{a}\,. \tag{53}$$

One may also introduce *inner product of trivectors with bivectors and trivectors.* They are direct generalizations of Eqs. (25)

$$\hat{T}\cdot(\mathbf{a}\wedge\mathbf{b}) = (\hat{T}\cdot\mathbf{a})\cdot\mathbf{b}\,,\quad (\mathbf{a}\wedge\mathbf{b})\cdot\hat{T} = \mathbf{a}\cdot(\mathbf{b}\cdot\hat{T})\,, \tag{54}$$

$$\hat{T}\cdot(\mathbf{a}\wedge\hat{\mathbf{B}}) = (\hat{T}\cdot\mathbf{a})\cdot\hat{\mathbf{B}}\,,\quad (\mathbf{a}\wedge\hat{\mathbf{B}})\cdot\hat{T} = \mathbf{a}\cdot(\hat{\mathbf{B}}\cdot\hat{T})\,. \tag{55}$$

These expressions do not depend on the choice of factors in the outer products (see Problem 14).

As an example, let us calculate the inner square of the trivector $\widehat{T} = \mathbf{a} \wedge \widehat{\mathbf{B}}$ with $\mathbf{a} \perp \widehat{\mathbf{B}}$:

$$\begin{aligned}\widehat{T} \cdot \widehat{T} &= (\mathbf{a} \wedge \widehat{\mathbf{B}}) \cdot (\mathbf{a} \wedge \widehat{\mathbf{B}}) = [(\mathbf{a} \wedge \widehat{\mathbf{B}}) \cdot \mathbf{a}] \cdot \widehat{\mathbf{B}} = [\mathbf{a} \cdot (\mathbf{a} \wedge \widehat{\mathbf{B}})] \cdot \widehat{\mathbf{B}} \\ &= [(\mathbf{a} \cdot \mathbf{a})\widehat{\mathbf{B}}] \cdot \widehat{\mathbf{B}} = |\,\mathbf{a}\,|^2 \widehat{\mathbf{B}} \cdot \widehat{\mathbf{B}} = -|\,\mathbf{a}\,|^2 |\,\widehat{\mathbf{B}}\,|^2 = -|\,\widehat{T}\,|^2 .\end{aligned}$$

We used Eqs. (50) and (29) in these transitions. We see from this that the inner square of a trivector is nonpositive, like the inner square of a volutor. Thus, for the unit trivector we obtain $\widehat{I} \cdot \widehat{I} = -1$. For the trivector (43), we obtain

$$\widehat{I} \cdot \widehat{T} = t\,\widehat{I} \cdot \widehat{I} = -t . \tag{56}$$

It is also possible to attach a tensor of the third order to any trivector $\widehat{T}$ through an equation similar to (27):

$$T_{ijk} = -(\mathbf{e}_i \wedge \mathbf{e}_j \wedge \mathbf{e}_k) \cdot \widehat{T} . \tag{57}$$

This is also an antisymmetric tensor — that is, transposition of any two of its indices changes its sign. Out of its 27 components, only six are different from zero, namely, the six possible permutations of the numbers 1, 2 and 3, and only one of them is independent because, by virtue of (41) and (56), they are proportional to the pseudoscalar t:

$$T_{ijk} = \varepsilon_{ijk} t .$$

Employing (51) and (21), we obtain from (54)

$$\begin{aligned}&(\mathbf{e}_1 \wedge \mathbf{e}_2 \wedge \mathbf{e}_3) \cdot (\mathbf{e}_1 \wedge \mathbf{e}_2) = (\mathbf{e}_2 \wedge \mathbf{e}_3) \cdot \mathbf{e}_2 = -\mathbf{e}_3 , \\ &(\mathbf{e}_1 \wedge \mathbf{e}_2 \wedge \mathbf{e}_3) \cdot (\mathbf{e}_2 \wedge \mathbf{e}_3) = -\mathbf{e}_1 , \\ &(\mathbf{e}_1 \wedge \mathbf{e}_2 \wedge \mathbf{e}_3) \cdot (\mathbf{e}_3 \wedge \mathbf{e}_1) = -\mathbf{e}_2 .\end{aligned}$$

These three equations can be expressed as a single one:

$$\widehat{I} \cdot (\mathbf{e}_i \wedge \mathbf{e}_j) = -\varepsilon_{ijk} \mathbf{e}_k .$$

Therefore, for any trivector $\widehat{T} = t\,\widehat{I}$ and bivector B, we obtain

$$\begin{aligned}\widehat{T} \cdot \widehat{\mathbf{B}} &= t\,\widehat{I} \cdot (B_{12}\mathbf{e}_1 \wedge \mathbf{e}_2 + B_{23}\mathbf{e}_2 \wedge \mathbf{e}_3 + B_{31}\mathbf{e}_3 \wedge \mathbf{e}_1) \\ &= -t(B_3\mathbf{e}_3 + B_1\mathbf{e}_1 + B_2\mathbf{e}_2) ,\end{aligned}$$

which may be written as

$$\widehat{T} \cdot \widehat{\mathbf{B}} = -t\mathbf{B} \,. \tag{58}$$

We ascertain here that the inner product of a trivector with a bivector is connected with the ordinary multiplication of a pseudoscalar and a pseudovector — both associated, via the Hodge map, with the trivector and bivector.

§5. Multivectors and Grassmann Algebra

Vectors, bivectors and trivectors are covered by the common name of *multivectors*. Specifically, a vector is a multivector of the first rank; a bivector, of the seocnd rank; and a trivector, of the third rank. We shall also call scalars multivectors of the zeroth rank. In this way we have defined the *rank of multivectors*. Multivectors of k-th rank will be called *k-vectors*.

We are already acquainted with the outer products of vectors with vectors and bivectors. We could also introduce an outer product of a vector with a trivector; however, it must be zero in three-dimensional space, because the magnitude of the result — called a *quadrivector* — should be a four-dimensional volume. Moreover, in the equation $|\mathbf{a} \wedge \widehat{T}| = |\mathbf{a}||\widehat{T}|\sin\alpha$, where α is the angle between $\mathbf{a}$ and $\widehat{T}$, we must put $\alpha = 0$ since all vectors $\mathbf{a}$ are parallel to all trivectors $\widehat{T}$.

All the inner products defined by Eqs. (17), (21), (49), (51), (54) and (55) are expressed through the scalar product of vectors. Therefore, they can be introduced in euclidean and pseudoeuclidean spaces (that is, in linear spaces with the scalar product not necessarily positive definite). From now on we shall also call the scalar product of two vectors the *inner product of two vectors*.

Yet the ordinary multiplication of a vector by a scalar is out of the sequence of the named inner and outer products. For the sake of agreement with the properties of the products, we assume that the multiplication of vectors by scalars is their outer product, whilst the inner product of a vector and a scalar is zero.

The commutativity or anticommutativity of the inner and outer products of vectors with multivectors changes from one to the other when the rank of the multivector increases by one, which we show in Table 1. These properties can be expressed in the following two equations:

$$\mathbf{a} \wedge M_k = (-1)^k M_k \wedge \mathbf{a} \,, \quad \mathbf{a} \cdot M_k = (-1)^{k+1} M_k \cdot \mathbf{a}$$

where M_k is a k-vector. One may verify that the double multiplication of the same kind by the same vector gives zero:

$$\mathbf{a} \wedge (\mathbf{a} \wedge M) = 0 \quad \text{and} \quad \mathbf{a} \cdot (\mathbf{a} \cdot M) = 0 \tag{59}$$

where M is any multivector (see Problem 16).

Table 1

Product of vector $\mathbf{a}$ by ...	Inner product	Outer product
scalar	$\mathbf{a}\cdot\lambda = -\lambda\cdot\mathbf{a} = 0$	$\mathbf{a}\wedge\lambda = \lambda\wedge\mathbf{a}$
vector	$\mathbf{a}\cdot\mathbf{b} = \mathbf{b}\cdot\mathbf{a}$	$\mathbf{a}\wedge\mathbf{b} = -\mathbf{b}\wedge\mathbf{a}$
bivector	$\mathbf{a}\cdot\widehat{\mathbf{B}} = -\widehat{\mathbf{B}}\cdot\mathbf{a}$	$\mathbf{a}\wedge\widehat{\mathbf{B}} = \widehat{\mathbf{B}}\wedge\mathbf{a}$
trivector	$\mathbf{a}\cdot\widehat{T} = \widehat{T}\cdot\mathbf{a}$	$\mathbf{a}\wedge\widehat{T} = -\widehat{T}\wedge\mathbf{a} = 0$

The connection of multivectors with exterior forms is worth mentioning. Recall that the *exterior form of k-th degree* (or *k-form*) is the multilinear and antisymmetric mapping ω_k of k vector variables from a linear space into reals: $\mathbf{v}_1, \mathbf{v}_2, \ldots, \mathbf{v}_k \to \omega_k(\mathbf{v}_1, \mathbf{v}_2, \ldots, \mathbf{v}_k)$. Antisymmetric means that it changes sign when any two arguments are transposed. In our case of three-dimensional space, only forms of the first, second and third degree may be non-zero.

Each vector $\mathbf{a}$ determines a 1-form by the scalar product

$$\omega_1(\mathbf{v}) = \mathbf{a}\cdot\mathbf{v} .$$

And if $\mathbf{a} \neq 0$ then $\omega_1 \neq 0$, so the kernel of the mapping $\mathbf{a} \to \omega_1$ is zero.

Each bivector $\widehat{\mathbf{B}}$ determines a 2-form by the inner product with the volutor $\mathbf{v}_1 \wedge \mathbf{v}_2$:

$$\omega_2(\mathbf{v}_1, \mathbf{v}_2) = \widehat{\mathbf{B}}\cdot(\mathbf{v}_1 \wedge \mathbf{v}_2) .$$

Linearity with respect to both arguments $\mathbf{v}_1$ and $\mathbf{v}_2$ follows from the linearity of the products involved, and antisymmetry follows from the anticommutativity of the outer product. If $\widehat{\mathbf{B}} \neq 0$ then $\omega_2 \neq 0$, so the kernel of the mapping $\widehat{\mathbf{B}} \to \omega_2$ is zero.

Each trivector $\widehat{T}$ determines a 3-form by the inner product with the trivector $\mathbf{v}_1 \wedge \mathbf{v}_2 \wedge \mathbf{v}_3$:

$$\omega_3(\mathbf{v}_1, \mathbf{v}_2, \mathbf{v}_3) = \widehat{T}\cdot(\mathbf{v}_1 \wedge \mathbf{v}_2 \wedge \mathbf{v}_3) .$$

The linearity and antisymmetry follow here just as previously. The kernel of the mapping $\widehat{\mathbf{T}} \to \omega_3$ is also zero.

In this way, we have assigned a k-form to each k-vector, and this assignment is linear. The reverse statement, that a multivector corresponds to each form, follows from comparing the dimensions of the respective linear spaces (of forms and multivectors) which turn out to be equal, and from the triviality of the described kernels. Hence the obtained mapping of multivectors into exterior forms is one-to-one. Of course, it exists only if the underlying vector space is euclidean or pseudoeuclidean. In the language of coordinates and the metric tensor, the described mapping consists of lowering the indices. All considerations in this book are performed in euclidean or pseudoeuclidean spaces, so one need not use both multivectors and exterior forms and the use of multivectors is sufficient.

Now we introduce a new linear space as the direct sum of the linear spaces of scalars, vectors, bivectors and trivectors. It thus has the dimension 1+3+3+1 = 8 if we start from the three-dimensional space of vectors. We write the general element of this space as the sum of multivectors of all ranks:

$$X = \lambda + \mathbf{a} + \widehat{\mathbf{B}} + \widehat{T} \,. \tag{60}$$

In honour of the English mathematician, William Clifford (1845–1879), who considered such sums, we shall call it a *cliffor.*[1)]

Linear operations for these objects are defined as follows:

$$\begin{aligned} \mu X &= \mu\lambda + \mu\mathbf{a} + \mu\widehat{\mathbf{B}} + \mu\widehat{T} \,, \\ X_1 + X_2 &= (\lambda_1 + \lambda_2) + (\mathbf{a}_1 + \mathbf{a}_2) + (\widehat{\mathbf{B}}_1 + \widehat{\mathbf{B}}_2) + (\widehat{T}_1 + \widehat{T}_2) \,. \end{aligned} \tag{61}$$

The outer product of a vector $\mathbf{b}$ with the cliffor (60) is defined by

$$\mathbf{b} \wedge X = \mathbf{b} \wedge \lambda + \mathbf{b} \wedge \mathbf{a} + \mathbf{b} \wedge \widehat{\mathbf{B}} + \mathbf{b} \wedge \widehat{T}$$

so as to fulfill the distributivity. The outer product of a k-vector $\mathbf{b}_1 \wedge \mathbf{b}_2 \wedge \ldots \wedge \mathbf{b}_k$ and the cliffor (60) is defined by

$$(\mathbf{b}_1 \wedge \ldots \wedge \mathbf{b}_k) \wedge X = \mathbf{b}_1 \wedge [\mathbf{b}_2 \wedge \ldots \wedge (\mathbf{b}_k \wedge X) \ldots] \,,$$

which, apart from distributivity, also ensures associativity. We may treat Eq. (61) also as the outer product of a scalar with a cliffor. We define the outer product of two cliffors X and $Y = \sum_{j=0}^{3} M_j$ as

$$Y \wedge X = \sum_{j=0}^{3} M_j \wedge X \,.$$

[1)]The form has been changed to render it similar to such words as vector, tensor, spinor. Many names occur in the literature for such composition: *Clifford number* [4,5], *c-number*, [5], *Clifford aggregate* [13,14], *tensor type* [15] and *multivector* [6,7,16]. In the last case, one has to introduce also the term *pure multivector* or *homogeneous multivector* for k-vector with definite rank k.

It is associative and distributive, but generally it is neither commutative nor anticommutative.

The linear space of cliffors endowed with outer multiplication is called *Grassmann algebra* in honour of the German mathematician, Hermann Grassmann (1809–1877), who first studied outer products.

A cliffor X is called *homogeneous* if three terms in combination (60) are zero, that is, at most one term is different from zero. Each multivector M_k can be mapped into a homogeneous cliffor in which the only nonzero term is M_k. This mapping of multivectors into homogeneous cliffors is one-to-one, so we may identify the two sets. Note that in this convention, the zero cliffor is identified with the homogeneous cliffor of each rank, hence it is simultaneously a scalar, vector, bivector and trivector. Therefore, zero has no direction as a multivector and has not been given any multivector marks.

The outer product of two homogeneous cliffors is commutative or anticommutative, which may be expressed in the following universal equation

$$M_k \wedge M_l = (-1)^{kl} M_l \wedge M_k \ . \tag{62}$$

In particular, scalars and bivectors commute with all multivectors and, therefore, with all cliffors. Let us call a combination of a scalar and bivector *even cliffor*. It follows from our previous observation that even cliffors commute with all cliffors in the outer product. The sum and the outer product of even cliffors is another even cliffor, so the set of even cliffors is a subalgebra of the Grassmann algebra.

If $\lambda = 0$ and $B = 0$ in Eq. (60), such a cliffor is called *odd*. Odd cliffors anticommute with respect to the outer product.

The expression $|X|$, defined by the equality

$$|X|^2 = \lambda^2 + |\,\mathbf{a}\,|^2 + |\,\widehat{\mathbf{B}}\,|^2 + |\,\widehat{T}\,|^2 \tag{63}$$

is called the *magnitude* of the cliffor (60). One may check that it has the properties: (i) $|X| = 0$ if, and only if, $X = 0$; (ii) $|\mu X| = |\mu||X|$ for any scalar μ: and (iii) $|X_1 + X_2| \leq |X_1| + |X_2|$. Hence it has the properties of a norm in the linear space of cliffors.

For the multivector $M_k = \mathbf{b}_1 \wedge \mathbf{b}_2 \wedge \ldots \wedge \mathbf{b}_k$, we define its *reverse* $M_k^\dagger = \mathbf{b}_k \wedge \mathbf{b}_{k-1} \wedge \ldots \wedge \mathbf{b}_2 \wedge \mathbf{b}_1$. It follows that

$$M_0^\dagger = M_0 \,, \quad M_1^\dagger = M_1 \,, \quad M_2^\dagger = -M_2 \,, \quad M_3^\dagger = -M_3 \ . \tag{64}$$

For the cliffor $X = \sum_{k=0}^{3} M_k$, we define $X^\dagger = \sum_{k=0}^{3} M_k^\dagger$. Therefore, for X represented in (60), we obtain $X^\dagger = \lambda + \mathbf{a} - \widehat{\mathbf{B}} - \widehat{T}$. From the very definition of the mapping $X \to X^\dagger$, called ***reversion***, we get

$$(X \wedge Y)^\dagger = Y^\dagger \wedge X^\dagger \, .$$

§6. Clifford Algebra

We introduce a new kind of product of two vectors

$$\mathbf{ab} = \mathbf{a} \cdot \mathbf{b} + \mathbf{a} \wedge \mathbf{b} \tag{65}$$

and call it a *Clifford product.* We write it just as a juxtaposition of the symbols $\mathbf{a}$ and $\mathbf{b}$, without any symbol between them. It is ***distributive*** with respect to the addition of both factors, which follows from the same property of the inner and outer product. Moreover, it is *homogeneous*: $\lambda(\mathbf{ab}) = (\lambda\mathbf{a})\mathbf{b} = \mathbf{a}(\lambda\mathbf{b})$ for real λ. Of course, Eq. (65) indicates that the Clifford product of two vectors is an even cliffor.

Note that for $\mathbf{a} \parallel \mathbf{b}$, product (65) is commutative (since then, only the first term survives), whereas for $\mathbf{a} \perp \mathbf{b}$ it is anticommutative. Besides, for each $\mathbf{a}$, there is

$$\mathbf{a}^2 = \mathbf{a}\,\mathbf{a} = \mathbf{a} \cdot \mathbf{a} = |\,\mathbf{a}\,|^2 \, . \tag{66}$$

We likewise introduce the Clifford product of a vector with a bivector

$$\mathbf{a}\,\widehat{\mathbf{B}} = \mathbf{a} \cdot \widehat{\mathbf{B}} + \mathbf{a} \wedge \widehat{\mathbf{B}} \, , \qquad \widehat{\mathbf{B}}\,\mathbf{a} = \widehat{\mathbf{B}} \cdot \mathbf{a} + \widehat{\mathbf{B}} \wedge \mathbf{a} \, , \tag{67}$$

and with a trivector

$$\mathbf{a}\,\widehat{T} = \mathbf{a} \cdot \widehat{T} + \mathbf{a} \wedge \widehat{T} \, , \qquad \widehat{T}\,\mathbf{a} = \widehat{T} \cdot \mathbf{a} + \widehat{T} \wedge \mathbf{a} \, . \tag{68}$$

These products are also homogeneous and distributive. The expressions $\mathbf{a}\,\widehat{\mathbf{B}}$ and $\widehat{\mathbf{B}}\,\mathbf{a}$ are odd cliffors and the second terms in (68) are zero in 3-dimensional space.

The Clifford product of vector $\mathbf{b}$ with a cliffor (60) is defined so as to ensure distributivity:

$$\mathbf{b}X = \mathbf{b}\lambda + \mathbf{b}\mathbf{a} + \mathbf{b}\widehat{\mathbf{B}} + \mathbf{b}\widehat{T} \, , \quad X\mathbf{b} = \lambda\mathbf{b} + \mathbf{a}\mathbf{b} + \widehat{\mathbf{B}}\mathbf{b} + \widehat{T}\mathbf{b} \, .$$

The ordinary products of vectors by scalars are first terms on the right-hand sides of both equations. The introduced Clifford product has the properties

$$(\mathbf{a}\,\mathbf{b})\mathbf{c} = \mathbf{a}(\mathbf{b}\,\mathbf{c}) \, , \quad (\mathbf{a}\,\widehat{\mathbf{B}})\mathbf{c} = \mathbf{a}(\widehat{\mathbf{B}}\,\mathbf{c}) \, , \quad (\mathbf{a}\,\widehat{T})\mathbf{c} = \mathbf{a}(\widehat{T}\,\mathbf{c}) \tag{69}$$

(see Problem 18), which could be called associativity for the products with two vectors at extremities. Note that when factors $\mathbf{b}_1, \ldots, \mathbf{b}_k$, forming the multivector $M_k = \mathbf{b}_1 \wedge \mathbf{b}_2 \wedge \ldots \wedge \mathbf{b}_k$, are mutually perpendicular, we may write $M_k = \mathbf{b}_1\mathbf{b}_2 \ldots \mathbf{b}_k$ with Clifford products in place of the outer products, since all the inner products give zero.

Intending to ensure associativity, we now define the Clifford product of the arbitrary multivector $M_k = \mathbf{b}_1\mathbf{b}_2 \ldots \mathbf{b}_k$ represented by perpendicular factors, with cliffor X:

$$M_k X = \mathbf{b}_1(\mathbf{b}_2 \ldots (\mathbf{b}_k X) \ldots),$$
$$X M_k = (\ldots((X\mathbf{b}_1)\mathbf{b}_2) \ldots)\mathbf{b}_k .$$

Then the *Clifford product of two cliffors* $Y = \sum_{k=0}^{3} M_k$ and X is defined as

$$YX = \sum_{k=0}^{3} M_k X, \qquad XY = \sum_{k=0}^{3} X M_k .$$

A linear space of cliffors with such a product is called *Clifford algebra.* Of course, it is *associative.* The Clifford algebra of the three-dimensional Euclidean space is referred to as *Pauli algebra.*

Let us calculate, as an example, the Clifford product of two volutors $\widehat{A} = \mathbf{a}_1\mathbf{a}_2$ and $\widehat{B} = \mathbf{b}_1\mathbf{b}_2$, where $\mathbf{a}_1 \perp \mathbf{a}_2$, $\mathbf{b}_1 \perp \mathbf{b}_2$. The last assumption allows us to write $\mathbf{a}_1\mathbf{a}_2 = \mathbf{a}_1 \wedge \mathbf{a}_2$ and $\mathbf{b}_1\mathbf{b}_2 = \mathbf{b}_1 \wedge \mathbf{b}_2$ interchangeably. Now

$$\begin{aligned}
\widehat{\mathbf{A}}\,\widehat{\mathbf{B}} &= (\mathbf{a}_1\mathbf{a}_2)(\mathbf{b}_1\mathbf{b}_2) = \mathbf{a}_1[\mathbf{a}_2(\mathbf{b}_1 \wedge \mathbf{b}_2)] \\
&= \mathbf{a}_1[(\mathbf{a}_2 \cdot \mathbf{b}_1)\mathbf{b}_2 - (\mathbf{a}_2 \cdot \mathbf{b}_2)\mathbf{b}_1 + \mathbf{a}_2 \wedge \mathbf{b}_1 \wedge \mathbf{b}_2] \\
&= (\mathbf{a}_2 \cdot \mathbf{b}_1)(\mathbf{a}_1 \cdot \mathbf{b}_2) - (\mathbf{a}_2 \cdot \mathbf{b}_2)(\mathbf{a}_1 \cdot \mathbf{b}_1) + (\mathbf{a}_2 \cdot \mathbf{b}_1)\mathbf{a}_1 \wedge \mathbf{b}_2 \\
&\quad - (\mathbf{a}_2 \cdot \mathbf{b}_2)\mathbf{a}_1 \wedge \mathbf{b}_1 - (\mathbf{a}_1 \cdot \mathbf{b}_1)\mathbf{a}_2 \wedge \mathbf{b}_2 \\
&\quad + (\mathbf{a}_1 \cdot \mathbf{b}_2)\mathbf{a}_2 \wedge \mathbf{b}_1 + \mathbf{a}_1 \wedge \mathbf{a}_2 \wedge \mathbf{b}_1 \wedge \mathbf{b}_2 .
\end{aligned}$$

We obtain a combination of a scalar, bivector and quadrivector. The scalar part is merely $\widehat{\mathbf{A}} \cdot \widehat{\mathbf{B}}$, the quadrivector part is by definition $\widehat{\mathbf{A}} \wedge \widehat{\mathbf{B}}$. The remaining bivector terms form the mingled product, in accordance with (31). So we may write the result more concisely as

$$\widehat{\mathbf{A}}\,\widehat{\mathbf{B}} = \widehat{\mathbf{A}} \cdot \widehat{\mathbf{B}} + \widehat{\mathbf{A}} \dot{\wedge} \widehat{\mathbf{B}} + \widehat{\mathbf{A}} \wedge \widehat{\mathbf{B}} . \qquad (70)$$

Of course, the last term vanishes in 3-dimensional space. In the particular case when $\widehat{\mathbf{A}} = \widehat{\mathbf{B}}$, the terms $\widehat{\mathbf{B}} \dot{\wedge} \widehat{\mathbf{B}}$ and $\widehat{\mathbf{B}} \wedge \widehat{\mathbf{B}}$ are zero and

$$\widehat{\mathbf{B}}^2 = \widehat{\mathbf{B}} \cdot \widehat{\mathbf{B}} = -|\widehat{\mathbf{B}}|^2 \qquad (71)$$

remains. We see that the Clifford square of a volutor is a nonpositive real number — we have here a geometrical model of imaginary numbers.

Equation (70) shows that the Clifford product of two bivectors is an even cliffor. The scalars in the Clifford product do not change the rank of a multivector. Therefore, we may conclude from the two last observations that the product of even cliffors is another even cliffor; and this means that the set of even cliffors forms a subalgebra of the Clifford algebra.

It follows from the anticommutativity of the outer product of vectors that $\hat{\mathbf{A}} \wedge \hat{\mathbf{B}} = \hat{\mathbf{B}} \wedge \hat{\mathbf{A}}$. We know from Sec. 2 that $\hat{\mathbf{A}} \cdot \hat{\mathbf{B}} = \hat{\mathbf{B}} \cdot \hat{\mathbf{A}}$ and $\hat{\mathbf{A}} \dot{\wedge} \hat{\mathbf{B}} = -\hat{\mathbf{B}} \dot{\wedge} \hat{\mathbf{A}}$; hence the following identities:

$$\begin{aligned} \frac{1}{2}(\hat{\mathbf{A}}\hat{\mathbf{B}} + \hat{\mathbf{B}}\hat{\mathbf{A}}) &= \hat{\mathbf{A}} \cdot \hat{\mathbf{B}} + \hat{\mathbf{A}} \wedge \hat{\mathbf{B}} \,, \\ \frac{1}{2}(\hat{\mathbf{A}}\hat{\mathbf{B}} - \hat{\mathbf{B}}\hat{\mathbf{A}}) &= \hat{\mathbf{A}} \dot{\wedge} \hat{\mathbf{B}} \,, \end{aligned} \tag{72}$$

are satisfied.

It is worth noticing that the Clifford product of two trivectors is equal to their inner product; namely, for $\hat{T} = t\hat{I}$ and $\hat{Z} = z\hat{I}$ we have $\hat{T}\hat{Z} = tz\hat{I}^2 = -tz$. On the other hand, due to $\hat{I} \cdot \hat{I} = -1$, we have $\hat{T} \cdot \hat{Z} = -tz$. Hence the identity

$$\hat{T}\hat{Z} = \hat{T} \cdot \hat{Z} \tag{73}$$

is valid in three-dimensional space. In particular, the Clifford square of a trivector is a nonpositive number

$$\hat{T}^2 = -t^2 = -|\hat{T}|^2 \,. \tag{74}$$

On the grounds of the identities $\mathbf{a}\hat{T} = \mathbf{a} \cdot \hat{T}$ and $\hat{T}\mathbf{a} = \hat{T} \cdot \mathbf{a}$ (see Problem 19), one may conclude that trivectors commute with vectors in three-dimensional space, hence with all multivectors and hence with all cliffors. Let $\hat{I}$ be a unit trivector; then, of course, $\hat{I}^2 = -1$. This property, along with the commutativity of $\hat{I}$ with all cliffors, allows us to treat $\hat{I}$ as the imaginary unit $i = \sqrt{-1}$.

The unit trivector $I = \mathbf{e}_1\mathbf{e}_2\mathbf{e}_3$[1)] for the right-handed orthonormal basis $\{\mathbf{e}_1, \mathbf{e}_2, \mathbf{e}_3\}$ is helpful as an algebraic model of both Hodge maps, since

$$I\mathbf{e}_1 = \mathbf{e}_1\mathbf{e}_2\mathbf{e}_3\mathbf{e}_1 = \mathbf{e}_2\mathbf{e}_3\mathbf{e}_1\mathbf{e}_1 = \mathbf{e}_2\mathbf{e}_3 \,,$$

[1)] The letter I will be used in this book only for the unit trivector, therefore we shall omit the caret above it.

and similarly

$$I\mathbf{e}_2 = \mathbf{e}_3\mathbf{e}_1\,, \quad I\mathbf{e}_3 = \mathbf{e}_1\mathbf{e}_2$$

which is a reconstruction of relations (14). This can be comprised in the single equality $*\mathbf{e}_i = I\,\mathbf{e}_i$. Hence, the general equation

$$\hat{\mathbf{B}} = *\,\mathbf{B} = I\mathbf{B} \tag{75}$$

for the Hodge map of pseudovectors. This is simply another version of Eq. (52). On the other hand, we have Eq. (44) for the Hodge map of pseudoscalars, so jointly

$$\hat{\mathbf{B}} = *\mathbf{B} = I\mathbf{B}\,, \quad \hat{T} = *t = It\,. \tag{76}$$

We obtain reverse equations from the property $I^2 = -1$:

$$\mathbf{B} = -I\hat{\mathbf{B}}\,, \quad t = -I\hat{T}\,. \tag{77}$$

Making use of the commutativity or anticommutativity of the products collected in Table 1, we may reproduce the inner and outer products of vectors with multivectors from Eqs. (65), (67) and (68)

$$\mathbf{a}\cdot\mathbf{b} = \frac{1}{2}(\mathbf{ab}+\mathbf{ba})\,, \qquad \mathbf{a}\wedge\mathbf{b} = \frac{1}{2}(\mathbf{ab}-\mathbf{ba})\,, \tag{78}$$

$$\mathbf{a}\cdot\hat{\mathbf{B}} = \frac{1}{2}(\mathbf{a}\hat{\mathbf{B}}-\hat{\mathbf{B}}\mathbf{a})\,, \qquad \mathbf{a}\wedge\hat{\mathbf{B}} = \frac{1}{2}(\mathbf{a}\hat{\mathbf{B}}+\hat{\mathbf{B}}\mathbf{a})\,, \tag{79}$$

$$\mathbf{a}\cdot\hat{T} = \frac{1}{2}(\mathbf{a}\hat{T}+\hat{T}\mathbf{a})\,, \qquad \mathbf{a}\wedge\hat{T} = \frac{1}{2}(\mathbf{a}\hat{T}-\hat{T}\mathbf{a})\,. \tag{80}$$

The last expression is zero in three-dimensional space.

We already know the properties of multivectors expressed in Eqs. (66), (71) and (74), which may be summarized by the observation that the squares of homogeneous cliffors are scalars. Employing Eq. (64) as well, we may write

$$\mathbf{a}\,\mathbf{a}^\dagger = \mathbf{a}^\dagger\mathbf{a} = |\,\mathbf{a}\,|^2, \quad \hat{\mathbf{B}}\hat{\mathbf{B}}^\dagger = \hat{\mathbf{B}}^\dagger\hat{\mathbf{B}} = |\,\hat{\mathbf{B}}\,|^2, \quad \hat{T}\hat{T}^\dagger = \hat{T}^\dagger\hat{T} = |\,\hat{T}\,|^2,$$

or together as

$$M_k M_k^\dagger = M_k^\dagger M_k = |\,M_k\,|^2 \tag{81}$$

for k-vector M_k.

One may verify (see Problem 20) that the Clifford product has the property $(XY)^\dagger = Y^\dagger X^\dagger$ for arbitrary cliffors X and Y. Moreover, the identity

$$(X^\dagger X)_s = (XX^\dagger)_s = |X|^2 \tag{82}$$

may be established (see Problem 21), where the subscript s denotes the scalar part of a given cliffor. This is a generalization of Eq. (81).

Let us apply (82) to the product MX where M is an arbitrary multivector

$$|MX|^2 = ((MX)^\dagger MX)_s = (X^\dagger M^\dagger MX)_s \,.$$

By virtue of (81), $M^\dagger M$ is a scalar, so it commutes with $X^\dagger$ and transforms the scalar part into a scalar part

$$|MX|^2 = (M^\dagger MX^\dagger X)_s = M^\dagger M(X^\dagger X)_s = |M|^2|X|^2 \,.$$

Hence

$$|MX| = |M||X| \,. \tag{83a}$$

Similarly one derives the complementary equation

$$|XM| = |X||M| \,. \tag{83b}$$

Unfortunately, Eqs. (83) are not generally valid when the multivector M is replaced by a cliffor, as we see in the following example. Let $\mathbf{n}$ be a unit vector, then $|1+\mathbf{n}| = \sqrt{2}$, $|1-\mathbf{n}| = \sqrt{2}$, whereas $|(1+\mathbf{n})(1-\mathbf{n})| = |1-\mathbf{n}^2| = |0| = 0$. The example suggests the inequality $|YX| \leq |Y||X|$ for any cliffors Y and X. This, however, is not true as the next example proves. Let $Y = 1+\mathbf{n}$, $X = 1+\mathbf{n}$, then $|X| = |Y| = \sqrt{2}$, whereas $|YX| = |(1+\mathbf{n})^2| = |2(1+\mathbf{n})| = 2|1+\mathbf{n}| = 2\sqrt{2}$ which is greater than $|Y||X| = 2$.

The magnitude of a cliffor defined by (63) in the eight-dimensional linear space of cliffors is the same as the norm in Euclidean space R^8. Therefore, using terms of mathematical analysis, we may ascertain that the set of cliffors is a complete normed space — that is, *Banach space*. However, the set of cliffors has the structure of Clifford algebra, hence the question arises whether it, equipped with the norm (63), is a normed or Banach algebra. According to Ref. 8, the condition $|X_1X_2| \leq |X_1||X_2|$ should be satisfied for any elements X_1, X_2 of the algebra. As we have seen above, this condition, generally, is not fulfilled.

Yet one may introduce another norm in Clifford algebra. Each cliffor Y can be treated as a linear operator acting according to the prescription $Y : X \rightarrow YX$ for all cliffors X, and the following *operator norm*

$$\|Y\| = \sup_{|X|=1} |YX| \tag{84a}$$

can be defined. This expression has all the properties of the norm and additionally satisfies

$$\|Y_1 Y_2\| \leq \|Y_1\| \|Y_2\| \,, \tag{84b}$$

(see Theorem 10.2 in Ref. 8). Thus we may claim that the Clifford algebra with norm (84a) is a *normed algebra.*

For the homogeneous cliffors $X = M$ with the property $|M| = 1$, the identity $|YM| = |Y|$ holds by virtue of (83b), so the least upper bound at the right-hand side of (84a) cannot be less than $|Y|$ and hence

$$\|Y\| \geq |Y| \tag{85}$$

for each cliffor Y. Of course it follows from (83a) that $\|M\| = |M|$ for homogeneous cliffors $Y = M$.

In associative algebra, an inverse element is very useful. We call the cliffor X *invertible* if a cliffor Y exists such that

$$XY = YX = 1 \,. \tag{86}$$

Then Y is called the *inverse* of X and is denoted X^{-1}. Not all nonzero cliffors are invertible. The non-invertible cliffors include those nonzero cliffors X which, with another nonzero cliffor Z, give zero as a product: $XZ = 0$. To see this, multiply the equality from the left by Y satisfying (86) — this would give $Z = 0$. Two cliffors $X = 1 + \mathbf{n}$, $Z = 1 - \mathbf{n}$, where $\mathbf{n}$ is a unit vector, provide an example of such a situation.

Equation (81) shows that all nonzero multivectors are invertible and allows us to write

$$M_k^{-1} = \frac{M_k^\dagger}{|M_k|^2} \,. \tag{87}$$

Moreover, nonzero cliffors of the form $X = \mathbf{a}_1 \mathbf{a}_2 \ldots \mathbf{a}_{k-1} \mathbf{a}_k$ are invertible since one can easily find that $X^{-1} = \mathbf{a}_k^{-1} \mathbf{a}_{k-1}^{-1} \ldots \mathbf{a}_2^{-1} \mathbf{a}_1^{-1}$. In particular, when all vectors $\mathbf{a}_i$ are of unit magnitude, then $X^{-1} = X^\dagger = \mathbf{a}_k \mathbf{a}_{k-1} \ldots \mathbf{a}_2 \mathbf{a}_1$.

So let us take two unit vectors $\mathbf{u}$ and $\mathbf{v}$. Then

$$\mathbf{v} \cdot \mathbf{u} = \cos\theta \,, \qquad \mathbf{v} \wedge \mathbf{u} = \widehat{\mathbf{1}} \sin\theta \,,$$

where θ is the angle between $\mathbf{u}$ and $\mathbf{v}$, and $\widehat{\mathbf{1}}$ is the unit volutor with the direction of $\mathbf{v} \wedge \mathbf{u}$. In such a case, we may write

$$U = \mathbf{v}\mathbf{u} = \mathbf{v} \cdot \mathbf{u} + \mathbf{v} \wedge \mathbf{u} = \cos\theta + \widehat{\mathbf{1}} \sin\theta \,. \tag{88}$$

Similarly,

$$U^{-1} = U^{\dagger} = \mathbf{uv} = \cos\theta - \widehat{\mathbf{i}}\sin\theta \ . \tag{89}$$

Of course, both cliffors U and U^{-1} are of unit magnitude — in other words, $|U| = |U^{-1}| = 1$. In this way, we see that the Clifford product of two unit vectors gives a unit even cliffor.

We have already mentioned that the set of even cliffors forms a subalgebra in the Clifford algebra. In consists — as we know — of linear combinations of scalars and bivectors; therefore, it forms a four-dimensional linear space. We may choose the elements 1, $\mathbf{e}_1 \wedge \mathbf{e}_2$, $\mathbf{e}_2 \wedge \mathbf{e}_3$, $\mathbf{e}_3 \wedge \mathbf{e}_1$ as the basis in it, where $\{\mathbf{e}_1, \mathbf{e}_2, \mathbf{e}_3\}$ is the right-handed orthonormal basis in the underlying vector space.

Let us look more closely at the algebraic properties of the chosen basis. Denoting $\widehat{\mathbf{i}} = \mathbf{e}_1 \wedge \mathbf{e}_2$, $\widehat{\mathbf{j}} = \mathbf{e}_2 \wedge \mathbf{e}_3$, and $\widehat{k} = \mathbf{e}_1 \wedge \mathbf{e}_3$, we may also write $\widehat{\mathbf{i}} = \mathbf{e}_1\mathbf{e}_2$, $\widehat{\mathbf{j}} = \mathbf{e}_2\mathbf{e}_3$, $\widehat{\mathbf{k}} = \mathbf{e}_1\mathbf{e}_3$. The squares of these elements are $\widehat{\mathbf{i}}^2 = \widehat{\mathbf{j}}^2 = \widehat{\mathbf{k}}^2 = -1$. The other products are

$$\widehat{\mathbf{i}}\widehat{\mathbf{j}} = \mathbf{e}_1\mathbf{e}_2\mathbf{e}_2\mathbf{e}_3 = \mathbf{e}_1\mathbf{e}_3 = \widehat{\mathbf{k}} \ ,$$
$$\widehat{\mathbf{j}}\widehat{\mathbf{i}} = \mathbf{e}_2\mathbf{e}_3\mathbf{e}_1\mathbf{e}_2 = \mathbf{e}_3\mathbf{e}_1 = -\widehat{\mathbf{k}}$$

which means anticommutativity. Similarly,

$$\widehat{\mathbf{j}}\widehat{\mathbf{k}} = -\widehat{\mathbf{k}}\widehat{\mathbf{j}} = \widehat{\mathbf{i}}, \qquad \widehat{\mathbf{k}}\widehat{\mathbf{i}} = -\widehat{\mathbf{i}}\widehat{\mathbf{k}} = \widehat{\mathbf{j}} \ .$$

Thus, the product of two elements out of $\widehat{\mathbf{i}}$, $\widehat{\mathbf{j}}$ and $\widehat{\mathbf{k}}$ gives the third one, with the plus sign when the cyclic order $\widehat{\mathbf{i}} \to \widehat{\mathbf{j}} \to \widehat{\mathbf{k}} \to \widehat{\mathbf{i}}$ is preserved, or the minus sign if the order is reversed. Finally, let us calculate the product of three basic volutors:

$$\widehat{\mathbf{i}}\widehat{\mathbf{j}}\widehat{\mathbf{k}} = \mathbf{e}_1\mathbf{e}_2\mathbf{e}_2\mathbf{e}_3\mathbf{e}_1\mathbf{e}_3 = -\mathbf{e}_1\mathbf{e}_1\mathbf{e}_3\mathbf{e}_3 = -1 \ .$$

An algebra with the basis elements $1, \imath, \jmath, k$ with the above-mentioned properties is called a *quaternionic field*, and its elements are called *quaternions*. We have thus shown that the subalgebra of even cliffors in three-dimensional space is isomorphic to the quaternionic field.

The unit quaternions, to which the cliffors of the form (88) correspond, are well suited for rotations. We shall discuss this question in Sec. 8.

§7. Functions

We shall now consider functions defined on the Clifford algebra and with values in it. The power function $X \to X^m$ is one of the simplest. We observe, by virtue of Eqs. (66), (71) and (74), that, for the respective multivectors, it yields

$$\mathbf{a}^m = \begin{cases} |\,\mathbf{a}\,|^m & \text{for even } m \\ |\,\mathbf{a}\,|^{m-1}\mathbf{a} & \text{for odd } m \end{cases},$$

$$\widehat{\mathbf{B}}^m = \begin{cases} (-1)^{m/2}|\,\widehat{\mathbf{B}}\,|^m & \text{for even } m \\ (-1)^{(m-1)/2}|\,\widehat{\mathbf{B}}\,|^{m-1}\widehat{\mathbf{B}} & \text{for odd } m \end{cases},$$

$$\widehat{T}^m = \begin{cases} (-1)^{m/2}|\,\widehat{T}\,|^m & \text{for even } m \\ (-1)^{(m-1)/2}|\,\widehat{T}\,|^{m-1}\widehat{T} & \text{for odd } m \end{cases}.$$

In particular, for the unit trivector, we obtain

$$I^{2m} = (-1)^m , \qquad I^{2m+1} = (-1)^m I . \tag{90}$$

The *exponential function* $X \to e^X$ is defined by means of the power series

$$e^X = \sum_{m=0}^{\infty} \frac{X^m}{m!} = 1 + X + \frac{X^2}{2!} + \frac{X^3}{3!} + \dots \tag{91}$$

This series is absolutely convergent since, combining the inequalities (85) and (84), we obtain $|\,X^m\,| \leq \|X^m\| \leq \|\,X\,\|^m$, hence

$$\Big|\sum_{m=0}^{\infty} \frac{X^m}{m!}\Big| \leq \sum_{m=0}^{\infty} \Big|\frac{X^m}{m!}\Big| \leq \sum_{m=0}^{\infty} \frac{\|\,X\,\|^m}{m!}$$

and the series on the right-hand side, consisting of real terms, is convergent. The set of cliffors is a Banach space, so the sum of this series is a certain cliffor.

Lemma: If $XY = YX$, then $e^X e^Y = e^{X+Y}$.

The proof is the same as given in standard calculus texts for the exponential functions of a complex variable, and is based on the development of the Newton binomial for commuting elements.

We introduce the *hyperbolic sine* and *cosine functions* by means of the equations

$$\sinh X = \frac{e^X - e^{-X}}{2} , \qquad \cosh X = \frac{e^X + e^{-X}}{2} .$$

They are the odd and even parts, respectively, of the exponential function

$$e^X = \cosh X + \sinh X .$$

We define the trigonometric functions *sine* and *cosine* by the well known expansions in the power series

$$\sin X = \sum_{n=0}^{\infty} (-1)^n \frac{X^{2n+1}}{(2n+1)!}, \qquad \cos X = \sum_{n=0}^{\infty} (-1)^n \frac{X^{2n}}{(2n)!}. \tag{92}$$

If J is a cliffor with the properties $J^2 = -1, JX = XJ$, then, by substitution of these formulas, we find that

$$\cosh JX = \cos X, \qquad \sinh JX = J \sin X \tag{93}$$

and

$$e^{JX} = \cos X + J \sin X. \tag{94}$$

One may equally easily check the equations

$$\cos JX = \cosh X, \qquad \sin JX = J \sinh X.$$

If so, when I is the unit trivector and α is a scalar, we obtain

$$e^{\alpha I} = \cos \alpha + I \sin \alpha. \tag{95}$$

Similarly, for any vector a,

$$e^{\mathbf{a} I} = \cos \mathbf{a} + I \sin \mathbf{a}.$$

We obtain from (92) and the expression for $\mathbf{a}^m$: $\cos \mathbf{a} = \cos |\mathbf{a}|$, $\sin \mathbf{a} = (\mathbf{a}/|\mathbf{a}|) \sin |\mathbf{a}|$. Therefore, after denoting $\mathbf{a} = \alpha \mathbf{n}$ for a unit vector $\mathbf{n}$ and $\alpha = |\mathbf{a}|$, we have

$$e^{\alpha \mathbf{n} I} = \cos \alpha + I \mathbf{n} \sin \alpha. \tag{96}$$

Each bivector in three-dimensional space can be written as a product of some vector and I, hence (96) is an expression for the exponential function with an arbitrary bivector exponent. Equation (95) is an expression for the exponential function with an arbitrary trivector exponent. Notice that functions (95) and (96) are periodic functions with respect to the scalar variable.

Using the formula for $\mathbf{a}^m$, we obtain from (91) for the vector exponents

$$e^{\mathbf{a}} = \cosh \mathbf{a} + \sinh \mathbf{a} = \cosh |\mathbf{a}| + \frac{\mathbf{a}}{|\mathbf{a}|} \sinh |\mathbf{a}|$$

and, after representing $\mathbf{a} = \alpha \mathbf{n}$,

$$e^{\alpha \mathbf{n}} = \cosh \alpha + \mathbf{n} \sinh \alpha. \tag{97}$$

For the scalar exponent, Eq. (91) gives the ordinary exponential function of the real variable.

Making use of (94), we write cliffor (88) of the previous section as

$$U = \mathbf{v}\mathbf{u} = e^{\hat{\mathbf{i}}\theta} .$$

By virtue of the Lemma, this equality allows to write

$$U^2 = e^{\hat{\mathbf{i}}2\theta} = \cos 2\theta + \hat{\mathbf{i}} \sin 2\theta .$$

All the functions considered in this section have the property $f(X)^\dagger = f(X^\dagger)$ (see Problem 23).

§8. Projections, Reflexions and Rotations

If $\mathbf{v}$ is a unit vector, then for any other vector $\mathbf{a}$, the inner product $\mathbf{a} \cdot \mathbf{v}$ gives the scalar value of the projection of $\mathbf{a}$ onto the direction of $\mathbf{v}$. When we multiply this scalar by $\mathbf{v}$, we obtain the vector $(\mathbf{a} \cdot \mathbf{v})\mathbf{v}$, which is called the *projection* of $\mathbf{a}$ onto the straight line of $\mathbf{v}$. Notice that the scalar $\mathbf{a} \cdot \mathbf{v}$ depends on the orientation of $\mathbf{v}$, whereas the vector $(\mathbf{a} \cdot \mathbf{v})\mathbf{v}$ does not. We introduce the notation

$$\mathbf{a}_{\parallel} = (\mathbf{a} \cdot \mathbf{v})\mathbf{v} . \tag{98}$$

This vector is also called a *component of* $\mathbf{a}$ *parallel to* $\mathbf{v}$. Then the *perpendicular component* is found as $\mathbf{a}_{\perp} = \mathbf{a} - \mathbf{a}_{\parallel}$.

Let us compute $\mathbf{a} \wedge \mathbf{v}$:

$$\mathbf{a} \wedge \mathbf{v} = (\mathbf{a}_{\parallel} + \mathbf{a}_{\perp}) \wedge \mathbf{v} = \mathbf{a}_{\parallel} \wedge \mathbf{v} + \mathbf{a}_{\perp} \wedge \mathbf{v} = \mathbf{a}_{\perp} \wedge \mathbf{v} .$$

We have omitted the term $\mathbf{a}_{\parallel} \wedge \mathbf{v}$, since the outer product of parallel vectors is zero. Now, by virtue of (21),

$$(\mathbf{a} \wedge \mathbf{v}) \cdot \mathbf{v} = (\mathbf{a}_{\perp} \wedge \mathbf{v}) \cdot \mathbf{v} = \mathbf{a}_{\perp}(\mathbf{v} \cdot \mathbf{v}) = \mathbf{a}_{\perp}$$

where we have used $\mathbf{v}^2 = 1$. Vector $\mathbf{v}$ is parallel to volutor $\mathbf{a} \wedge \mathbf{v}$, so $(\mathbf{a} \wedge \mathbf{v}) \wedge \mathbf{v} = 0$. In that case, we may replace the inner product in the last formula with the Clifford product

$$\mathbf{a}_{\perp} = (\mathbf{a} \wedge \mathbf{v})\mathbf{v} . \tag{99}$$

Notice the similarity to (98). Equations (98) and (99) can be generalized to any multivector in place of $\mathbf{a}$ (see Problem 22).

Now we consider the mapping of vectors $\mathbf{a} \to \mathbf{a}' = \mathbf{vav}$ for a fixed unit vector $\mathbf{v}$. Use the representation $\mathbf{a} = \mathbf{a}_{||} + \mathbf{a}_\perp$ and the fact that $\mathbf{a}_{||}$ commutes with $\mathbf{v}$, whilst $\mathbf{a}_\perp$ anticommutes with $\mathbf{v}$

$$\mathbf{vav} = \mathbf{va}_{||}\mathbf{v} + \mathbf{va}_\perp\mathbf{v} = \mathbf{a}_{||} - \mathbf{a}_\perp = \mathbf{a}' \ .$$

We ascertain that the mapping $\mathbf{a} \to \mathbf{a}'$ is inversion with respect to the straight line on which the vector $\mathbf{v}$ lies.

A general multivector sandwiched between two vectors $\mathbf{v}$ changes into the product of the mapped vectors forming the multivector. For example, a bivector $\mathbf{a} \wedge \mathbf{b}$ transforms into

$$\begin{aligned}\mathbf{v}(\mathbf{a} \wedge \mathbf{b})\mathbf{v} &= \frac{1}{2}\mathbf{v}(\mathbf{ab} - \mathbf{ba})\mathbf{v} = \frac{1}{2}[(\mathbf{vav})(\mathbf{vbv}) - (\mathbf{vbv})(\mathbf{vav})] \\ &= \frac{1}{2}(\mathbf{a}'\mathbf{b}' - \mathbf{b}'\mathbf{a}') = \mathbf{a}' \wedge \mathbf{b}' \ .\end{aligned}$$

We used Eq. (78) twice in the transitions. In this way, one shows that any multivector transforms under the described inversion according to the formula

$$M' = \mathbf{v}M\mathbf{v} \ . \tag{100}$$

By virtue of the linearity of the transformation, we may carry this formula over to cliffors: $X' = \mathbf{v}X\mathbf{v}$.

When the mapping $\mathbf{a} \to \mathbf{a}' = \mathbf{vav}$ is followed by the next one $\mathbf{a}' \to \mathbf{a}'' = \mathbf{uau}$ with another unit vector $\mathbf{u}$, the composite mapping

$$\mathbf{a} \to \mathbf{a}'' = \mathbf{uvavu} = (\mathbf{vu})^\dagger \mathbf{avu} \tag{101}$$

is formed, that is

$$\mathbf{a}'' = U^{-1}\mathbf{a}U \tag{102}$$

where $U = \cos\theta + \widehat{\mathbf{i}}\sin\theta$ as in (88).

In order to see the geometric meaning of this mapping, it is worthwhile to break $\mathbf{a}$ down into its components: $\mathbf{a}_{||}$ parallel to $\widehat{\mathbf{i}}$ and $\mathbf{a}_\perp$ perpendicular to $\widehat{\mathbf{i}}$. Then $\mathbf{a}_{||}\widehat{\mathbf{i}} = \mathbf{a}_{||} \cdot \widehat{\mathbf{i}}$ and $\mathbf{a}_{||}$ anticommutes with $\widehat{\mathbf{i}}$, which gives

$$U^{-1}\mathbf{a}_{||} = (\cos\theta - \widehat{\mathbf{i}}\sin\theta)\mathbf{a}_{||} = \mathbf{a}_{||}(\cos\theta + \widehat{\mathbf{i}}\sin\theta) = \mathbf{a}_{||}U \ ;$$

whereas $\mathbf{a}_\perp\widehat{\mathbf{i}} = \mathbf{a}_\perp \wedge \widehat{\mathbf{i}}$ and $\mathbf{a}_\perp$ commutes with $\widehat{\mathbf{i}}$, which implies

$$U^{-1}\mathbf{a}_\perp = (\cos\theta - \widehat{\mathbf{i}}\sin\theta)\mathbf{a}_\perp = \mathbf{a}_\perp(\cos\theta - \widehat{\mathbf{i}}\sin\theta)\mathbf{a}_\perp U^{-1} \ .$$

Therefore, mapping (102) yields

$$\mathbf{a}'' = U^{-1}(\mathbf{a}_{\|} + \mathbf{a}_{\perp})U = U^{-1}\mathbf{a}_{\|}U + U^{-1}\mathbf{a}_{\perp}U = \mathbf{a}_{\|}UU + \mathbf{a}_{\perp}U^{-1}U\ ,$$
$$\mathbf{a}'' = \mathbf{a}_{\|}U^2 + \mathbf{a}_{\perp}\ . \tag{103}$$

We may transform the first term on the right-hand side:

$$\mathbf{a}_{\|}U^2 = \mathbf{a}_{\|}(\cos 2\theta + \hat{\mathbf{i}}\sin 2\theta) = \mathbf{a}_{\|}\cos 2\theta + \mathbf{a}_{\|}\hat{\mathbf{i}}\sin 2\theta\ .$$

We have already noticed that $\mathbf{a}_{\|}\mathbf{i} = \mathbf{a}_{\|}\cdot\mathbf{i}$; the vector $\mathbf{a}_{\|}\cdot\hat{\mathbf{i}}$ has the magnitude $|\,\mathbf{a}_{\|}\,|$ and — as we know from Fig. 13 — is the image of $\mathbf{a}_{\|}$ rotated by the angle $\pi/2$ in the sense of $\hat{\mathbf{i}}$. In this way, the expression $\mathbf{a}_{\|}U^2$ is the sum of $\mathbf{a}_{\|}$ with the coefficient $\cos 2\theta$ and another vector of the same length but perpendicular to $\mathbf{a}_{\|}$ and having the coefficient $\sin 2\theta$. This means that $\mathbf{a}_{\|}U^2$ is $\mathbf{a}_{\|}$ rotated in the plane of $\hat{\mathbf{i}}$ by the angle 2θ (see Fig. 22).

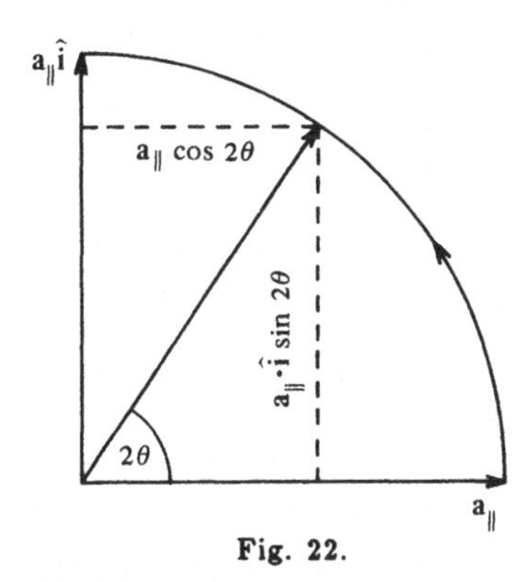

Fig. 22.

It can be seen from (103) that the component $\mathbf{a}_{\perp}$ is left invariant, hence we may consider that $\mathbf{a}''$ is the image of $\mathbf{a}$ rotated around an axis perpendicular to $\hat{\imath}$ (see Fig. 23). In this way, we have shown that mapping (101) is simply rotation by an angle two times larger than the angle between the vectors $\mathbf{v}$ and $\mathbf{u}$ in the plane determined by them and in the direction from $\mathbf{v}$ to $\mathbf{u}$. By the way, we have also shown that any rotation of vectors can be expressed as a composition of two inversions with respect to certain straight lines lying in the plane of rotation (see Fig. 24).

Similarly, a rotation of an arbitrary cliffor by angle 2θ in direction $\hat{\mathbf{i}}$ can be written as

$$X \to X'' = U^{-1}XU$$

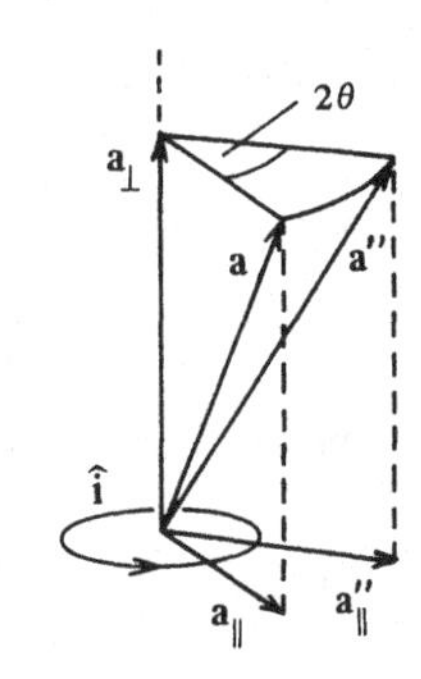

Fig. 23.

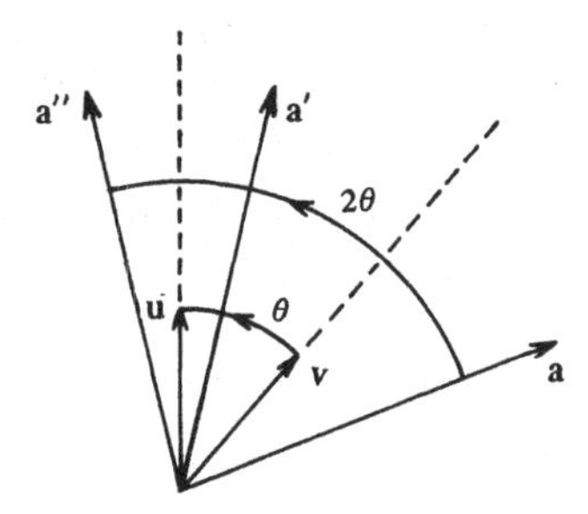

Fig. 24.

where $U = \cos\theta + \widehat{\imath}\sin\theta$. Having used formula (97), we may also write this as

$$X'' = e^{-\widehat{\mathbf{i}}\theta} X e^{\widehat{\mathbf{i}}\theta} = e^{-\widehat{\boldsymbol{\theta}}} X e^{\widehat{\boldsymbol{\theta}}} \tag{104}$$

where $\widehat{\boldsymbol{\theta}} = \widehat{\mathbf{i}}\,\theta$ is the volutor of the rotation angle.

Summing up the considerations of this section, we can state that the unit even cliffor $U = e^{\widehat{\boldsymbol{\theta}}}$, which is the Clifford product of the two unit vectors **v** and **u**, determines through (104) the rotation of cliffors by angle $2\widehat{\boldsymbol{\theta}}$ where $\widehat{\boldsymbol{\theta}}$ is the angle between **u** and **v**, directed from **v** to **u**. It obviously follows from this that $U^{-1} = U^\dagger$ describes the opposite rotation. We stress that multivectors of different ranks transform according to a common formula. The situation is different from that in tensor calculus.

Also notice the following property of transformation (104): $(X'')_s = X_s$; that is, the scalar part of a cliffor is left invariant during the rotation. Moreover,

$$\begin{aligned}(X'')^\dagger X'' &= (e^{-\widehat{\boldsymbol{\theta}}} X e^{\widehat{\boldsymbol{\theta}}})^\dagger e^{-\widehat{\boldsymbol{\theta}}} X e^{\widehat{\boldsymbol{\theta}}} = e^{-\widehat{\boldsymbol{\theta}}} X^\dagger e^{\widehat{\boldsymbol{\theta}}} e^{-\widehat{\boldsymbol{\theta}}} X e^{\widehat{\boldsymbol{\theta}}} \\ &= e^{-\widehat{\boldsymbol{\theta}}} X^\dagger X e^{\widehat{\boldsymbol{\theta}}} = (X^\dagger X)'' .\end{aligned}$$

This, in conjunction with (82), means that

$$|X''| = |X| \,. \tag{105}$$

One may also represent rotations by matrices. It is sufficient to rotate basis vectors $\mathbf{e}_i \to \mathbf{e}'_i = U^{-1}\mathbf{e}_i U$ and use the fact that $\mathbf{e}'_i$ can be expanded in the basis $\{\mathbf{e}_1, \mathbf{e}_2, \mathbf{e}_3\}$:

$$\mathbf{e}'_i = R_{ik}\mathbf{e}_k \,. \tag{106}$$

In this way, each rotation determines a matrix $R = \{R_{ik}\}$ called the *rotation matrix*. We assume the reader knows from linear algebra that it is an *orthogonal matrix* which means

$$R^T = R^{-1} \tag{107}$$

(the superscript T denotes transposition of the matrix). If we take an arbitrary vector $\mathbf{r} = x_i\mathbf{e}_i$ and rotate it $\mathbf{r} \to \mathbf{r}' = U^{-1}\mathbf{r}U$, we obtain, due to (106)

$$\mathbf{r}' = U^{-1}x_i\mathbf{e}_i U = x_i\mathbf{e}'_i = x_i R_{ik}\mathbf{e}_k \,.$$

We write this relation symbolically as

$$\mathbf{r}' = R\mathbf{r} \,. \tag{108}$$

This means that the rotated vector $\mathbf{r}'$ has, in the initial basis, the components $x'_k = x_i R_{ik} = (R^T)_{ki}x_i$. By virtue of (107), we may write this as

$$x'_k = (R^{-1})_{ki}x_i \,. \tag{109}$$

§9. Functions of a Scalar Variable

If a cliffor $F(t)$ corresponds to each value of a scalar variable t, we say that $t \to F(t)$ is *cliffor-valued function of a scalar variable*. We call the function F *continuous* at t_0 if $\lim_{t\to t_0} |F(t) - F(t_0)| = 0$. We write then $F(t_0) = \lim_{t\to t_0} F(t)$.

A *derivative* of F at the point t_0, denoted as $\dot{F}(t_0)$ or $\frac{dF}{dt}(t_0)$, is defined by the formula

$$\dot{F}(t_0) = \lim_{\Delta t\to 0} \frac{F(t_0 + \Delta t) - F(t_0)}{\Delta t} \,.$$

A function which has a derivative at each point is called *differentiable.* The derivatives of the sum and product of two differentiable cliffor-valued functions F and G satisfy the identities

$$\frac{d}{dt}(F+G) = \dot{F} + \dot{G} ,$$
$$\frac{d}{dt}(FG) = \dot{F}G + F\dot{G} . \tag{110}$$

The proof of the second equality differs from the proof of the well-known theorem of differential calculus in that one has to preserve the order of the differentiated factors. Specifically, we get

$$\frac{dF^2}{dt} = \dot{F}F + F\dot{F} . \tag{111}$$

The right-hand side equals the classical result $2F\dot{F}$ only if F commutes with $\dot{F}$.

In order to understand the meaning of this modification, let us consider a function $\mathbf{v}$ which may denote, e.g., the velocity of a particle. Equation (111) then gives

$$\frac{d\mathbf{v}^2}{dt} = \dot{\mathbf{v}}\mathbf{v} + \mathbf{v}\dot{\mathbf{v}} = 2\mathbf{v}\cdot\dot{\mathbf{v}} . \tag{112}$$

On the other hand, $\mathbf{v}^2 = v^2$, where $v = |\mathbf{v}|$ is a scalar, so

$$\frac{d\mathbf{v}^2}{dt} = \frac{dv^2}{dt} = 2v\dot{v} . \tag{113}$$

Introduce a unit vector $\mathbf{n}$ by the formula $\mathbf{v} = v\mathbf{n}$. When the parameter t changes, the vector $\mathbf{v}$ may change its magnitude v and direction $\mathbf{n}$. If the magnitude v is constant, Eq. (113) gives $\frac{d\mathbf{v}^2}{dt} = 0$, hence, due to (112), $\mathbf{v}\cdot\dot{\mathbf{v}} = 0$ and we see that a vector changing only the direction is perpendicular to its derivative. When — on the other hand — the direction n is constant, then $\dot{\mathbf{v}} = \dot{v}\mathbf{n}$ and Eq. (112) reduces to

$$\frac{d\mathbf{v}^2}{dt} = 2v\dot{v} = 2\mathbf{v}\dot{\mathbf{v}} = 2\mathbf{v}\cdot\dot{\mathbf{v}} .$$

Thus, expression (111) need not be equal to $2F\dot{F}$ if F changes its direction, but it is equal to $2F\dot{F}$ when F changes only its magnitude.

Lemma. The vector-valued function $t \to \mathbf{v}(t)$ has a constant magnitude if and only if a bivector-valued function $t \to \widehat{\mathbf{\Omega}}(t)$ exists such that

$$\dot{\mathbf{v}} = \mathbf{v}\cdot\widehat{\mathbf{\Omega}} . \tag{114}$$

Proof of the sufficient condition. Let $\dot{\mathbf{v}} = \mathbf{v} \cdot \widehat{\mathbf{\Omega}}$, then $\mathbf{v} \cdot \dot{\mathbf{v}} = \mathbf{v} \cdot (\mathbf{v} \cdot \widehat{\mathbf{\Omega}}) = 0$ by virtue of (59). In that case, due to (112), $\frac{d\mathbf{v}^2}{dt} = 0$ and, due to (113), $\dot{v} = 0$, that is $|\mathbf{v}| = \text{const}$.

Proof of the necessary condition. Let v be constant. For $\mathbf{v} = 0$, of course, $\widehat{\mathbf{\Omega}} = 0$ satisfies (114). So assuming $\mathbf{v} \neq 0$, we may consider $\mathbf{v}^{-1}$ in the sense of Sec. 6. For constant v, we have $\mathbf{v} \cdot \dot{\mathbf{v}} = 0$, due to (113) and (112). By virtue of (17), we then obtain $\mathbf{v} \cdot (\mathbf{v}^{-1} \wedge \dot{\mathbf{v}}) = \dot{\mathbf{v}}$, which means that (114) is satisfied by $\widehat{\mathbf{\Omega}} = \mathbf{v}^{-1} \wedge \dot{\mathbf{v}}$. Is it the only bivector satisfying (114)? Represent

$$\widehat{\mathbf{\Omega}}' = \mathbf{v}^{-1} \wedge \dot{\mathbf{v}} + \widehat{\mathbf{B}} \,.$$

Of course, the condition $\mathbf{v} \cdot \widehat{\mathbf{B}} = 0$ must be satisfied, which means that $\widehat{\mathbf{B}}$ is perpendicular to $\mathbf{v}$. Hence the bivectorial solution of Eq. (114) is not unique.■

Let us now discuss the exponential function $F : t \to e^{At}$ for a certain constant cliffor A. We obtain from definition (91):

$$\frac{d}{dt}e^{At} = \sum_{m=0}^{\infty} \frac{d}{dt}\frac{A^m t^m}{m!} = \sum_{m=0}^{\infty} \frac{mA^m t^{m-1}}{m!} = \sum_{m=1}^{\infty} \frac{A^m t^{m-1}}{(m-1)!}$$
$$= A\sum_{m=1}^{\infty} \frac{A^{m-1} t^{m-1}}{(m-1)!} = A\sum_{n=0}^{\infty} \frac{A^n t^n}{n!} \,,$$

that is,

$$\frac{d}{dt}e^{At} = A\, e^{At} \,. \tag{115}$$

Since A commutes with this exponential function, we may also write

$$\frac{d}{dt}e^{At} = e^{At} A \,. \tag{116}$$

That means, for this function, that $\frac{dF}{dt} = A\,F = F\,A$.

By virtue of Eq. (104) and the Lemma, the uniform rotation of a vector $\mathbf{v}$ with constant angular velocity $\widehat{\omega}$ may be expressed as the function

$$t \to \mathbf{v}(t) = e^{-\frac{1}{2}\widehat{\omega}t}\mathbf{v}\, e^{\frac{1}{2}\widehat{\omega}t} \,. \tag{117}$$

Its derivative, according to (110), is

$$\frac{d}{dt}\mathbf{v}(t) = \frac{d}{dt}\left(e^{-\frac{1}{2}\widehat{\omega}t}\right)\mathbf{v}\, e^{\frac{1}{2}\widehat{\omega}t} + e^{-\frac{1}{2}\widehat{\omega}t}\mathbf{v}\frac{d}{dt}e^{\frac{1}{2}\widehat{\omega}t} \,.$$

Now we use (115) and (116):

$$\dot{\mathbf{v}}(t) = -\frac{1}{2}\widehat{\omega}\, e^{-\frac{1}{2}\widehat{\omega}t}\mathbf{v}\, e^{\frac{1}{2}\widehat{\omega}t} + e^{-\frac{1}{2}\widehat{\omega}t}\mathbf{v}\, e^{\frac{1}{2}\widehat{\omega}t}\frac{1}{2}\widehat{\omega}$$
$$= -\frac{1}{2}\widehat{\omega}\mathbf{v}(t) + \frac{1}{2}\mathbf{v}(t)\widehat{\omega}\,.$$

Hence, by virtue of the first Eq. (79),

$$\dot{\mathbf{v}}(t) = \mathbf{v}(t) \cdot \widehat{\omega}\,. \tag{118}$$

We obtain a special case of formula (114) with the constant bivector $\widehat{\Omega} = \widehat{\omega}$.

Similarly, for the uniform rotation of a volutor $\widehat{\mathbf{M}}$,

$$t \to \widehat{\mathbf{M}}(t) = e^{-\frac{1}{2}\widehat{\omega}t}\widehat{\mathbf{M}}\, e^{\frac{1}{2}\widehat{\omega}t}$$

the same calculations give

$$\dot{\widehat{\mathbf{M}}}(t) = -\frac{1}{2}\widehat{\omega}\,\widehat{\mathbf{M}}(t) + \frac{1}{2}\widehat{\mathbf{M}}(t)\widehat{\omega}\,.$$

Now we use (72):

$$\dot{\widehat{\mathbf{M}}}(t) = \widehat{\mathbf{M}}(t)\dot{\wedge}\widehat{\omega}\,. \tag{119}$$

It is a differential equation satisfied by the uniformly rotating volutor.

§10. Nabla Operator

The following expression, built of partial derivatives and vectors of an orthonormal basis in Cartesian coordinates, is called a *nabla operator*:

$$\nabla = \mathbf{e}_1\frac{\partial}{\partial x_1} + \mathbf{e}_2\frac{\partial}{\partial x_2} + \mathbf{e}_3\frac{\partial}{\partial x_3}\,.$$

It has meaning only when acting on differentiable functions of three variables x_1, x_2, x_3, which stand for the Cartesian coordinates x, y and z. We call k-vector-valued functions *k-vector fields*. We may obtain different fields by action of the nabla on various fields.

1. When acting on a scalar field φ, the nabla gives only a vector field

$$\nabla\varphi = \mathbf{e}_1\frac{\partial\varphi}{\partial x_1} + \mathbf{e}_2\frac{\partial\varphi}{\partial x_2} + \mathbf{e}_3\frac{\partial\varphi}{\partial x_3}\,.$$

It is called *gradient* and denoted gradφ. We may also write it in our notation as $\nabla \wedge \varphi$.

2. When acting on a vector field $\mathbf{A}$, the nabla may give:
(i) a scalar field

$$\nabla \cdot \mathbf{A} = \frac{\partial A_1}{\partial x_1} + \frac{\partial A_2}{\partial x_2} + \frac{\partial A_3}{\partial x_3}$$

called *divergence* and denoted div$\mathbf{A}$. We shall call it the *inner derivative* of field $\mathbf{A}$;
(ii) a bivector field

$$\nabla \wedge \mathbf{A} = \left(\frac{\partial A_2}{\partial x_1} - \frac{\partial A_1}{\partial x_2}\right)\mathbf{e}_1 \wedge \mathbf{e}_2 + \left(\frac{\partial A_3}{\partial x_2} - \frac{\partial A_2}{\partial x_3}\right)\mathbf{e}_2 \wedge \mathbf{e}_3 + \left(\frac{\partial A_1}{\partial x_3} - \frac{\partial A_3}{\partial x_1}\right)\mathbf{e}_3 \wedge \mathbf{e}_1$$

which we call the *outer derivative* of $\mathbf{A}$. Notice that the bivector $\nabla \wedge \mathbf{A}$ is dual to the pseudovector $\nabla \times \mathbf{A}$, known as the *curl* of $\mathbf{A}$. We may thus write

$$\nabla \wedge \mathbf{A} = I(\nabla \times \mathbf{A}) \ . \tag{120}$$

3. When acting on a bivector field $\widehat{\mathbf{B}}$, the nabla may give:
(i) a vector field

$$\begin{aligned}\nabla \cdot \widehat{\mathbf{B}} &= \left(\frac{\partial B_{31}}{\partial x_3} + \frac{\partial B_{21}}{\partial x_2}\right)\mathbf{e}_1 + \left(\frac{\partial B_{12}}{\partial x_1} + \frac{\partial B_{32}}{\partial x_3}\right)\mathbf{e}_2 + \left(\frac{\partial B_{23}}{\partial x_2} + \frac{\partial B_{13}}{\partial x_1}\right)\mathbf{e}_3 \\ &= \frac{\partial B_{kl}}{\partial x_k}\mathbf{e}_l\end{aligned}$$

(the summation convention is understood), which we call the *inner derivative* of $\widehat{\mathbf{B}}$. In analogy to (24), the identity

$$\nabla \cdot \widehat{\mathbf{B}} = -\nabla \times \mathbf{B} \tag{121}$$

holds, which is visible after rewriting the previous formula in the form

$$\nabla \cdot \widehat{\mathbf{B}} = \left(\frac{\partial B_2}{\partial x_3} - \frac{\partial B_3}{\partial x_2}\right)\mathbf{e}_1 + \left(\frac{\partial B_3}{\partial x_1} - \frac{\partial B_1}{\partial x_3}\right)\mathbf{e}_2 + \left(\frac{\partial B_1}{\partial x_2} - \frac{\partial B_2}{\partial x_1}\right)\mathbf{e}_3 \ ;$$

(ii) a trivector field

$$\nabla \wedge \widehat{\mathbf{B}} = \left(\frac{\partial B_{12}}{\partial x_3} + \frac{\partial B_{23}}{\partial x_1} + \frac{\partial B_{31}}{\partial x_2}\right)\mathbf{e}_1 \wedge \mathbf{e}_2 \wedge \mathbf{e}_3$$

which we call the *outer derivative* of $\widehat{\mathbf{B}}$. Notice that the equality $\nabla \wedge \widehat{\mathbf{B}} = I \operatorname{div} \mathbf{B}$ is satisfied — it may also be written as

$$\nabla \wedge (I\mathbf{B}) = I(\nabla \cdot \mathbf{B}) \ . \tag{122}$$

4. By action of the nabla on a trivector field $\widehat{T} = tI$ in three-dimensional space, one may obtain only the bivector field

$$\nabla \cdot \widehat{T} = \frac{\partial t}{\partial x_1}\mathbf{e}_2 \wedge \mathbf{e}_3 + \frac{\partial t}{\partial x_2}\mathbf{e}_3 \wedge \mathbf{e}_1 + \frac{\partial t}{\partial x_3}\mathbf{e}_1 \wedge \mathbf{e}_2$$

which should be called the *inner derivative* of $\widehat{T}$. We may easily see the equality $\widehat{T} = I\,\mathrm{grad}\,t$, which may also be written as

$$\nabla \cdot (I\,t) = I\,(\nabla \wedge t)\ . \tag{123}$$

One may also consider the outer derivative $\nabla \wedge \widehat{T}$ but we have to take it as equal to zero. Similarly $\nabla \cdot \varphi = 0$ for the scalar field φ.

Equations (120)–(123) show that all new derivatives of a multivector field in 3-dimensional space can be reduced to three operators of classical vector calculus: gradient, curl and divergence. This explains why physicists have for so long dispensed with inner and outer derivatives. It was done, of course, at the expense of introducing pseudoscalar and pseudovector fields.

A remark to readers familiar with differential geometry: We mentioned in Sec. 5 the one-to-one correspondence between multivectors and exterior forms. Now we should add that within this relation, exterior differentials d and interior differentials $\delta = *d*$ of the differential forms correspond to the outer and inner derivatives of the multivector fields. More precisely: if a differential k-form ω is related to the k-vector field M, then the $(k+1)$-form $d\omega$ is related to the $(k+1)$-vector field $\nabla \wedge M$, whilst the $(k-1)$-form $\delta\omega$ is related to the $(k-1)$-vector field $\nabla \cdot M$.

The identities

$$\nabla \wedge (\nabla \wedge M) = 0 \quad \text{and} \quad \nabla \cdot (\nabla \cdot M) = 0 \tag{124}$$

for an arbitrary twice-differentiable multivector field M may be verified for the nabla in the same way as Eqs. (59) for vectors (see Problem 25). It follows from this and from (120) that

$$0 = \nabla \wedge (\nabla \wedge \varphi) = I[\nabla \times (\nabla \varphi)]\ ,$$

i.e., curl (grad φ) = 0. Moreover, it follows from (124), (120) and (122) that

$$0 = \nabla \wedge (\nabla \wedge \mathbf{A}) = I[\nabla \cdot (\nabla \times \mathbf{A})]\ ,$$

i.e., div (curl $\mathbf{A}$) = 0.

While contemplating identities (124), a natural question arises: Is there a multivector field N such that for the multivector field M with the property

$\nabla \wedge M = 0$, the equality $M = \nabla \wedge N$ occurs? The answer to this question is in some cases positive. Before formulating it, we give a definition. Now, an open region Ω in R^3 is called *contractible to a point* $\mathbf{x}_0 \in \Omega$ if there exists a continuous mapping $h : \Omega \times \langle 0, 1\rangle \to \Omega$ with continuous partial derivatives such that for each $\mathbf{x} \in \Omega$, the conditions $h(\mathbf{x}, 0) = \mathbf{x}_0$, $h(\mathbf{x}, 1) = \mathbf{x}$ hold. In other words: any point of Ω can be connected with $\mathbf{x}_0$ by a continuous curve lying in Ω with continuous tangent vector, and the family of such curves can be defined in a continuous and differentiable manner.

Poincaré Lemma. If Ω is a region contractible to a point, then for each k-vector field M defined in Ω with the property $\nabla \wedge M = 0$, a $(k-1)$-vector field N exists such that $M = \nabla \wedge N$.

The proof of this Lemma in terms of the differential forms can be found in Ref. 9.

When M is vector field $\mathbf{F}$, the condition $\nabla \wedge M = 0$ means curl $\mathbf{F} = 0$. Then the Poincaré Lemma refers to the existence of a scalar field U such that $\mathbf{F} = \text{grad}U$. This version of the Lemma is well-known in mechanics and corresponds to the fact that the conservative force field has a potential.

We may also define the outer and inner derivatives of the arbitrary differentiable cliffor field $X = \sum_{k=0}^{3} M_k$ according to

$$\nabla \wedge X = \sum_{k=0}^{3} \nabla \wedge M_k \quad \text{and} \quad \nabla \cdot X = \sum_{k=0}^{3} \nabla \cdot M_k \ .$$

As one may check, they are linear operators. Then the following counterparts of Eqs. (124) are fulfilled for the cliffor fields

$$\nabla \wedge (\nabla \wedge X) = 0 \quad \text{and} \quad \nabla \cdot (\nabla \cdot X) = 0 \ . \tag{125}$$

Now we define the *Clifford derivative* of the cliffor field

$$\nabla X = \nabla \cdot X + \nabla \wedge X \ .$$

By simple inspection with the use of the commutativity of the partial derivatives for twice differentiable cliffor fields, we may assure ourselves that the Clifford square of the nabla operator is

$$\nabla^2 = \nabla\nabla = \nabla \cdot \nabla = \frac{\partial^2}{(\partial x_1)^2} + \frac{\partial^2}{(\partial x_2)^2} + \frac{\partial^2}{(\partial x_3)^2} \ .$$

We denote it Δ, and call it the *Laplace operator* of *Laplacian*. This result also means that $\nabla \wedge \nabla = \frac{1}{2}(\nabla\nabla - \nabla\nabla) = 0$ as an operator acting on twice-differentiable cliffor field. The Laplacian is a scalar operator, hence when acting on a multivector field, it does not change the rank of the field.

Let us calculate the Clifford square of the nabla using associativity in the Clifford algebra:

$$\begin{aligned}\nabla^2 X &= \nabla(\nabla X) = \nabla(\nabla \cdot X + \nabla \wedge X)\\ &= \nabla \cdot (\nabla \cdot X) + \nabla \cdot (\nabla \wedge X) + \nabla \wedge (\nabla \cdot X) + \nabla \wedge (\nabla \wedge X)\,.\end{aligned}$$

The first and last term vanish by virtue of (125). We have thus proven the useful identity:

$$\nabla \cdot (\nabla \wedge X) + \nabla \wedge (\nabla \cdot X) = \Delta X \tag{126}$$

valid for the arbitrary twice-differentiable cliffor field X.

The equality $\Delta = \nabla^2$ can be expressed in the words: the Clifford derivative is a square root of the Laplacian. This is a novelty in comparison with classical vector calculus, in which none of the operators — gradient, curl nor divergence — has such a property. One may meet the equality $\Delta = \nabla^2$ in classical vector calculus but it is valid only for scalar fields, where it is a superposition of gradient with divergence; therefore, not the square of a single operation.

We shall also need the following cliffor operator

$$\nabla + \frac{1}{u}\frac{\partial}{\partial t} \tag{127}$$

acting on cliffor fields depending additionally on time t, where u is a constant with the dimension of velocity. Its Clifford product with another similar operator $\nabla - \frac{1}{u}\frac{\partial}{\partial t}$ gives

$$\left(\nabla + \frac{1}{u}\frac{\partial}{\partial t}\right)\left(\nabla - \frac{1}{u}\frac{\partial}{\partial t}\right) = \nabla^2 - \frac{1}{u^2}\frac{\partial^2}{\partial t^2}\,. \tag{128}$$

The operator on the right-hand side is called the *d'Alembert operator* or *Dalambertian*, and is denoted as $\Box$. The product in reversed order yields the same result:

$$\left(\nabla - \frac{1}{u}\frac{\partial}{\partial t}\right)\left(\nabla + \frac{1}{u}\frac{\partial}{\partial t}\right) = \Delta - \frac{1}{u^2}\frac{\partial^2}{\partial t^2} = \Box\,. \tag{129}$$

The Dalambertian, like the Laplacian, is a scalar operator.

Now we shall discuss the behaviour of the nabla operator under rotations. We first need to consider the behaviour of partial derivatives. Let f be an arbitrary cliffor-valued function of position $\mathbf{r} \to f(\mathbf{r})$, which may also be treated as a function of the three Cartesian coordinates $f(x_1, x_2, x_3)$. Our principal assumption is that for a rotation written for basis vectors according to (106) as $\mathbf{e}'_i = U^{-1}\mathbf{e}_i$, $U = R_{ik}\mathbf{e}_k$, the rotation of the function $f \to f'$ is given by the equality $f'(\mathbf{r}') = U^{-1}f(\mathbf{r})U$. Since the position $\mathbf{r}$ is arbitrary, we may substitute $\mathbf{r} \to R^{-1}\mathbf{r}$ (this implies substitution $\mathbf{r}' \to \mathbf{r}$) and obtain

$$f'(\mathbf{r}) = U^{-1}f(R^{-1}\mathbf{r})U \,. \tag{130}$$

By virtue of (109) we write $f'(x_1, x_2, x_3) = U^{-1}f(R_{1j}x_j, R_{2j}x_j, R_{3j}x_j)U$. The cliffor U does not depend on position, hence the partial derivatives are

$$\frac{\partial f'}{\partial x_k} = U^{-1}\frac{\partial f}{\partial x_i}\frac{\partial(R_{ij}x_j)}{\partial x_k}U = R_{ik}U^{-1}\frac{\partial f}{\partial x_i}U \,.$$

Now we may go on to the nabla operator acting on the rotated function

$$\nabla f' = \mathbf{e}_k\frac{\partial f'}{\partial x_k} = \mathbf{e}_k R_{ik}U^{-1}\frac{\partial f}{\partial x_i}U \,.$$

We use (106)

$$\nabla f' = \mathbf{e}'_i U^{-1}\frac{\partial f}{\partial x_i}U = U^{-1}\mathbf{e}_i U U^{-1}\frac{\partial f}{\partial x_i}U = U^{-1}\mathbf{e}_i\frac{\partial f}{\partial x_i}U \,.$$

In this way, we have obtained

$$\nabla f' = (\nabla f)' \,. \tag{131}$$

After adding the arguments, it can be written as

$$\nabla f'(\mathbf{r}) = U^{-1}(\nabla f)(R^{-1}\mathbf{r})U \,.$$

§11. Integration

We shall use the common name *manifold* for curves, surfaces or spatial regions in three-dimensional space. In the case of curves and surfaces, we assume that the functions occuring in their parametric equations are *piecewise smooth* — that is, arbitrary many times differentiable on finite number of open regions which cover almost all manifold. The set of all vectors tangent to the manifold at a given point (of the smooth region) is called *tangent space* and its dimension is called the *dimension of the manifold.* It is obvious that the dimension of a curve is one; of surface, two; and of a spatial region, three.

We choose, at each point of a smooth segment of a curve, a unit tangent vector in such a way that the mapping of the point into that vector is continuous on the segment. In this way, the *orientation* of the segment is given.

It is equivalent to putting an arrow on that segment of the curve. At the points where two smooth segments merge, the arrows on both sides should be compatible, that is, when one is incoming to the point, the other should be outgoing from it. In this way, an *orientation* is given to piecewise smooth curve.

Similarly, assigning continuously a unit tangent volutor to each point of a smooth fragment of a surface gives an orientation to that fragment. At the points where two smooth fragments merge, the unit volutors on both sides of the mergence line should be compatible, like the faces of the polyhedron shown in Fig. 18 (compare Problem 1). In this way, an *orientation* is given to piecewise smooth surface.[1)]

The orientation of a spatial region is one unit trivector; usually, we shall choose the right-handed unit trivector I.

Let n be the multivector function defining the orientation of a given manifold V; let v be the positive measure for the manifold (for a curve it is length, for a surface it is area and for a spatial region it is volume). Let f be a cliffor-valued field defined on some region of space containing V. We introduce the expression $dV = ndv$, called a *directed measure*, and define two types of integrals

$$\int_V dV \cdot f = \int_V dv(n \cdot f) \tag{132}$$

and

$$\int_V dV \wedge f = \int_V dv(n \wedge f) \tag{133}$$

where, on the right-hand side, we have the inner product $n \cdot f$ or the outer product $n \wedge f$ at respective points of the manifold, and the integral is to be understood as the Lebesgue or Riemann integral of the cliffor function over the real measure v. We call Eq. (132) *adherence* of the field f to the manifold V and Eq. (133) *protrudence* of f from V.

In particular, when the manifold V is a curve C and the cliffor field f is a force field $\mathbf{F}$, then $dv = dl$, $dV = \mathbf{n}\,dl = d\mathbf{l}$ and the integral (132) — i.e., $\int_C d\mathbf{l} \cdot \mathbf{F}$ — is the work done along C by the force field.

By dint of the Hodge map, many of these integrals may be connected with integrals of (pseudo)scalar and (pseudo)vector fields known from classical vector calculus. For instance, we have for the protrudence of a vector field **a**

[1)]Such an orientation does not exist for all surfaces in three-dimensional space. When an orientation does not exist for a given surface, it is called a non-orientable surface — a famous example is the Möbius strip. These will not be treated in this book.

from a curve C, due to Eqs. (13) and (75),

$$\int_C d\mathbf{l} \wedge \mathbf{a} = I \int_C d\mathbf{l} \times \mathbf{a} \, .$$

For the adherence of a bivector field $\widehat{\mathbf{B}}$ to a curve C, due to (24), we may write

$$\int_C d\mathbf{l} \cdot \widehat{\mathbf{B}} = - \int_C d\mathbf{l} \times \mathbf{B} \, ,$$

where $\widehat{\mathbf{B}} = I\mathbf{B}$. For the protrudence of $\widehat{\mathbf{B}}$ from C, we use Eq. (46)

$$\int_C d\mathbf{l} \wedge \widehat{\mathbf{B}} = I \int_C d\mathbf{l} \cdot \mathbf{B} \, . \tag{134}$$

Finally, the adherence of a trivector field $\widehat{T} = It$ to a curve C may be transformed with the use of (53):

$$\int_C d\mathbf{l} \cdot \widehat{T} = I \int_C t \, d\mathbf{l} \, .$$

One may act similarly with surface integrals. So by virtue of (24), we have for a surface S

$$\int_S d\widehat{\mathbf{s}} \cdot \mathbf{a} = - \int_S d\mathbf{s} \times \mathbf{a}$$

where $d\widehat{\mathbf{s}} = I \, d\mathbf{s}$ and $d\mathbf{s}$ is the traditionally used vector element of integration normal to the surface. Moreover, due to (46), we have

$$\int_S d\widehat{\mathbf{s}} \wedge \mathbf{a} = I \int_S d\mathbf{s} \cdot \mathbf{a} \tag{135}$$

and, due to (30),

$$\int_S d\widehat{\mathbf{s}} \cdot \widehat{\mathbf{B}} = - \int_S d\mathbf{s} \cdot \mathbf{B} \, . \tag{136}$$

Finally, by virtue of (58), we obtain

$$\int_S d\widehat{\mathbf{s}} \cdot \widehat{T} = - \int_S t \, d\mathbf{s} \, .$$

To wind up, we give similar formulae for the volume integrals over a region Ω, where only adherences may occur. By virtue of (53), (58) and (56), respectively, we have

$$\int_\Omega d\widehat{V} \cdot \mathbf{a} = I \int_\Omega dv \, \mathbf{a} \, , \quad \int_\Omega d\widehat{V} \cdot \widehat{\mathbf{B}} = - \int_\Omega dv \mathbf{B} \, , \quad \int_\Omega d\widehat{V} \cdot \widehat{T} = - \int_\Omega dv \, t \, ,$$

where $d\widehat{V} = Idv$.

We see that all curvilinear, surface and volume integrals of the two types may be reduced to classical integrals of the (pseudo)scalar and (pseudo)vector fields with known products between them. This clarifies why physicists dispensed with multivector integrals in three-dimensional space.

It is also proper to find counterparts of important theorems connecting curvilinear, surface and volume integrals.

The *Stokes theorem* applies to a surface S bounded by a directed closed curve (loop) C, where the orientation of C is related to the orientation of the normal vector ds by the rule of right-hand screw (see Fig. 25). Then, according to the theorem, the following identity

$$\int_C d\mathbf{l} \cdot \mathbf{A} = \int_S d\mathbf{s} \cdot (\nabla \times \mathbf{A})$$

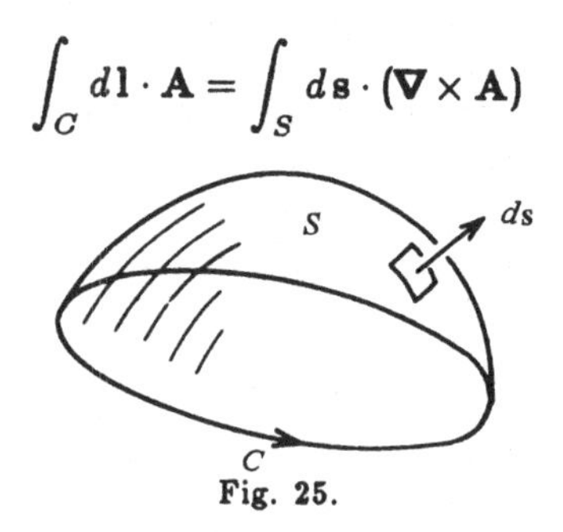

Fig. 25.

holds for a differentiable vector or pseudovector field $\mathbf{A}$. We introduce the notation $C = \partial S$ which should be read: The loop C is a boundary of the surface S. Then the last equality is written as

$$\int_{\partial S} d\mathbf{l} \cdot \mathbf{A} = \int_S d\mathbf{s} \cdot (\nabla \times \mathbf{A}) . \tag{137}$$

One should inscribe it with the use of $d\widehat{\mathbf{s}}$ in place of $d\mathbf{s}$. The volutor $d\widehat{\mathbf{s}}$ is more intimately connected with the surface than the vector normal to it. Moreover, the two-dimensional orientation of S is automatically connected with the orientation of its boundary ∂S, as the orientation of the volutor from Fig. 1 is connected to the orientation of its boundary. And one need not adduce any screw.

Further procedure depends on whether the field $\mathbf{A}$ is a vector or pseudovector field. If $\mathbf{A}$ is a pseudovector $\mathbf{H}$, we apply (121) to $\nabla \times \mathbf{H}$ and obtain

$$\int_{\partial S} d\mathbf{l} \cdot \mathbf{H} = -\int_S d\mathbf{s} \cdot (\nabla \cdot \widehat{\mathbf{H}}) .$$

We multiply both sides by I and apply Eqs. (134) and (135)

$$\int_{\partial S} d\mathbf{l} \wedge \widehat{\mathbf{H}} = -\int_S d\widehat{\mathbf{s}} \wedge (\boldsymbol{\nabla} \cdot \widehat{\mathbf{H}}) \,. \tag{138}$$

If, instead, $\mathbf{A}$ is a vector $\mathbf{E}$, one need only change the right hand side of (137). We know that $\boldsymbol{\nabla} \wedge \mathbf{E}$ is dual to $\boldsymbol{\nabla} \times \mathbf{E}$ and $d\widehat{\mathbf{s}}$ is dual to $d\mathbf{s}$, so by virtue of (30), we have $d\mathbf{s} \cdot (\boldsymbol{\nabla} \times \mathbf{E}) = -d\widehat{\mathbf{s}} \cdot (\boldsymbol{\nabla} \wedge \mathbf{E})$. Therefore, Eq. (137) may be written as

$$\int_{\partial S} d\mathbf{l} \cdot \mathbf{E} = -\int_S d\widehat{\mathbf{s}} \cdot (\boldsymbol{\nabla} \wedge \mathbf{E}) \,. \tag{139}$$

Equations (138) and (139) are the counterparts of the Stokes theorem we were looking for.

The *Gauss theorem* concerns a spatial region Ω bounded by a closed surface $\partial\Omega$ when the normal vector $d\mathbf{s}$ is directed outwards to $\partial\Omega$. (The volutor $d\widehat{\mathbf{s}}$ dual to $d\mathbf{s}$ then has counterclockwise orientation if the surface is looked at from the outside.) According to the Gauss theorem, the identity

$$\int_{\partial\Omega} d\mathbf{s} \cdot \mathbf{A} = \int_\Omega dv(\boldsymbol{\nabla} \cdot \mathbf{A}) \tag{140}$$

holds for a differentiable vector or pseudovector field $\mathbf{A}$. In order to introduce $d\widehat{\mathbf{s}}$ in place of $d\mathbf{s}$, one again needs to consider two cases. When $\mathbf{A}$ is a pseudovector $\mathbf{B}$, then we write $d\mathbf{s} \cdot \mathbf{B} = -d\widehat{\mathbf{s}} \cdot \widehat{\mathbf{B}}$ and apply (122): $\boldsymbol{\nabla} \cdot \mathbf{B} = -I(\boldsymbol{\nabla} \wedge \widehat{\mathbf{B}})$. After these changes, Eq. (140) takes the form

$$\int_{\partial\Omega} d\widehat{\mathbf{s}} \cdot \widehat{\mathbf{B}} = I \int_\Omega dv(\boldsymbol{\nabla} \wedge \widehat{\mathbf{B}}) \,.$$

The counterclockwise orientation of $d\mathbf{s}$ at the boundary $\partial\Omega$ fits to the right-handed orientation of the region Ω (compare Figs. 17 and 18), hence we introduce the directed volume element $d\widehat{V} = I\,dv$ and write the last equality as

$$\int_{\partial\Omega} d\widehat{\mathbf{s}} \cdot \widehat{\mathbf{B}} = \int_\Omega d\widehat{V}(\boldsymbol{\nabla} \wedge \widehat{\mathbf{B}}) \,.$$

The Clifford product of two trivectors may be replaced by their inner product (see Eq. (73)), therefore

$$\int_{\partial\Omega} d\widehat{\mathbf{s}} \cdot \widehat{\mathbf{B}} = \int_\Omega d\widehat{V} \cdot (\boldsymbol{\nabla} \wedge \widehat{\mathbf{B}}) \,. \tag{141}$$

When $\mathbf{A}$ is a vector $\mathbf{D}$, we multiply (140) by I and apply identity (135) to the left-hand side:

$$\int_{\partial\Omega} d\hat{\mathbf{s}} \wedge \mathbf{D} = I \int_{\Omega} dv(\nabla \cdot \mathbf{D}) = \int_{\Omega} d\hat{V}(\nabla \cdot \mathbf{D}) = \int_{\Omega} d\hat{V} \wedge (\nabla \cdot \mathbf{D}) \,. \quad (142)$$

Formulas (141) and (142) are the counterparts of the Gauss theorem that we looked for.

Equations (139) and (141) connect the adherences of k-vector fields to k-dimensional manifolds and yield scalars as the result. They have as their equivalent in differential geometry the *generalized Stokes theorem*:

$$\int_{\partial V} \omega = \int_{V} d\omega$$

where V is a region or a surface, its boundary ∂V is a surface or a curve, and ω is 2-form or 1-form respectively.

Equations (138) and (142) concern the protrudences of k-vector fields from $(3-k)$-dimensional manifolds, and the results of integration are trivectors. In terms of differential geometry, they should be written as

$$* \int_{\partial V} *\omega = * \int_{V} d * \omega \,.$$

§12. Cliffor-valued Distributions

12.1. Test Functions and Distributions

The distribution theory rests on so-called test functions — we start by introducing this notion. We consider functions φ of four variables $(\mathbf{r}, t) \in R^4$ (three spatial variables, one time variable) with values in the Pauli algebra. The closure of the set of points $(\mathbf{r}, t) \in R^4$ in which $\varphi(\mathbf{r}, t) \neq 0$ is called *support* of ϕ. Recall that the closed and bounded set in R^4 is called *compact.* We call φ a *test function* if it has a compact support and is differentiable arbitrarily many times. Test functions can be added and multiplied by scalars — this means that the set of all test functions $\mathcal{D}$ is a linear space — we call it the *space of test functions.* The convergence of sequences of its elements — is defined as follows: A sequence of test functions $\varphi_n \in \mathcal{D}$ is convergent to zero if

1. all functions φ_n are zero outside a common compact set
2. for an arbitrary partial derivative

$$\varphi_n^{(q)} = \frac{\partial^{q_1+q_2+q_3+q_4}}{\partial x_1^{q_1} \partial x_2^{q_2} \partial x_3^{q_3} \partial t^{q_4}} \varphi_n \quad (143)$$

the condition $\lim_{n\to\infty}\left\{\sup_{\mathbf{r},t}\|\varphi_n^{(q)}(\mathbf{r},t)\|\right\}=0$ is satisfied.

A linear and continuous mapping F from D into the Clifford algebra (a linear functional F) is called *distribution*. The value of F on a test function φ is denoted by (F,φ).

If f is a locally integrable function on R^4 — that is, integrable on each bounded set — the distribution is defined according to the formula

$$(F,\varphi)=\int_{R^4} f(\mathbf{r},t)\varphi(\mathbf{r},t)d^4v \tag{144}$$

where $d^4v = dv\,dt$. We omit the proof that it satisfies all the required assumptions. In this way, to any locally integrable function, a distribution corresponds under (144), called a *regular distribution*. An example of such a distribution is the expression

$$(F,\varphi)=\int_{R^4}\frac{1}{r}\varphi(\mathbf{r},t)d^4v$$

obtained from the locally integrable function

$$f(\mathbf{r},t)=\begin{cases}\frac{1}{r}=\frac{1}{|\mathbf{r}|} & \text{for } \mathbf{r}\neq 0\\ a & \text{for } \mathbf{r}=0\end{cases}$$

and for all t, with an arbitrary real a.[1)]

Distributions which are not representable in the form of (144) with a locally integrable function f, are called *singular*. An example of such distribution is the four-dimensional *Dirac delta* $(D,\varphi)=\varphi(0)$. Singular distributions are also often written in the form of (144), but instead of the function f,

---

[1)] The integrability of this function in any bounded region whose closure does not contain the point $\mathbf{r}=0$ is obvious. The integrability in a bounded set containing $\mathbf{r}=0$ can be checked on the example of the set $K(r_0)\times T$, where T is an interval and $K(r_0)$ is a ball with radius r_0 and center in origin. By changing the variables x_1, x_2, x_3 into the spherical ones r, ϑ, φ:

$$\int_{K(r_0)}\int_T\frac{1}{r}dv\,dt=|T|\int_0^{r_0}r^2dr\int_0^{\pi}\sin\vartheta d\vartheta\int_0^{2\pi}d\varphi\frac{1}{r}$$
$$=4\pi|T|\int_0^{r_0}r\,dr=2\pi r_0^2|T|\,.$$

something indefinite stands, which is understood rather symbolically. Let us call this a *generalized function.*[1] For the Dirac delta one writes

$$(D, \varphi) = \int_{R^4} \delta^4(\mathbf{r}, t)\varphi(\mathbf{r}, t) d^4v = \varphi(0) \ . \tag{145}$$

The fact that only the value of φ at the point zero contributes to this integral is described by the expression: The generalized Dirac delta function vanishes outside the zero

$$\delta^4(\mathbf{r}, t) = 0 \quad \text{for } (\mathbf{r}, t) \neq 0$$

and its value in zero must be infinite in order that the integral should be nonzero when $\varphi(0) \neq 0$.

One may also introduce distributions over spaces with smaller dimension. When the test functions depend on fewer variables, the corresponding distributions are defined as previously, with the obvious modifications. We write, for instance, the representatives of equality (144) in three dimensions as

$$(F, \varphi) = \int_{R^3} f(\mathbf{r})\varphi(\mathbf{r}) dv \ ,$$

in two dimensions as

$$(F, \varphi) = \iint\limits_{R \times R} f(x_1, x_2)\varphi(x_1, x_2) dx_1 dx_2$$

and in one dimension as

$$(F, \varphi) = \int_R f(x)\varphi(x) dx \ .$$

We do the same with formula (145) for the Dirac delta:

$$(D, \varphi) = \int_{R^3} \delta^3(\mathbf{r})\varphi(\mathbf{r}) dv = \varphi(0) \qquad \text{in three dimensions,}$$

$$(D, \varphi) = \iint\limits_{R \times R} \delta^2(x_1, x_2)\varphi(x_1, x_2) dx_1 dx_2 = \varphi(0) \quad \text{in two dimensions,}$$

$$(D, \varphi) = \int_R \delta(x)\varphi(x) dx = \varphi(0) \qquad \text{in one dimension.}$$

---

[1] In mathematical literature, the terms "distribution" and "generalized function" are synonymous. I have indulged here in their distinction. Now one may admit that distribution and its corresponding generalized function have different physical dimensions. For instance, the four-dimensional Dirac delta as a distribution, i.e. functional, is dimensionless (it maps a test function with some physical dimension into its value with the same dimension), whereas as a generalized function, it has the dimension (length)$^{-3}$ (time)$^{-1}$.

The one-dimensional Dirac delta has the property $a\delta(ax) = \delta(x)$ for arbitrary $a > 0$. The three-dimensional Dirac delta as a generalized function has an interpretation as the spatial density of something (e.g. mass or charge) of bigness one, concentrated in the origin of coordinates. In two dimensions, the Dirac delta is a surface density, and in one dimension a linear density. The identity $\delta^n(x_1, x_2, \ldots, x_n) = \delta(x_1)\delta(x_2)\ldots\delta(x_n)$ is satisfied, which is understood as the equality of the above integrals taken as iterated integrals for any test function.

The *translated Dirac delta* $(D_{\mathbf{a},\tau}, \varphi) = \varphi(\mathbf{a}, \tau)$ can also be introduced, for which the counterpart of Eq. (145) has the form

$$\int_{R^4} \delta^4(\mathbf{r} - \mathbf{a}, t - \tau)\varphi(\mathbf{r}, t)d^4v = \varphi(\mathbf{a}, \tau) \ . \tag{146}$$

This formula can be extended not only on test functions, but also on arbitrary continuous functions φ; we refer the reader to Sec. VI.3 of Ref. 10.[1] The Dirac delta also has the property

$$\delta^4(\mathbf{r} - \mathbf{r}', t - t') = \delta^4(\mathbf{r}' - \mathbf{r}, t' - t) \ . \tag{147}$$

Now we introduce differentiation of distributions: the q-th *derivative* of a distribution F is defined by

$$(F^{(q)}, \varphi) = (-1)^{|q|}(F, \varphi^{(q)}) \tag{148}$$

which, with the use of integral representation (144), has the form

$$\int_{R^4} f^{(q)}(\mathbf{r}, t)\varphi(\mathbf{r}, t)d^4v = (-1)^{|q|}\int_{R^4} f(\mathbf{r}, t)\varphi^{(q)}(\mathbf{r}, t)d^4v$$

where $\varphi^{(q)}$ is defined by (143) and $|q| = q_1 + q_2 + q_3 + q_4$. The reader should convince himself that this gives a continuous functional on $\mathcal{D}$ — that is, distribution; for this purpose, the specific convergence chosen for $\mathcal{D}$ is helpful. It follows from this that all distributions may be arbitrarily many times differentiated and distributions are always obtained. In particular, locally integrable functions have derivatives in the distribution sense. If such a function is differentiable and has bounded partial derivatives, then its distribution derivative

[1] One may even take a generalized function in place of φ, but then the integral on the left-hand side in (146) is understood more formally as the so-called *convolution*. The scope of this book does not allow us to explain the details.

corresponds to the ordinary derivative, which can be checked by using the rule of integration by parts.

We shall need the inner, outer and Clifford derivatives of distributions which are defined as follows

$$(\nabla \cdot F, \varphi) = \mathbf{e}_i \cdot \left(\frac{\partial F}{\partial x_i}, \varphi\right) = -\mathbf{e}_i \cdot \left(F, \frac{\partial \varphi}{\partial x_i}\right) ,$$

$$(\nabla \wedge F, \varphi) = \mathbf{e}_i \wedge \left(\frac{\partial F}{\partial x_i}, \varphi\right) = -\mathbf{e}_i \wedge \left(F, \frac{\partial \varphi}{\partial x_i}\right) ,$$

$$(\nabla F, \varphi) = \mathbf{e}_i \left(\frac{\partial F}{\partial x_i}, \varphi\right) = -\mathbf{e}_i \left(F, \frac{\partial \varphi}{\partial x_i}\right) .$$

(The summation convention holds for repeated indices.)

12.2. Fundamental Solutions of Differential Equations

We shall be interested in nonhomogeneous linear differential equations of the form

$$K\, f(\mathbf{r}, t) = j(\mathbf{r}, t) \tag{149}$$

where j and f are generalized cliffor-valued functions and K is a differential operator formed of the nabla and time derivative. In these equations, j is a given function and f is an unknown function. We say that a generalized function h is a *fundamental solution* of Eq. (149) if it satisfies

$$Kh(\mathbf{r}, t) = \delta^4(\mathbf{r}, t) . \tag{150}$$

This equation should be understood in the distribution sense, that is

$$\int_{R^4} [K\, h(\mathbf{r}, t)] \varphi(\mathbf{r}, t) d^4 v = \int_{R^4} \delta^4(\mathbf{r}, t) \varphi(\mathbf{r}, t) d^4 v$$

for each $\varphi \in \mathcal{D}$, or in another notation $(KH, \varphi) = \varphi(0)$. Here H denotes the distribution corresponding to the generalized function h. The fundamental solution is not unique: one may always add to it a generalized function f satisfying $Kf = 0$.

If the operator K does not contain a time derivative, we take as its fundamental solution a generalized function g which does not depend on time. Then, instead of (150), we take

$$K\, g(\mathbf{r}) = \delta^3(\mathbf{r}) . \tag{151}$$

Lemma 1. The cliffor field

$$f(\mathbf{r},t) = \int_{R^4} h(\mathbf{r}-\mathbf{r}',t-t')j(\mathbf{r}',t')d^4v' + f_0(\mathbf{r},t) \tag{152}$$

where h is the fundamental solution and f_0 satisfies $K\,f_0 = 0$, is a solution of Eq. (149).

Proof. The operator K does not act on primed variables — i.e. the integration variables — so one may put it under the integration sign in (152):

$$K\,f(\mathbf{r},t) = \int_{R^4} K\,h(\mathbf{r}-\mathbf{r}',t-t')j(\mathbf{r}',t')d^4v' \,.$$

We make use of (150) and (147)

$$K\,f(\mathbf{r},t) = \int_{R^4} \delta^4(\mathbf{r}-\mathbf{r}',t-t')j(\mathbf{r}',t')dv'$$
$$= \int_{R^4} \delta^4(\mathbf{r}'-\mathbf{r},t'-t)j(\mathbf{r}',t')d^4v'$$

and of (146)

$$K\,f(\mathbf{r},t) = j(\mathbf{r},t) \,.$$

In cases when the operator K does not contain a time derivative, the formula (152) should be replaced by

$$f(\mathbf{r}) = \int_{R^3} g(\mathbf{r}-\mathbf{r}')j(\mathbf{r}')dv' + f_0(\mathbf{r}) \,. \tag{153}$$

The proof is analogous, and rests on Eq. (151) in place of (150).

Lemma 2. The function

$$g(\mathbf{r}) = \begin{cases} \frac{1}{4\pi}\frac{\mathbf{r}}{r^3} & \text{for } r \neq 0 \\ 0 & \text{for } r = 0 \end{cases}$$

is the fundamental solution for the differential equation

$$\nabla f(\mathbf{r}) = j(\mathbf{r}) \,. \tag{154}$$

Proof. The function $\mathbf{r}/r^3$ for $r \neq 0$ may be represented as $-\nabla\frac{1}{r}$ (see Problem 33), hence

$$\nabla g(\mathbf{r}) = \nabla\frac{1}{4\pi}\left(-\nabla\frac{1}{r}\right) = -\frac{1}{4\pi}\nabla^2\frac{1}{r} = -\frac{1}{4\pi}\Delta\frac{1}{r} \,.$$

Now we refer to the identity on Pg. 47 in Vladimirov's book [11]:

$$\Delta \frac{1}{r} = -4\pi \delta^3(\mathbf{r}) \tag{155}$$

and obtain

$$\nabla g(\mathbf{r}) = \delta^3(\mathbf{r}) ,$$

that is, condition (151).■

Lemma 3. The generalized functions

$$D^{(-)}(\mathbf{r}, t) = -\frac{1}{4\pi r}\delta\left(t - \frac{r}{u}\right), \qquad D^{(+)}(\mathbf{r}, t) = -\frac{1}{4\pi r}\delta\left(t + \frac{r}{u}\right)$$

are fundamental solutions of the equation

$$\Box A = j . \tag{156}$$

Proof. We check this for one of the functions. We calculate the consecutive derivatives

$$\frac{\partial}{\partial x_i} D^{(-)}(\mathbf{r}, t) = -\frac{1}{4\pi}\left[\left(\frac{\partial}{\partial x_i}\frac{1}{r}\right)\delta\left(t - \frac{r}{u}\right) + \frac{1}{r}\delta'\left(t - \frac{r}{u}\right)\frac{\partial}{\partial x_i}\left(-\frac{r}{u}\right)\right] .$$

(The prime means here the derivative with respect to the whole argument of the one-dimensional Dirac delta.) Now

$$\begin{aligned}
&\frac{\partial^2}{\partial x_i^2} D^{(-)}(\mathbf{r}, t) \\
&= -\frac{1}{4\pi}\left\{\left(\frac{\partial^2}{\partial x_i^2}\frac{1}{r}\right)\delta\left(t - \frac{r}{u}\right) + 2\left(\frac{\partial}{\partial x_i}\frac{1}{r}\right)\delta'\left(t - \frac{r}{u}\right)\frac{\partial}{\partial x_i}\left(-\frac{r}{u}\right)\right. \\
&\qquad \left. + \frac{1}{r}\frac{\partial}{\partial x_i}\left[\delta'\left(t - \frac{r}{u}\right)\frac{\partial}{\partial x_i}\left(-\frac{r}{u}\right)\right]\right\} \\
&= -\frac{1}{4\pi}\left[\left(\frac{\partial^2}{\partial x_i^2}\frac{1}{r}\right)\delta\left(t - \frac{r}{u}\right) - \frac{2x_i}{r^3}\delta'\left(t - \frac{r}{u}\right)\left(-\frac{x_i}{ur}\right)\right. \\
&\qquad \left. + \frac{1}{r}\delta''\left(t - \frac{r}{u}\right)\left(-\frac{x_i}{ur}\right)^2 + \frac{1}{r}\delta'\left(t - \frac{r}{u}\right)\frac{x_i^2 - r^2}{ur^3}\right] \\
&= -\frac{1}{4\pi}\left[\left(\frac{\partial^2}{\partial x_i^2}\frac{1}{r}\right)\delta\left(t - \frac{r}{u}\right) + \frac{3x_i^2 - r^2}{ur^4}\delta'\left(t - \frac{r}{u}\right) + \frac{x_i^2}{u^2 r^3}\delta''\left(t - \frac{r}{u}\right)\right] .
\end{aligned}$$

After summation with respect to i, the middle term in the bracket vanishes and we obtain

$$\Delta D^{(-)}(\mathbf{r}, t) = -\frac{1}{4\pi}\left[\left(\Delta\frac{1}{r}\right)\delta\left(t - \frac{r}{u}\right) + \frac{1}{u^2 r}\delta''\left(t - \frac{r}{u}\right)\right] .$$

Moreover,

$$\frac{\partial^2}{\partial t^2} D^{(-)}(\mathbf{r}, t) = -\frac{1}{4\pi r} \delta''\left(t - \frac{r}{u}\right) .$$

Thus, the Dalambertian acting on $D^{(-)}$ gives

$$\Box D^{(-)}(\mathbf{r}, t) = -\frac{1}{4\pi} \left(\Delta \frac{1}{r}\right) \delta\left(t - \frac{r}{u}\right) .$$

Now we use identity (155)

$$\Box D^{(-)}(\mathbf{r}, t) = \delta^3(\mathbf{r}) \delta\left(t - \frac{r}{u}\right) .$$

We recall that this is the distribution equality. The presence of the first delta demands that we put $\mathbf{r} = 0$ in the other, so

$$\Box D^{(-)}(\mathbf{r}, t) = \delta^3(\mathbf{r}) \delta(t) = \delta^4(\mathbf{r}, t) .$$

The calculations are similar for the other solution $D^{(+)}$.■

The d'Alembert operator is linear, hence the following combination of the fundamental solutions

$$D(\mathbf{r}, t) = \alpha D^{(-)}(\mathbf{r}, t) + \beta D^{(+)}(\mathbf{r}, t) \tag{157}$$

with the coefficients α, β satisfying $\alpha + \beta = 1$ is also a fundamental solution.

Equation (156) may contain cliffor functions A and j; nevertheless, the fundamental solutions $D^{(+)}$ and $D^{(-)}$ are scalars. This is connected with the fact that $\Box$ is the scalar operator. However, the fundamental solution (157) may be cliffor-valued if one takes α and β as cliffors.

Lemma 4. The generalized functions

$$h^{(\pm)}(\mathbf{r}, t) = \left(\boldsymbol{\nabla} - \frac{1}{u}\frac{\partial}{\partial t}\right) D^{(\pm)}(\mathbf{r}, t)$$

are fundamental solutions of the equation

$$\boldsymbol{\nabla} f + \frac{1}{u}\frac{\partial f}{\partial t} = j . \tag{158}$$

Proof. We obtain, with the use of identity (128),

$$\left(\boldsymbol{\nabla} + \frac{1}{u}\frac{\partial}{\partial t}\right) h^{(\pm)} = \left(\boldsymbol{\nabla} + \frac{1}{u}\frac{\partial}{\partial t}\right)\left(\boldsymbol{\nabla} - \frac{1}{u}\frac{\partial}{\partial t}\right) D^{(\pm)} = \Box D^{(\pm)} = \delta^4 .■$$

Let us calculate these generalized functions explicitly,

$$h^{(\pm)}(\mathbf{r},t) = \nabla D^{(\pm)}(\mathbf{r},t) - \frac{1}{u}\frac{\partial}{\partial t}D^{(\pm)}(\mathbf{r},t)$$
$$= -\frac{1}{4\pi}\left[-\frac{\mathbf{r}}{r^3}\delta\left(t \pm \frac{r}{u}\right) \pm \frac{\mathbf{r}}{ur^2}\delta'\left(t \pm \frac{r}{u}\right)\right] + \frac{1}{u}\frac{1}{4\pi r}\delta'\left(t \pm \frac{r}{u}\right).$$

Hence,

$$h^{(\pm)}(\mathbf{r},t) = \frac{1}{4\pi}\left[\frac{\mathbf{r}}{r^3}\delta\left(t \pm \frac{r}{u}\right) + \frac{1}{ur}\left(1 \mp \frac{\mathbf{r}}{r}\right)\delta'\left(t \pm \frac{r}{u}\right)\right]. \tag{159}$$

We see that these cliffor-valued generalized functions are combinations of scalars and vectors. This is connected with the fact that the operator $\nabla + \frac{1}{u}\frac{\partial}{\partial t}$ standing in (158) contains the vector and scalar part.

The following fact is worth noticing. If the right-hand side of (158) does not depend on time, the solution obtained by means of (152), with h given by (159), is

$$f(\mathbf{r},t) = \frac{1}{4\pi}\int_{R^4}\left[\frac{\mathbf{r}-\mathbf{r}'}{|\mathbf{r}-\mathbf{r}'|^3}\delta\left(t - t' \pm \frac{|\mathbf{r}-\mathbf{r}'|}{u}\right)\right.$$
$$\left.+\frac{1}{u|\mathbf{r}-\mathbf{r}'|}\left(1 \mp \frac{\mathbf{r}-\mathbf{r}'}{|\mathbf{r}-\mathbf{r}'|}\right)\delta'\left(t - t' \pm \frac{|\mathbf{r}-\mathbf{r}'|}{u}\right)\right]j(\mathbf{r}')d^4v' + f_0(\mathbf{r},t).$$

In accordance with (148), we carry over the time derivative of second delta onto j and obtain zero, therefore

$$f(\mathbf{r},t) = \frac{1}{4\pi}\int_{R^3}\int_{R}\frac{\mathbf{r}-\mathbf{r}'}{|\mathbf{r}-\mathbf{r}'|^3}\delta\left(t - t' \pm \frac{|\mathbf{r}-\mathbf{r}'|}{u}\right)j(\mathbf{r}')dv'\,dt' + f_0(\mathbf{r},t).$$

The integration with the one-dimensional Dirac delta over the time variable gives one, hence

$$f(\mathbf{r},t) = \frac{1}{4\pi}\int_{R^3}\frac{\mathbf{r}-\mathbf{r}'}{|\mathbf{r}-\mathbf{r}'|^3}j(\mathbf{r}')dv' + f_0(\mathbf{r},t).$$

This is the solution of Eq. (154) computed according to (153) with the fundamental solution g of Lemma 2.

Problems

1. One may endow the faces of a polyhedron with orientations — in this way, they become volutors. We say that two neighbouring faces have *compatible orientations* if their common edge has opposite orientations as the boundary of the two faces. Show that the sum of volutors constituting the surface of the polyhedron is zero if all neighbouring faces have compatible orientations. How is this result connected with the vector sum of sides of a polygon?
2. Show that the Hodge map of pseudovectors into volutors is linear.
3. Demonstrate that the inner product of a vector with a volutor is distributive with respect to the addition of both factors and does not depend on the factorization of the volutor in the outer product.
4. Prove the formula $(\mathbf{a} \cdot \widehat{\mathbf{B}})_k = \sum_{j=1}^{3} a_j B_{jk}$.
5. Prove the identity $\mathbf{a} \cdot \widehat{\mathbf{B}} = -\mathbf{a} \times \mathbf{B}$. Illustrate it with a picture.
6. Prove the identity $\varepsilon_{klm}\varepsilon_{kij} = \delta_{li}\delta_{mj} - \delta_{lj}\delta_{im}$, making use of Problem 5.
7. Let the angular velocity of a particle be given by $\widehat{\omega} = (\mathbf{r} \wedge \mathbf{v})/r^2$, where $\mathbf{r}$ is the radius vector of the particle beginning on the rotation axis and perpendicular to it. Show that the linear velocity is $\mathbf{v} = \mathbf{r} \cdot \widehat{\omega}$. How do these formulae change if $\mathbf{r}$ is not perpendicular to the axis?
8. Demonstrate that the inner product of volutors is commutative and does not depend on their factorization in the outer product.
9. Let $\widehat{\mathbf{B}} = \mathbf{b}_1 \wedge \mathbf{b}_2$, $\widehat{\mathbf{C}} = \mathbf{c}_1 \wedge \mathbf{c}_2$. Show that $\widehat{\mathbf{C}} \cdot \widehat{\mathbf{B}} = -\det\{\mathbf{c}_i \cdot \mathbf{b}_j\}$. Show that the coordinates of any volutor $\widehat{\mathbf{B}}$ in an orthonormal basis $\{\mathbf{e}_1, \mathbf{e}_2, \mathbf{e}_3\}$ may be computed as $B_{ij} = -(\mathbf{e}_i \wedge \mathbf{e}_j) \cdot \widehat{\mathbf{B}}$.
10. Demonstrate that the mingled product of volutors does not depend on their factorization in the outer product and that it can be expressed by coordinates as follows
$$(\widehat{\mathbf{B}} \dot{\wedge} \widehat{\mathbf{C}})_{jk} = B_{ji}C_{ik} - B_{ki}C_{ij} \ .$$
11. Show in pictures that the three trivectors $\mathbf{a} \wedge (\mathbf{b} \wedge \mathbf{c})$, $\mathbf{b} \wedge (\mathbf{c} \wedge \mathbf{a})$, and $\mathbf{c} \wedge (\mathbf{a} \wedge \mathbf{b})$ are equal.
12. How may the sum of two trivectors be found by a geometrical construction?
13. Demonstrate that the inner product of a vector with a trivector is distributive with respect to addition of both factors and does not depend on the factorization of the trivector in the outer product.

14. Demonstrate that the inner product of a bivector with a trivector and of two trivectors does not depend on their factorization in the outer products.
15. Show that the vector $\hat{\mathbf{D}} \cdot \hat{T}$ is perpendicular to the volutor $\hat{\mathbf{D}}$.
16. Show that for any vector a and multivector M, the identities

$$\mathbf{a} \wedge (\mathbf{a} \wedge M) = 0 \qquad \text{and} \qquad \mathbf{a} \cdot (\mathbf{a} \cdot M) = 0$$

are satisfied. *Hint*: For the inner product, use the formula $\mathbf{a} \cdot (\mathbf{b} \wedge M_k) = (\mathbf{a} \cdot \mathbf{b}) M_k - \mathbf{b} \wedge (\mathbf{a} \cdot M_k)$ and the induction principle.
17. Let $\hat{T} = \mathbf{a}_1 \wedge \mathbf{a}_2 \wedge \mathbf{a}_3$, $\hat{V} = \mathbf{b}_1 \wedge \mathbf{b}_2 \wedge \mathbf{b}_3$. Show that $\hat{T} \cdot \hat{V} = -\det\{\mathbf{a}_i \cdot \mathbf{b}_j\}$.
18. Show the associativity of the products $\mathbf{abc}$, $\mathbf{a}\hat{\mathbf{B}}\mathbf{c}$, and $\mathbf{a}\hat{T}\mathbf{c}$.
19. Show that $\hat{T}M = \hat{T} \cdot M$ and $M\hat{T} = M \cdot \hat{T}$ for any trivector T and multivector M.
20. Demonstrate the property $(XY)^\dagger = Y^\dagger X^\dagger$ for any cliffors X, Y.
21. Prove the identity $|X|^2 = (XX^\dagger)_s = (X^\dagger X)_s$ for any cliffor X.
22. Show that for any multivector M and a unit vector $\mathbf{n}$ the expression $(M \cdot \mathbf{n})\mathbf{n}$ is the component of M parallel to $\mathbf{n}$, whereas $(M \wedge \mathbf{n})\mathbf{n}$ is the component of M perpendicular to $\mathbf{n}$.
23. Show that all cliffor valued functions on the Clifford algebra defined by the power series with real coefficients have the property $f(X^\dagger) = f(X)^\dagger$.
24. Let two cliffors X and Y satisfy $XY = YX$. Demonstrate that

$$\sin(X+Y) = \sin X \cos Y + \sin Y \cos X \,,$$
$$\cos(X+Y) = \cos X \cos Y - \sin X \sin Y \,.$$

25. Show that for a sufficiently regular multivector field M the identities $\nabla \wedge (\nabla \wedge M) = 0$, $\nabla \cdot (\nabla \cdot M) = 0$ hold. What does the expression "sufficiently regular" mean?
26. Prove the formula $\nabla \wedge (M_k \wedge X) = (\nabla \wedge M_k) \wedge X + (-1)^k M_k \wedge (\nabla \wedge X)$, where M_k is a k-vector field and X is a cliffor field.
27. Do the formulae

$$\operatorname{curl}(f\,\mathbf{A}) = f \operatorname{curl} \mathbf{A} + (\operatorname{grad} f) \times \mathbf{A} \,,$$
$$\operatorname{div}(\mathbf{A} \times \mathbf{B}) = \mathbf{B} \cdot \operatorname{curl} \mathbf{A} - \mathbf{A} \cdot \operatorname{curl} \mathbf{B}$$

have anything to do with the previous Problem?
28. Prove the formulae
(i) $\nabla \cdot (\mathbf{A} \cdot \hat{\mathbf{B}}) = (\nabla \wedge \mathbf{A}) \cdot \hat{\mathbf{B}} - \mathbf{A} \cdot (\nabla \cdot \hat{\mathbf{B}})$,
(ii) $\nabla \wedge (\mathbf{A} \cdot \mathbf{B}) = (\mathbf{A} \cdot \nabla)\mathbf{B} + (\mathbf{B} \cdot \nabla)\mathbf{A} - \mathbf{B} \cdot (\nabla \wedge \mathbf{A}) - \mathbf{A} \cdot (\nabla \wedge \mathbf{B})$,

(iii) $\boldsymbol{\nabla}\cdot(\mathbf{A}\wedge\mathbf{B}) = \mathbf{B}(\boldsymbol{\nabla}\cdot\mathbf{A}) - \mathbf{A}(\boldsymbol{\nabla}\cdot\mathbf{B}) + (\mathbf{A}\cdot\boldsymbol{\nabla})\mathbf{B} - (\mathbf{B}\cdot\boldsymbol{\nabla})\mathbf{A}$,
(iv) $\boldsymbol{\nabla}\wedge(\widehat{\mathbf{A}}\,\widehat{\mathbf{B}}) = (\widehat{\mathbf{A}}\wedge\boldsymbol{\nabla})\cdot\widehat{\mathbf{B}} + (\widehat{\mathbf{B}}\wedge\boldsymbol{\nabla})\cdot\widehat{\mathbf{A}} - \widehat{\mathbf{B}}\cdot(\boldsymbol{\nabla}\cdot\widehat{\mathbf{A}}) - \widehat{\mathbf{A}}\cdot(\boldsymbol{\nabla}\cdot\widehat{\mathbf{B}})$,
(v) $\boldsymbol{\nabla}\wedge(\mathbf{A}\cdot\widehat{\mathbf{B}}) = \widehat{\mathbf{B}}(\boldsymbol{\nabla}\cdot\mathbf{A}) - \mathbf{A}(\boldsymbol{\nabla}\wedge\widehat{\mathbf{B}}) + (\mathbf{A}\cdot\boldsymbol{\nabla})\widehat{\mathbf{B}} - (\widehat{\mathbf{B}}\wedge\boldsymbol{\nabla})\mathbf{A}$,
(vi) $\boldsymbol{\nabla}\cdot(\mathbf{A}\wedge\widehat{\mathbf{B}}) = (\mathbf{A}\cdot\boldsymbol{\nabla})\widehat{\mathbf{B}} + (\widehat{\mathbf{B}}\wedge\boldsymbol{\nabla})\mathbf{A} - \widehat{\mathbf{B}}\dot{\wedge}(\boldsymbol{\nabla}\wedge\mathbf{A}) - \mathbf{A}\wedge(\boldsymbol{\nabla}\cdot\widehat{\mathbf{B}})$.

29. Prove the identity $\boldsymbol{\nabla}(\mathbf{A}\,\widehat{\mathbf{B}}) = (\boldsymbol{\nabla}\mathbf{A})\widehat{\mathbf{B}} - \mathbf{A}(\boldsymbol{\nabla}\widehat{\mathbf{B}}) + 2(\mathbf{A}\cdot\boldsymbol{\nabla})\widehat{\mathbf{B}}$. *Hint*: Use Problems 26 and 28 (i), (v), (vi).
30. Prove the formula $\boldsymbol{\nabla}\cdot(fM) = f(\boldsymbol{\nabla}\cdot M) + (\boldsymbol{\nabla}f)\cdot M$, where f is a scalar field and M is a multivector field.
31. Let $\mathbf{a},\mathbf{b}$ be constant vectors. Show that for vector fields $\mathbf{A}(\mathbf{r}) = \mathbf{a}(\mathbf{b}\cdot\mathbf{r})$ and $\mathbf{A}'(\mathbf{r}) = -\mathbf{b}(\mathbf{a}\cdot\mathbf{r})$ the equality $\boldsymbol{\nabla}\wedge\mathbf{A} = \boldsymbol{\nabla}\wedge\mathbf{A}'$ holds.
32. Compute $\boldsymbol{\nabla}\wedge\mathbf{A}$ if
    (i) $\mathbf{A}(\mathbf{r}) = (\mathbf{n}\cdot\mathbf{r})(\mathbf{n}\cdot\widehat{\mathbf{B}})$ for $\mathbf{n}\|\widehat{\mathbf{B}}$, $\mathbf{n}^2 = 1$,
    (ii) $\mathbf{A}(\mathbf{r}) = \mathbf{e}_3\ln\left[(x_1^2+x_2^2)/a^2\right]$ for real $a \neq 0$. Sketch the fields $\mathbf{A}$ and $\boldsymbol{\nabla}\wedge\mathbf{A}$.
33. Check directly that $\boldsymbol{\nabla}\wedge\mathbf{r} = 0$. Is the field $\mathbf{A}(\mathbf{r}) = \mathbf{r}$ the gradient of a scalar field? What does the gradient look like for the scalar field $\varphi(\mathbf{r}) = \varphi(r)$ with spherical symmetry?
34. Compute $\boldsymbol{\nabla}(\mathbf{r}\cdot\mathbf{a})$, $\boldsymbol{\nabla}\left(\frac{\mathbf{r}\cdot\mathbf{a}}{r^n}\right)$, $\boldsymbol{\nabla}\wedge(\mathbf{r}\cdot\widehat{\mathbf{B}})$, $\boldsymbol{\nabla}\wedge\left(\frac{\mathbf{r}\cdot\widehat{\mathbf{B}}}{r^n}\right)$, $(\mathbf{a}\cdot\boldsymbol{\nabla})\mathbf{r}$, $\boldsymbol{\nabla}\wedge(\mathbf{a}\wedge\mathbf{r})$, $\boldsymbol{\nabla}\wedge(\widehat{\mathbf{B}}\wedge\mathbf{r})$, $\boldsymbol{\nabla}\cdot(\mathbf{a}\wedge\mathbf{r})$, $\boldsymbol{\nabla}\cdot(\widehat{\mathbf{B}}\wedge\mathbf{r})$, $\boldsymbol{\nabla}\cdot(\widehat{\mathbf{B}}\cdot\mathbf{r})$, where $\mathbf{a}$ is a constant vector and $\widehat{\mathbf{B}}$ is a constant bivector.
35. Compute the Clifford derivatives $\boldsymbol{\nabla}\mathbf{r}$, $\boldsymbol{\nabla}(\mathbf{a}\wedge\mathbf{r})$, $\boldsymbol{\nabla}(\mathbf{a}\mathbf{r})$, $\boldsymbol{\nabla}(\mathbf{r}\mathbf{a})$, where $\mathbf{a}$ is a constant vector.
36. Find the Clifford derivatives $\boldsymbol{\nabla}\mathbf{r}^k$ and $\boldsymbol{\nabla}\mathbf{r}_{\mathbf{n}}^k$ where k is an integer and $\mathbf{r}_{\mathbf{n}} = (\mathbf{r}\cdot\mathbf{n})\mathbf{n}$ for some unit vector $\mathbf{n}$.
37. Express the Clifford derivatives $\boldsymbol{\nabla}(\mathbf{A}\cdot\mathbf{B})$, $\boldsymbol{\nabla}(\mathbf{A}\wedge\mathbf{B})$, $\boldsymbol{\nabla}(\mathbf{A}\mathbf{B})$ by the derivatives of single factors. *Hint*: Use Problems 26 and 28.
38. Is the volutor field $\mathbf{a}\wedge\mathbf{r}$ an outer derivative of a vector field? If the answer is positive, find it.
39. Show that $\Delta\frac{1}{r} = 0$ for $r \neq 0$.
40. Find $\boldsymbol{\nabla}x_j$ where x_j are Cartesian coordinates of the radius vector $\mathbf{r}$.
41. Let ξ_1, ξ_2, ξ_3 be curvilinear coordinates in three-dimensional space. Show that under a change of variables $\xi_j' = f_j(\xi_1, \xi_2, \xi_3)$, the vectors $\boldsymbol{\nabla}\xi_i$ transform in the same manner as differentials $d\xi_i$. Notice that $\boldsymbol{\nabla}\xi_i$ are tangent to the lines of coordinates when ξ_i are orthogonal coordinates.
42. Show that the equality $\int_{\partial V} d\mathbf{s}f = \int_V dv\boldsymbol{\nabla}f$ for a scalar field f follows from the Gauss theorem. *Hint*: Take the scalar product of both sides with a constant vector $\mathbf{a}$.
43. Using the Stokes theorem, show that the equality $\int_C d\mathbf{l}\cdot\mathbf{r}\varphi(r) = 0$ holds

for any spherically symmetric scalar function φ and any closed curve C.

44. Show that the identity $\int_{\partial S} d\mathbf{l}\varphi(r) = -\int_{S} d\mathbf{s} \times \mathbf{r}\frac{1}{r}\frac{d\varphi}{dr}$ for a spherically symmetric scalar function φ follows from the Stokes theorem. *Hint*: Take the scalar product of both sides with a constant vector $\mathbf{a}$ and use Problem 33.

Chapter 1
ELECTROMAGNETIC FIELD

§1. The Magnetic Field is a Bivector Field

In order to analyse the nature of the magnetic field, let us consider the motion of a charged particle in the electromagnetic field. The so-called *Lorentz force* arising from the electromagnetic field acts on such a particle such that

$$\mathbf{F} = q\mathbf{E} + q\mathbf{v} \times \mathbf{B}$$

where $\mathbf{E}$ is the electric field, $\mathbf{B}$ the magnetic induction, $\mathbf{v}$ is the velocity of the particle and q, its electric charge. It follows from this equation that different forces act on stationary particles and on moving particles.

The magnetic field is responsible for part of the force depending on the velocity of charge. This is the principal manifestation of the magnetic field, since all other magnetic actions can be derived from it. In the past, physicists considered that the magnetic field acted on magnetic poles, also called magnetic charges, but later such views were abandoned.[1)]

[1)] For some time past, scientists have pondered over magnetic charges known under the name *Dirac monopoles* in elementary particle theory. An intensive search for these charges had been carried out in many laboratories. Even if the monopoles would have been discovered, it is doubtful whether they could be used for determining or measuring the magnetic field.

What, in that case, are magnets? Magnets are systems of vortical electric currents. Even the elementary magnets connected with atomic spins are produced by the vortical movements of electrons. Therefore we take the standpoint that both the electric and magnetic field can be defined and determined through the forces by which they act on *electric* charges. Furthermore, this takes place by the intermediary of the Lorentz force.

We may determine the electric field through the Lorentz force $\mathbf{F}_{el} = q\mathbf{E}$ acting on a resting test charge. By definition, the electric field is then the ratio of force to charge:

$$\mathbf{E} = q^{-1}\mathbf{F}_{el} \,. \tag{1}$$

For a collection of charges with different signs and values located at the same point, the forces $\mathbf{F}_{el}$ have different magnitudes and orientations but always lie on the same straight line — this is a feature of the electric field at that point.

For the determination of the magnetic field, we need to have a moving charge. After subtracting the electric part from the Lorentz force the magnetic force remains as

$$\mathbf{F} - \mathbf{F}_{el} = \mathbf{F}_{\mathrm{mag}} = q\mathbf{v} \times \mathbf{B}. \tag{2}$$

We see from this formula that for a collection of charges with different velocities, neither the magnitudes nor the directions of the forces are the same. For various directions of $\mathbf{v}$, all the directions of $\mathbf{F}_{\mathrm{mag}}$ are possible which lie on a plane perpendicular to $\mathbf{B}$. We already know notions which have two-dimensional directions as their relevant features, therefore we claim that the *magnetic field in its essence is a bivector quantity.*

In an isotropic medium characterized by the magnetic permeability μ, the magnetic induction $d\mathbf{B}$ produced by the vectorial element $d\mathbf{l}$ of the length of the current J, by virtue of the Biot-Savart law, equals

$$d\mathbf{B} = \frac{\mu J}{4\pi}\,\frac{d\mathbf{l} \times \mathbf{r}}{r^3}$$

where $\mathbf{r}$ is the vector connecting the element of the current with the point in which the field is produced. The magnetic induction is the vector product of two polar vectors, hence it is an axial vector.

Bearing in mind that each axial vector is connected via the Hodge map with a bivector, let us assume that the magnetic induction B and the magnetic field H, proportional to B, *are bivectors*. Then Eq. (2) will be written as

$$\mathbf{F}_{\mathrm{mag}} = -q\mathbf{v}\cdot\hat{\mathbf{B}} \,, \tag{3}$$

by virtue of Eq. (0.24). We illustrate Eq. (2) in Fig. 26a and Eq. (3) in Fig. 26b. Of course Eq. (3) also shows that the vector $\mathbf{F}_{\text{mag}}$ for fixed $\widehat{\mathbf{B}}$ and different $\mathbf{v}$ always lies in a plane which is a characteristic of the magnetic field.

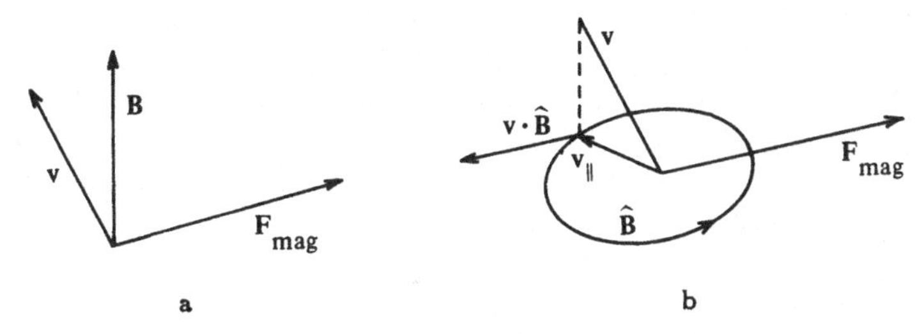

Fig. 26.

If $\mathbf{v}\|\widehat{\mathbf{B}}$, then $\mathbf{v}\cdot\widehat{\mathbf{B}} = \mathbf{v}\widehat{\mathbf{B}}$ with the Clifford product between the factors and $\mathbf{F}_{\text{mag}} = -q\mathbf{v}\widehat{\mathbf{B}}$. From this, we obtain

$$\widehat{\mathbf{B}} = -q^{-1}\mathbf{v}^{-1}\mathbf{F}_{\text{mag}} = -q^{-1}\mathbf{v}^{-1}\wedge\mathbf{F}_{\text{mag}} = q^{-1}\mathbf{F}_{\text{mag}}\wedge\mathbf{v}^{-1}$$

where we use the fact that $\mathbf{F}_{\text{mag}}\perp\mathbf{v}$ and $\mathbf{F}_{\text{mag}}\perp\mathbf{v}^{-1}$. This equation can be used to determine the bivector of magnetic induction when the test charge q performs a planar motion (but not along a straight line) in the magnetic field. This is analogous to Eq. (1). Of course it is valid only for small $\mathbf{v}$ when the field can be treated as uniform.

Once we have agreed that the magnetic field is to be described as a bivector field, we shall look for a counterpart for the force lines known for the vector field. First, we remember that: *force lines for the electric field* are directed curves to which the field $\mathbf{E}$ is tangent at each point and has compatible orientation. Hence the following definition enforces itself: *Force surfaces for the magnetic field* are directed surfaces to which the field $\widehat{\mathbf{B}}$ is tangent at each point and has compatible orientation. The notion known up to now as "force lines of the magnetic field " can still be used in the mathematical sense since, for a smooth (pseudo) vector field $\mathbf{B}$ or $\mathbf{H}$ (they are parallel with compatible orientations when μ is scalar), tangent curves may be found. However, its name is misleading — it originated in times when magnetic charges were used. (If the magnetic charges existed, forces would act on them parallel to $\mathbf{H}$.) Henceforth, we propose the expression *vectorial lines of the magnetic field* in place of "force lines of the magnetic field".

Let us now compose a quantity giving account of the electric and magnetic field jointly. In the Clifford algebra formalism we may add multivectors of different ranks, so we could merely add electric quantities to the magnetic ones. We should, however, follow the demand that the added terms have the same physical dimension. Let us take $\mathbf{e} = \sqrt{\varepsilon}\,\mathbf{E}$ instead of the electric field $\mathbf{E}$ (ε is the electric permittivity) and $\widehat{\mathbf{b}} = \frac{1}{\sqrt{\mu}}\widehat{\mathbf{B}}$ instead of the magnetic induction. Both quantities $\mathbf{e}$ and $\widehat{\mathbf{b}}$ have the dimension $\sqrt{\mathrm{J/m^3}} = \mathrm{s^{-1}}\sqrt{\mathrm{kg/m}}$ in the SI system of units. Their squares are equal to the doubled energy density of the electric and magnetic fields, respectively, which will be shown in Sec. 2.6. The quantities $\mathbf{e}$ and $\mathbf{b}$ may also be expressed through the electric induction $\mathbf{D}$ and magnetic field $\widehat{\mathbf{H}}$: $\mathbf{e} = \frac{1}{\sqrt{\varepsilon}}\mathbf{D}$, $\widehat{\mathbf{b}} = \sqrt{\mu}\,\widehat{\mathbf{H}}$. We form the following cliffor quantity

$$f = \mathbf{e} + \widehat{\mathbf{b}} \tag{4}$$

which we call *electromagnetic cliffor*. We may always extract out of this combination the electric field as the vector part and the magnetic field as the bivector part. This procedure is similar to the selection from a vector of a component parallel to a chosen direction.

Using Eq. (0.75), we may also write $f = \mathbf{e} + I\mathbf{b}$. In this way, the electromagnetic cliffor becomes a sort of complex vector, the real part of which is the electric part and the imaginary part the magnetic part. By virtue of Eq. (0.64), we have $f^\dagger = \mathbf{e} - \widehat{\mathbf{b}} = \mathbf{e} - I\mathbf{b}$, hence the reversion operation $f \to f^\dagger$ is equivalent to the complex conjugation of the "complex" vector f.

Let us calculate the square of the cliffor f

$$f^2 = (\mathbf{e}+\widehat{\mathbf{b}})(\mathbf{e}+\widehat{\mathbf{b}}) = \mathbf{e}^2 + \mathbf{e}\widehat{\mathbf{b}} + \widehat{\mathbf{b}}\mathbf{e} + \widehat{\mathbf{b}}^2 = \mathbf{e}^2 + 2\mathbf{e}\wedge\widehat{\mathbf{b}} + \widehat{\mathbf{b}}^2 \,,$$
$$f^2 = |\mathbf{e}|^2 - |\widehat{\mathbf{b}}|^2 + 2\mathbf{e}\wedge\widehat{\mathbf{b}} \,. \tag{5}$$

It is composed, as we can see, of the scalar $|\mathbf{e}|^2 - |\widehat{\mathbf{b}}|^2$ and the trivector $2\mathbf{e}\wedge\widehat{\mathbf{b}} = 2I(\mathbf{e}\cdot\mathbf{b})$. We shall also need the Clifford product of f with $f^\dagger$

$$f\,f^\dagger = (\mathbf{e}+\widehat{\mathbf{b}})(\mathbf{e}-\widehat{\mathbf{b}}) = \mathbf{e}^2 + \widehat{\mathbf{b}}\mathbf{e} - \mathbf{e}\widehat{\mathbf{b}} - \widehat{\mathbf{b}}^2 = |\mathbf{e}|^2 + |\widehat{\mathbf{b}}|^2 + 2\widehat{\mathbf{b}}\cdot\mathbf{e} \,,$$
$$f\,f^\dagger = |f|^2 + 2\widehat{\mathbf{b}}\cdot\mathbf{e} \,. \tag{6}$$

We see that $f\,f^\dagger$ is a combination of the scalar $|f|^2$ and the vector $2\widehat{\mathbf{b}}\cdot\mathbf{e} = 2\mathbf{e}\times\mathbf{b}$.

§2. Integral Maxwell Equations

To express the Faraday law of general physics, we need the notion of *magnetic flux* through a surface S: $\int_S d\mathbf{s}\cdot\mathbf{B}$ and the notion of *circulation* of the electric field around the loop $\partial S : \int_{\partial S} d\mathbf{l}\cdot\mathbf{E}$. It is known that the latter is the *electromotive force* around the loop ∂S understood as a circuit. The orientation of the vector $d\mathbf{s}$ is related to the orientation of the loop ∂S by the Stokes Theorem discussed in Sec. 0.11. The *Faraday law* for a stationary surface S is as follows: The time rate of change of magnetic flux is opposite to the electromotive force around the boundary ∂S of the surface

$$\frac{d}{dt}\int_S d\mathbf{s}\cdot\mathbf{B} = -\int_{\partial S} d\mathbf{l}\cdot\mathbf{E}\,.$$

We are going to inscribe this law using bivectors in place of pseudovectors $d\mathbf{s}$ and $\mathbf{B}$. Now, due to Eq. (0.136) we have $\int_S d\mathbf{s}\cdot\mathbf{B} = -\int_S d\widehat{\mathbf{s}}\cdot\widehat{\mathbf{B}}$ which means that the magnetic flux is equal (with the minus sign) to the adherence of $\widehat{\mathbf{B}}$ to the surface S. Thus, the Faraday law can be rewritten as

$$\frac{d}{dt}\int_S d\widehat{\mathbf{s}}\cdot\widehat{\mathbf{B}} = \int_{\partial S} d\mathbf{l}\cdot\mathbf{E}\,. \tag{7}$$

This can be stated as follows: the adherence of the field $\dot{\widehat{\mathbf{B}}} = \frac{d\widehat{\mathbf{B}}}{dt}$ to a stationary surface is equal to the adherence of the field $\mathbf{E}$ to its boundary.

Now we pass to the generalization of the *Ampere-Oersted law* found by Maxwell. It states that, for a stationary surface S, the circulation of $\mathbf{H}$ around the loop ∂S is equal to the sum of the current J flowing through S and of the so-called *displacement current*, i.e., the time derivative of the integral of $\mathbf{D}$ over S

$$\int_{\partial S} d\mathbf{l}\cdot\mathbf{H} = J + \frac{d}{dt}\int_S d\mathbf{s}\cdot\mathbf{D}\,.$$

In order to inscribe this law using bivectors in place of pseudovectors $d\mathbf{s}$ and $\mathbf{H}$ we need to use Eqs. (0.134) and (0.135) after multiplying both sides by the unit trivector I

$$\int_{\partial S} d\mathbf{l}\wedge\widehat{\mathbf{H}} = IJ + \frac{d}{dt}\int_S d\widehat{\mathbf{s}}\wedge\mathbf{D}\,. \tag{8}$$

The Ampere-Oersted law, written in our notation, relates the protrudence of $\dot{\mathbf{D}}$ from S with the protrudence of $\widehat{\mathbf{H}}$ from ∂S and with the current J flowing through S.

The *Gauss law* has the form

$$\int_{\partial\Omega} d\mathbf{s}\cdot\mathbf{D} = Q$$

and states that the flux of electric induction through the closed surface $\partial\Omega$ which is the boundary of some region Ω is equal to the electric charge Q

contained in Ω. We intend to use the volutor $d\hat{\mathbf{s}}$ instead of $d\mathbf{s}$. To this end, we multiply both sides by I and use Eq. (0.135)

$$\int_{\partial\Omega} d\hat{\mathbf{s}} \wedge \mathbf{D} = IQ \,. \tag{9}$$

This is the new form of the Gauss law. It relates the protrudence of $\mathbf{D}$ from $\partial\Omega$ with the charge Q contained in Ω. Notice that the closed surface $\partial\Omega$ has counterclockwise orientation when looked at from the outside. This follows from the traditional covention that $d\mathbf{s}$ is directed outwards.

The *magnetic Gauss law* remains as

$$\int_{\partial\Omega} d\mathbf{s} \cdot \mathbf{B} = 0 \,,$$

stating that the magnetic flux through the closed surface is always zero — this corresponds to the lack of magnetic charges in nature.[1)] By virtue of (0.136), we obtain

$$\int_{\partial\Omega} d\hat{\mathbf{s}} \cdot \hat{\mathbf{B}} = 0 \,. \tag{10}$$

This equation shows that the adherence of $\hat{\mathbf{B}}$ to the closed surface is zero.

Equations (7)–(10) have the common name of *integral Maxwell equations*. We shall derive differential equations corresponding to them.

§3. Differential Maxwell Equations

When the electric field $\mathbf{E}$ is differentiable, we may apply Stokes Theorem (0.139) to the right-hand side of (7) such that

$$\frac{d}{dt}\int_S d\hat{\mathbf{s}} \cdot \hat{\mathbf{B}} = -\int_S d\hat{\mathbf{s}} \cdot (\nabla \wedge \mathbf{E}) \,.$$

The surface S is stationary, so we may put the time derivative under the integration sign

$$\int_S d\hat{\mathbf{s}} \cdot \dot{\hat{\mathbf{B}}} + \int_S d\hat{\mathbf{s}} \cdot (\nabla \wedge \mathbf{E}) = 0 \,.$$

The surface of integration is the same in both integrals, hence we may write one integral:

$$\int_S d\hat{\mathbf{s}} \cdot (\dot{\hat{\mathbf{B}}} + \nabla \wedge \mathbf{E}) = 0 \,. \tag{11}$$

---

1)See Footnote on Pg.71.

If the fields $\dot{\widehat{\mathbf{B}}}$ and $\nabla \wedge \mathbf{E}$ are continuous, this equality, valid for any surface S, implies that the integrand vanishes, i.e.

$$\dot{\widehat{\mathbf{B}}} + \nabla \wedge \mathbf{E} = 0 \,. \tag{12}$$

We now transform similarly Eq. (8). The electric current J is expressed as the surface integral of the vector $\mathbf{j}$ of *current density*: $J = \int_S d\mathbf{s} \cdot \mathbf{j}$. Making use of Eq. (0.135), we may write $IJ = \int_S d\widehat{\mathbf{s}} \wedge \mathbf{j}$ and Eq. (8) takes the form

$$\int_{\partial S} d\mathbf{l} \wedge \widehat{\mathbf{H}} = \int_S d\widehat{\mathbf{s}} \wedge \mathbf{j} + \int_S d\widehat{\mathbf{s}} \wedge \dot{\mathbf{D}} \,,$$

where we have also used the stationary quality of the surface S. If the field $\widehat{\mathbf{H}}$ is differentiable, we may apply Stokes Theorem (0.138) to the left-hand side

$$-\int_S d\widehat{\mathbf{s}} \wedge (\nabla \cdot \widehat{\mathbf{H}}) = \int_S d\widehat{\mathbf{s}} \wedge \mathbf{j} + \int_S d\widehat{\mathbf{s}} \wedge \dot{\mathbf{D}} \,.$$

All the integrals are carried over the same surface, thus

$$\int_S d\widehat{\mathbf{s}} \wedge (\nabla \cdot \widehat{\mathbf{H}} + \dot{\mathbf{D}} + \mathbf{j}) = 0 \,. \tag{13}$$

If the fields $\nabla \cdot \widehat{\mathbf{H}}$, $\dot{\mathbf{D}}$ and $\mathbf{j}$ are continuous, the arbitrariness of S implies that the integrand vanishes:

$$\nabla \cdot \widehat{\mathbf{H}} + \dot{\mathbf{D}} + \mathbf{j} = 0 \,. \tag{14}$$

We pass now to Eq. (9). The electric charge Q is expressed as the volume integral of the *charge density* ρ: $Q = \int_\Omega dv\rho$. After introducing the directed volume $d\widehat{V} = Idv$, we write Eq. (9) in the form

$$\int_{\partial\Omega} d\widehat{\mathbf{s}} \wedge \mathbf{D} = \int_\Omega d\widehat{\mathbf{V}}\rho \,.$$

If the field $\mathbf{D}$ is differentiable, we use Eq. (0.142) and obtain

$$\int_\Omega d\widehat{\mathbf{V}}\,(\nabla \cdot \mathbf{D} - \rho) = 0 \,.$$

If the fields $\nabla \cdot \mathbf{D}$ and ρ are continuous, the arbitrariness of Ω implies

$$\nabla \cdot \mathbf{D} - \rho = 0 \,. \tag{15}$$

Finally, we consider Eq. (10). The identity (0.141) is applicable here after assuming the differentiability of $\widehat{\mathbf{B}}$. It gives $\int_\Omega d\widehat{\mathbf{V}} \cdot (\nabla \wedge \widehat{\mathbf{B}}) = 0$ and, if the field $\nabla \wedge \widehat{\mathbf{B}}$ is continuous,

$$\nabla \wedge \widehat{\mathbf{B}} = 0 \, . \tag{16}$$

Equations (12), (14), (15) and (16) are called *differential Maxwell equations*. They were derived under the assumption of the continuity of the expressions $\dot{\widehat{\mathbf{B}}}$, $\nabla \wedge \mathbf{E}$, $\nabla \cdot \widehat{\mathbf{H}}$, $\dot{\mathbf{D}}$, $\mathbf{j}$, $\nabla \cdot \mathbf{D}$, ρ and $\nabla \wedge \widehat{\mathbf{B}}$ — therefore they seem to be less general than the integral ones. However, in case the named expressions are dicontinuous, the differential Maxwell equations are assumed to be valid as distribution equations — in this way, they become equivalent to the integral Maxwell equations.

The differential Maxwell equations can be arranged in two groups. The first one containing outer derivatives

$$\nabla \wedge \mathbf{E} + \frac{\partial \widehat{\mathbf{B}}}{\partial t} = 0 \tag{17a}$$

$$\nabla \wedge \widehat{\mathbf{B}} = 0 \tag{17b}$$

has zeros on the right-hand side. These equations are homogeneous with respect to field quantities. The second group, containing inner derivatives

$$\nabla \cdot \mathbf{H} + \frac{\partial \mathbf{D}}{\partial t} = -\mathbf{j} \tag{17c}$$

$$\nabla \cdot \mathbf{D} = \rho \tag{17d}$$

includes the sources of the electromagnetic field, that is charge or current densities, on the right-hand side. Therefore these equations are inhomogeneous with respect to field quantities. Notice that each equation has a different multivector rank of terms that it contains.

The equations reveal the similarity between the quantities $\widehat{\mathbf{H}}$ and $\mathbf{D}$, which depend directly on sources, and between $\widehat{\mathbf{B}}$ and $\mathbf{E}$, which do not depend immediately on sources, but — as we know from the expression for the Lorentz force — are important for the action of the field on the electric charges. This is called the *Lorentz-Abraham analogy* in the older textbooks. In this connection, one can meet special terminology: *intensity quantities* for $\mathbf{E}$ and $\widehat{\mathbf{B}}$, and *magnitude quantities* for $\mathbf{D}$ and $\widehat{\mathbf{H}}$ (see Ref. 2, Sec. 9).

We are going to show how all the Maxwell equations can be contained in a single equation using Clifford algebra. Inserting $\mathbf{e}$ and $\widehat{\mathbf{b}}$ into Eqs. (17), we obtain

$$\nabla \wedge \left(\frac{1}{\sqrt{\varepsilon}}\mathbf{e}\right) + \frac{\partial}{\partial t}(\sqrt{\mu}\,\widehat{\mathbf{b}}) = 0\,, \qquad \nabla \wedge (\sqrt{\mu}\,\widehat{\mathbf{b}}) = 0\,,$$

$$\nabla \cdot \left(\frac{1}{\sqrt{\mu}}\widehat{\mathbf{b}}\right) + \frac{\partial}{\partial t}(\sqrt{\varepsilon}\,\mathbf{e}) = -\mathbf{j}\,, \qquad \nabla \cdot (\sqrt{\varepsilon}\,\mathbf{e}) = \rho\,.$$

To simplify derivations, let us introduce the notation $\sqrt{\varepsilon} = \eta$, $\sqrt{\mu} = \tau$. After differentiation of the products in the brackets with the use of Problems 0.26 and 0.30, we obtain

$$\left(\nabla\frac{1}{\eta}\right) \wedge \mathbf{e} + \frac{1}{\eta}\nabla \wedge \mathbf{e} + \frac{\partial \tau}{\partial t}\widehat{\mathbf{b}} + \tau\frac{\partial \widehat{\mathbf{b}}}{\partial t} = 0\,, \tag{18}$$

$$(\nabla\tau) \wedge \widehat{\mathbf{b}} + \tau\nabla \wedge \widehat{\mathbf{b}} = 0\,, \tag{19}$$

$$\left(\nabla\frac{1}{\tau}\right) \cdot \widehat{\mathbf{b}} + \frac{1}{\tau}\nabla \cdot \widehat{\mathbf{b}} + \frac{\partial \eta}{\partial t}\mathbf{e} + \eta\frac{\partial \mathbf{e}}{\partial t} = -\mathbf{j}\,, \tag{20}$$

$$(\nabla\eta) \cdot \mathbf{e} + \eta\nabla \cdot \mathbf{e} = \rho\,. \tag{21}$$

We multiply (18) by η and (19) by $\frac{1}{\tau}$ and add them. After using cliffor quantity (4) and the identity $0 = \nabla\left(\eta\frac{1}{\eta}\right) = \frac{1}{\eta}\nabla\eta + \eta\nabla\frac{1}{\eta}$, we obtain

$$\nabla \wedge f + \eta\tau\frac{\partial \widehat{\mathbf{b}}}{\partial t} - \frac{1}{\eta}(\nabla\eta) \wedge \mathbf{e} + \frac{1}{\tau}(\nabla\tau) \wedge \widehat{\mathbf{b}} + \eta\frac{\partial \tau}{\partial t}\widehat{\mathbf{b}} = 0\,.$$

Similarly, multiplication of (20) by τ and (21) by $\frac{1}{\eta}$ and adding yields

$$\nabla \cdot f + \eta\tau\frac{\partial \mathbf{e}}{\partial t} - \frac{1}{\tau}(\nabla\tau) \cdot \widehat{\mathbf{b}} + \frac{1}{\eta}(\nabla\eta) \cdot \mathbf{e} + \tau\frac{\partial \eta}{\partial t}\mathbf{e} = \frac{1}{\eta}\rho - \tau\mathbf{j}\,.$$

We add the last two equations

$$\nabla f + \eta\tau\frac{\partial f}{\partial t} + \frac{1}{\eta}(\mathbf{e}\wedge\nabla\eta + \mathbf{e}\cdot\nabla\eta) + \frac{1}{\tau}(\widehat{\mathbf{b}}\cdot\nabla\tau + \widehat{\mathbf{b}}\wedge\nabla\tau) + \tau\frac{\partial \eta}{\partial t}\mathbf{e} + \eta\frac{\partial \tau}{\partial t}\widehat{\mathbf{b}} = \frac{1}{\eta}\rho - \tau\mathbf{j}\,.$$

We recognize Clifford products in the brackets, therefore we have (on eliminating the provisional notation η and τ):

$$\nabla f + \sqrt{\varepsilon\mu}\frac{\partial f}{\partial t} + \mathbf{e}\frac{1}{\sqrt{\varepsilon}}\left(\nabla\sqrt{\varepsilon} + \sqrt{\varepsilon\mu}\,\frac{\partial\sqrt{\varepsilon}}{\partial t}\right)$$

$$+\,\widehat{\mathbf{b}}\frac{1}{\sqrt{\mu}}\left(\nabla\sqrt{\mu} + \sqrt{\varepsilon\mu}\,\frac{\partial\sqrt{\mu}}{\partial t}\right) = \frac{1}{\sqrt{\varepsilon}}\rho - \sqrt{\mu}\,\mathbf{j}\,.$$

Let us introduce the cliffor differential operator $\mathcal{D} = \nabla + \sqrt{\varepsilon\mu}\frac{\partial}{\partial t}$. Then our equation takes the form

$$\mathcal{D}f + \mathbf{e}\frac{1}{\sqrt{\varepsilon}}\mathcal{D}\sqrt{\varepsilon} + \widehat{\mathbf{b}}\frac{1}{\sqrt{\mu}}\mathcal{D}\sqrt{\mu} = \frac{1}{\sqrt{\varepsilon}}\rho - \sqrt{\mu}\mathbf{j} \, .$$

We see that there is some value in introducing one more cliffor $\tilde{j} = \frac{1}{\sqrt{\varepsilon}}\rho - \sqrt{\mu}\mathbf{j}$, which may be called the ***density of electric sources.*** Then our equation can eventually be written as

$$\mathcal{D}f + \mathbf{e}\mathcal{D}\ln\sqrt{\varepsilon} + \widehat{\mathbf{b}}\mathcal{D}\ln\sqrt{\mu} = \tilde{j} \, . \tag{22}$$

This is the sought-after form of the Maxwell equations contained in a single formula. It contains a combination of multivectors of all ranks. One may reconstruct individual Maxwell equations (18) – (21) selecting only multivectors of specific ranks. In this way, the single equation is equivalent to the four equations of (17). We shall call it the ***Maxwell equation.***

The equation has a remarkable property. If f_1 and f_2 are electromagnetic fields with the sources $\tilde{j}_1$ and $\tilde{j}_2$ respectively, their sum $f_1 + f_2$ is the field with the sources $\tilde{j}_1 + \tilde{j}_2$. This property is called the ***superposition principle*** and is important for solving complicated problems. It permits us to find separately fields produced by independent sources and then to add the results for the sum of sources. It will be used, for instance, in Sec. 6 for finding the magnetic field illustrated on Figs. 35 and 36. In regions devoid of charges and currents, the superposition principle states that the sum of fields being solutions of the Maxwell equation is again a solution of the equation. In optics, it constitutes a mathematical basis for the explanation of the diffraction of the wave as a result of interference.

The relatively developed left-hand side of Eq. (22) stems from the differentiation of products like $\sqrt{\varepsilon}\,\mathbf{e}$ in which both factors can be, in general, functions of time and position. If the material medium is uniform in space and constant in time, the equation becomes simpler

$$\mathcal{D}f = \tilde{j} \, . \tag{23}$$

In the regions where charge and current densities are zero, it takes the form

$$\mathcal{D}f = \nabla f + \sqrt{\varepsilon\mu}\frac{\partial f}{\partial t} = 0 \, . \tag{24}$$

It is then called the ***homogeneous Maxwell equation.***

Let us apply the operator $\mathcal{D}^* = \nabla - \sqrt{\varepsilon\mu}\frac{\partial}{\partial t}$ to it. By virtue of Eq. (0.129) and the substitution $u = 1/\sqrt{\varepsilon\mu}$, we obtain

$$\Box f = \Delta f - \varepsilon\mu\frac{\partial^2 f}{\partial t^2} = 0 . \tag{25}$$

This equality is called the ***wave equation.*** Since the d'Alembertian $\Box$ is a scalar operator, the separation of the vector and bivector parts of (25) gives two equations

$$\Box \mathbf{e} = 0 , \qquad \Box \widehat{\mathbf{b}} = 0$$

which means that the electric and magnetic fields separately satisfy the wave equation. They, however, do not satisfy Eq. (24) separately.

After remembering that the basic vector coefficients $\mathbf{e}_i$ fulfil the algebraic relation $\mathbf{e}_i\mathbf{e}_j + \mathbf{e}_j\mathbf{e}_i = 2\delta_{ij}$, the similarity of our Maxwell equation (24) to the Dirac equation from quantum mechanics becomes visible. The Dirac equation is called a first order wave equation. As we shall see in Chap. 4, Eq. (24) also has a wave solution, therefore it also deserves such a name.

It is worth noticing another property of the Maxwell equation. Let us take both sides of Eq. (22) at the point $(R^{-1}\mathbf{r}, t)$, where R denotes the rotation matrix as in (0.108), and multiply by U^{-1} and U from respective sides:

$$U^{-1}(\mathcal{D} f + \mathbf{e}\,\mathcal{D}\ln\sqrt{\varepsilon} + \widehat{\mathbf{b}}\,\mathcal{D}\ln\sqrt{\mu})U = U^{-1}\tilde{j}U .$$

The scalar part of the operator $\mathcal{D}$ is invariant under rotations, hence Eq. (0.131) implies $\mathcal{D} f' = (\mathcal{D} f)'$ where prime denotes the rotation of function according to Eq. (0.130). This identity applied to the previous equation gives

$$\mathcal{D} f' + \mathbf{e}'\,\mathcal{D}\ln\sqrt{\varepsilon'} + \widehat{\mathbf{b}}'\,\mathcal{D}\ln\sqrt{\mu'} = \tilde{j}' .$$

We see that the rotated functions again satisfy the Maxwell equation. This property is called the ***symmetry of the Maxwell equation under rotations.***

The homogeneous equation (24) has one more symmetry. We know that the unit trivector I commutes with all cliffors. As a constant quantity, it also commutes with derivatives, hence after multiplication of (24) by I, we obtain

$$\nabla(If) + \sqrt{\varepsilon\mu}\,\frac{\partial}{\partial t}\,(If) = 0 .$$

We see that if f satisfies the homogeneous Maxwell equation, then so does If. Let us look more closely at If,

$$If = I(\mathbf{e} + I\mathbf{b}) = I\mathbf{e} + I^2\mathbf{b} = -\mathbf{b} + I\mathbf{e} .$$

Thus If, as the sum of vector and bivector, again describes the electromagnetic field but with interchanged electric and magnetic parts: $\mathbf{e} \to -\mathbf{b}$, $\mathbf{b} \to \mathbf{e}$. We remember that I realizes the Hodge map. Now we ascertain that the *Hodge map of the electromagnetic field constitutes a symmetry of the Maxwell equation.* This symmetry was noticed by Heaviside at the end of the last century and is also called *duality symmetry.*

The existence of this symmetry implies that the mapping

$$f \to e^{I\alpha} f \tag{26}$$

for real α is also a symmetry of the Maxwell equation, since after applying the operator $e^{I\alpha} = \cos\alpha + I\sin\alpha$ to Eq. (24) we obtain

$$\nabla(e^{I\alpha} f) + \sqrt{\varepsilon\mu}\,\frac{\partial}{\partial t}(e^{I\alpha} f) = 0\,.$$

This symmetry of the Maxwell equations was considered by Larmor and Reinich around 1925. The electromagnetic cliffor changes under mapping (26) as follows

$$\begin{aligned} f \to e^{I\alpha} f &= (\cos\alpha + I\sin\alpha)(\mathbf{e} + I\mathbf{b}) \\ &= \mathbf{e}\cos\alpha - \mathbf{b}\sin\alpha + I\mathbf{e}\sin\alpha + I\mathbf{b}\cos\alpha\,. \end{aligned} \tag{27}$$

This means that for the electric and magnetic fields separately,

$$\mathbf{e} \to \mathbf{e}\cos\alpha - \mathbf{b}\sin\alpha\,, \qquad \mathbf{b} \to \mathbf{b}\cos\alpha + \mathbf{e}\sin\alpha\,.$$

Let us call it the *Larmor-Reinich transformation.*

We now intend to derive a conclusion following from the inhomogeneous equations (17c,d). The inner derivative of (17c) is

$$\nabla\cdot(\nabla\cdot\mathbf{H}) + \nabla\cdot\frac{\partial\mathbf{D}}{\partial t} = -\nabla\cdot\mathbf{j}\,.$$

The first term vanishes by virtue of Eq. (0.125), hence

$$\nabla\cdot\frac{\partial\mathbf{D}}{\partial t} = -\nabla\cdot\mathbf{j}\,.$$

The time derivative of (17d) is

$$\frac{\partial}{\partial t}\nabla\cdot\mathbf{D} = \frac{\partial\rho}{\partial t}\,.$$

The left-hand sides of the last two equations are identical for sufficiently regular fields $\mathbf{D}$. In this way, we obtain

$$\frac{\partial \rho}{\partial t} + \nabla \cdot \mathbf{j} = 0 . \tag{28}$$

This is the condition relating charge and current densities ρ and $\mathbf{j}$. It is called the *continuity equation* for the electric charge, since after integration over a region Ω, it gives $\frac{d}{dt} \int_\Omega \rho \, dv = - \int_\Omega dv (\nabla \cdot \mathbf{j})$. We apply Gauss Theorem (0.140)

$$\frac{d}{dt} Q(\Omega) = - \int_{\partial\Omega} d\mathbf{s} \cdot \mathbf{j} .$$

Now we see distinctly the conservation of charge: the time rate of charge $Q(\Omega)$ contained in Ω is equal to the current flowing through the boundary $\partial\Omega$ into Ω.

§4. Boundary Conditions at an Interface

It happens quite often that the physical properties of a medium described by ε and μ sharply change on a surface Σ which is treated as an interface separating two distinct media. One may expect then that the fields $\mathbf{E}, \mathbf{D}, \widehat{\mathbf{B}}, \widehat{\mathbf{H}}$ are also discontinuous and the densities ρ and $\mathbf{j}$ are to be replaced by the *surface density of charge* σ and *linear density of surface currents* $\mathbf{i}$. In this case, we have equation $Q = \int_\Sigma ds\sigma$ for the charge collected on the interface Σ, and $J = \int_L (\mathbf{n} \times d\mathbf{l}) \cdot \mathbf{i}$ for the current flowing on Σ through a directed curve L with the length element $d\mathbf{l}$ and the normal vector $\mathbf{n}$ to Σ. By virtue of Eqs. (0.48) and (0.76), we may write

$$IJ = \int_L (\mathbf{n} \wedge d\mathbf{l}) \wedge \mathbf{i} . \tag{29}$$

We shall now find conditions for the discontinuities of the fields on the interface, imposed by the integral Maxwell equations. We choose S for Eq. (7) as the rectangle perpendicular to Σ, as shown in Fig. 27. We assume that S is stationary, therefore we may write (7) as

$$\int_{\partial S} d\mathbf{l} \cdot \mathbf{E} = \int_S d\widehat{\mathbf{s}} \cdot \dot{\widehat{\mathbf{B}}} .$$

We let the height $AD = BC$ tend to zero. Then the area of the rectangle also tends to zero. If we assume that the time derivative of magnetic induction is bounded, the integral on the right-hand side tends to zero. If, moreover, the

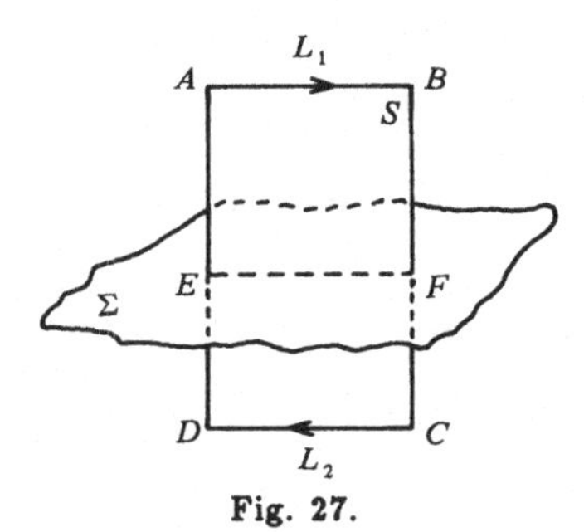

Fig. 27.

electric field is bounded, the linear integral over segments DA and BC also tends to zero. We thus obtain, in the limit,

$$\int_{L_1} d\mathbf{l} \cdot \mathbf{E}_1 + \int_{L_2} d\mathbf{l} \cdot \mathbf{E}_2 = 0$$

where the subscripts 1 and 2 refer to the media above and below the interface Σ. Both linear integrals are in fact performed over the same segment, only on opposite sides of Σ and with opposite directions, so we have

$$\int_{EF} d\mathbf{l} \cdot (\mathbf{E}_1 - \mathbf{E}_2) = 0 \ .$$

The segment EF is arbitrary on Σ, hence we infer that the tangent component of the integrand to the interface must vanish, which means that

$$\mathbf{E}_{1t} = \mathbf{E}_{2t} \ .$$

We arrive at the conclusion that the *tangent component of the electric field must be continuous.*

In order to obtain a condition for the normal component of the electric field, we choose Ω for Eq. (9) as the cylinder with bases S_1 and S_2 parallel to Σ, and with side surface denoted S_s. Σ intersects Ω along S_0 as shown in Fig. 28. We give to $\partial\Omega$ the orientation counterclockwise when looked at from the outside. The integral over the closed surface $\partial\Omega$ can be separated into three parts and Eq. (9) takes the form

$$\int_{S_1} d\hat{\mathbf{s}} \wedge \mathbf{D}_1 + \int_{S_2} d\hat{\mathbf{s}} \wedge \mathbf{D}_2 + \int_{S_s} d\hat{\mathbf{s}} \wedge \mathbf{D} = IQ(h) \ .$$

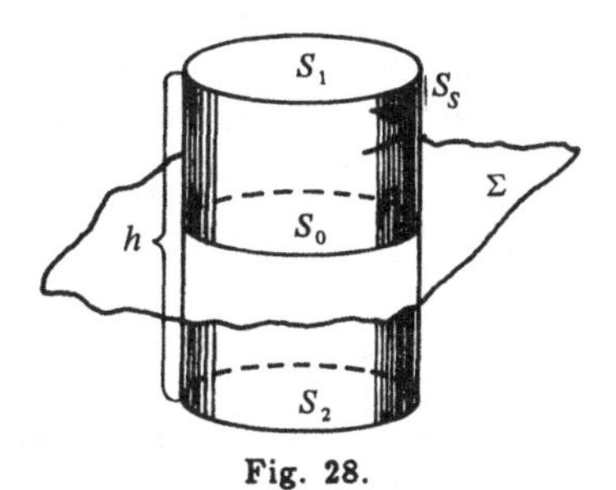

Fig. 28.

The charge contained in the cylinder changes together with its height h, so we have written the functional dependence $Q(h)$.

We let h tend to zero. If we assume that $\mathbf{D}$ is bounded, the integral over S_s tends to zero and we are left with

$$\int_{S_1} d\hat{\mathbf{s}} \wedge \mathbf{D}_1 + \int_{S_2} d\hat{\mathbf{s}} \wedge \mathbf{D}_2 = I \lim_{h \to 0} Q(h) \ .$$

The surfaces S_1 and S_2 coincide with S_0 in the limit, only the orientations are opposite. We assume that S_0 and S_1 have the same orientation, hence after substitution of $\lim_{h\to 0} Q(h) = \int_{S_0} ds\sigma$, we get

$$\int_{S_0} d\hat{\mathbf{s}} \wedge (\mathbf{D}_1 - \mathbf{D}_2) = I \int_{S_0} ds\, \sigma \ .$$

We introduce the unit bivector $\hat{\mathbf{n}}$ such that $d\hat{\mathbf{s}} = \hat{\mathbf{n}}\, ds$

$$\int_{S_0} ds[\hat{\mathbf{n}} \wedge (\mathbf{D}_1 - \mathbf{D}_2) - I\sigma] = 0 \ .$$

The circle S_0 is arbitrary on Σ, hence we infer that the integrand must vanish: $\hat{\mathbf{n}} \wedge (\mathbf{D}_1 - \mathbf{D}_2) = I\sigma$. We write $\hat{\mathbf{n}} = I\,\mathbf{n}$ for a normal vector $\mathbf{n}$ to Σ and use the identity (0.46): $I[(\mathbf{D}_1 - \mathbf{D}_2) \cdot \mathbf{n}] = I\sigma$. This means that $(\mathbf{D}_1 - \mathbf{D}_2) \cdot \mathbf{n} = \sigma$ or

$$D_{1n} - D_{2n} = \sigma$$

where D_{1n}, D_{2n} represent the coordinates of vectors $\mathbf{D}_1$, $\mathbf{D}_2$ along the direction $\mathbf{n}$ with its orientation towards the medium 1. After multiplying the last equality by $\mathbf{n}$, we obtain

$$\mathbf{D}_{1n} - \mathbf{D}_{2n} = \sigma\,\mathbf{n} \ . \tag{30}$$

Here $\mathbf{D}_{1n}$, $\mathbf{D}_{2n}$ are the components of $\mathbf{D}_1$, $\mathbf{D}_2$ orthogonal to Σ. This equation states that the *normal component of the electric induction has a leap proportional to the surface density of charge.*

In the case of lacking surface charges, we obtain the condition $\mathbf{D}_{1n} = \mathbf{D}_{2n}$, i.e.

$$\varepsilon_1 \mathbf{E}_{1n} = \varepsilon_2 \mathbf{E}_{2n} .$$

This shows that, for $\varepsilon_1 \neq \varepsilon_2$, the electric field must have a discontinuous normal component. This fact implies the occurrence of deflection of the force lines for the electric field on the interface between two different dielectric media (see Problem 4).

Now we consider a *recess*, that is, an empty space, in the dielectric medium without any surface charges. It follows from the continuity of $\mathbf{E}_t$ and $\mathbf{D}_n$ that the force acting on a test charge located in such a recess depends on its shape. For simplicity, we consider a recess in the shape of a cylinder with the axis parallel to $\mathbf{E}$. If the height of the cylinder is large in comparison with its diameter, the continuity of $\mathbf{E}_t$ far from the bases of the cylinder implies the relation

$$\mathbf{E}_{in} = \mathbf{E}$$

where $\mathbf{E}_{in}$ denotes the electric field inside the recess. If the height is small in comparison with the diameter, the continuity of $\mathbf{D}_n$ (far from the side surface of the cylinder) implies $\mathbf{D}_{in} = \mathbf{D}$, such that

$$\mathbf{E}_{in} = \frac{1}{\varepsilon_0} \mathbf{D}$$

where ε_0 is the electric permittivity of the vacuum. Both situations are illustrated in Fig. 29. The following rule stems from these facts: A measurement of a force acting on a test charge located in a long slit parallel to the electric force lines determines the electric field, whereas in a flattened slit perpendicular to the force lines it gives the electric induction (divided by ε_0) — this is called the *Kelvin rule.*

In order to find the continuity condition following from Eq. (10) we again choose the integration surface of Fig. 28 and Eq. (10) thus gives $\int_S d\widehat{\mathbf{s}} \cdot \widehat{\mathbf{B}} = 0$. In the limit $h \to 0$ we are left with

$$\int_{S_0} d\widehat{\mathbf{s}} \cdot (\widehat{\mathbf{B}}_1 - \widehat{\mathbf{B}}_2) = 0 .$$

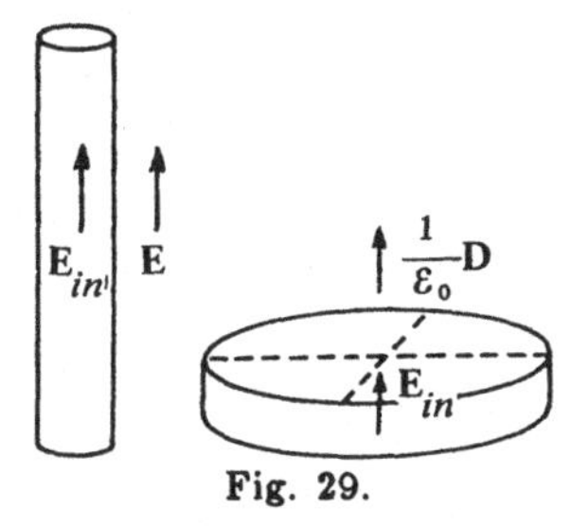

Fig. 29.

Since the circle S_0 on Σ is arbitrary, the integrand must have the tangent component to the interface equal to zero

$$\hat{\mathbf{B}}_{1t} = \hat{\mathbf{B}}_{2t} \ .$$

We obtain the conclusion that the *tangent component of the magnetic induction bivector must be continuous.*

We now pass to Eq. (8). We choose the integration surface similar to that shown in Fig. 27 but we additionally mark the normal vector $\mathbf{n}$ to the interface (see Fig. 30). This surface is stationary, therefore we obtain from (8)

$$\int_{\partial S} d\mathbf{l} \wedge \hat{\mathbf{H}} = IJ + \int_S d\hat{\mathbf{s}} \wedge \dot{\mathbf{D}} \ .$$

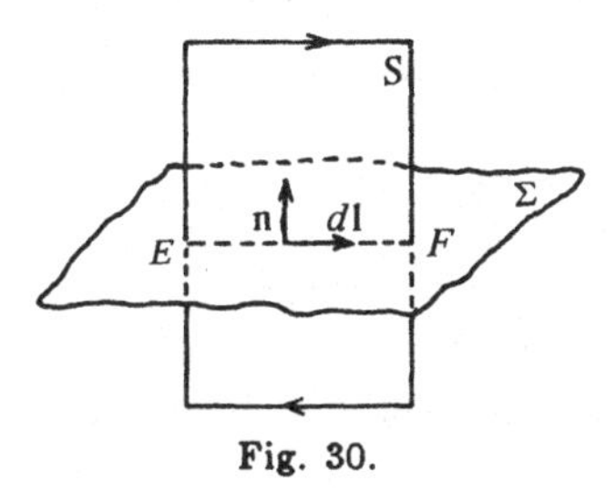

Fig. 30.

When the height h tends to zero and the derivative $\dot{\mathbf{D}}$ is bounded, the last integral tends to zero. If, moreover, $\hat{\mathbf{H}}$ is bounded, the integral over vertical segments of the boundary ∂S also tends to zero. Thus we obtain in the limit

$$\int_{EF} d\mathbf{l} \wedge \hat{\mathbf{H}}_1 - \int_{EF} d\mathbf{l} \wedge \hat{\mathbf{H}}_2 = I \lim_{h \to 0} J(h) \ , \tag{31}$$

where $J(h)$ is the current flowing through the rectangle S of the height h.

Let us look more closely at the surface integral giving the current

$$IJ(h) = \int_S d\hat{\mathbf{s}} \wedge \mathbf{j} = \int_{EF} \int_h (\mathbf{n} \wedge d\mathbf{l}) \wedge \mathbf{j}\, dh\,,$$

since $d\hat{\mathbf{s}} = \mathbf{n}\, dh \wedge dl$. If surface currents exist, the surface density $\mathbf{j}$ of the volume currents is singular and should be replaced by the linear density $\mathbf{i}$ of the surface currents: $\mathbf{i} = \lim_{h \to 0} \int_h \mathbf{j}\, dh$. Thus we obtain

$$I \lim_{h \to 0} J(h) = \int_{EF} (\mathbf{n} \wedge d\mathbf{l}) \wedge \mathbf{i} = \int_{EF} d\mathbf{l} \wedge (\mathbf{i} \wedge \mathbf{n})\,,$$

where we use the associativity and cyclical symmetry for the outer product. Now Eq. (31) takes the form

$$\int_{EF} d\mathbf{l} \wedge (\hat{\mathbf{H}}_1 - \hat{\mathbf{H}}_2 - \mathbf{i} \wedge \mathbf{n}) = 0\,.$$

Since the segment EF lying on Σ is otherwise arbitrary, we obtain the following condition for the normal components

$$\hat{\mathbf{H}}_{1n} - \hat{\mathbf{H}}_{2n} = \mathbf{i} \wedge \mathbf{n}\,,$$

because the outer product $d\mathbf{l} \wedge \hat{\mathbf{H}}$ "picks out" from $\hat{\mathbf{H}}$ only the component perpendicular to $d\mathbf{l}$. The volutor $\mathbf{i} \wedge \mathbf{n}$ is already normal to Σ. Using the perpendicularity $\mathbf{i} \perp \mathbf{n}$, we may also write the Clifford product $\mathbf{i}\mathbf{n}$ instead of $\mathbf{i} \wedge \mathbf{n}$:

$$\hat{\mathbf{H}}_{1n} - \hat{\mathbf{H}}_{2n} = \mathbf{i}\mathbf{n}\,. \tag{32}$$

We see from this that the *normal component of the magnetic field bivector has a leap proportional to the linear density of the surface currents.*

In the absence of surface currents, we have the condition $\hat{\mathbf{H}}_{1n} = \hat{\mathbf{H}}_{2n}$, i.e.

$$\frac{1}{\mu_1}\hat{\mathbf{B}}_{1n} = \frac{1}{\mu_2}\hat{\mathbf{B}}_{2n}\,.$$

This means that for $\mu_1 \neq \mu_2$, the magnetic induction (as the bivector) must have a discontinuous normal component.

When considering a recess in the magnetic medium, we discover — just as before — the corresponding *Kelvin-like rule:* A measurement of forces acting on test charges moving in a flat slit parallel to the magnetic force surfaces determines the magnetic induction, whereas for those moving in a long slit perpendicular to the force surfaces gives the magnetic field (multiplied by μ_0) (see Fig. 31). Comparison of this sentence with the Kelvin rule for the electric field confirms once more the similarity between $\mathbf{E}$ and $\hat{\mathbf{B}}$ on the one hand and between $\mathbf{D}$ and $\hat{\mathbf{H}}$ on the other.

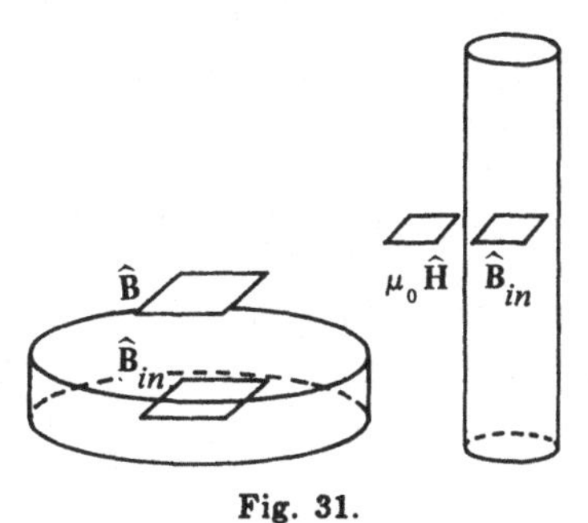

Fig. 31.

Let us make a comparison of all four conditions obtained:

$$\mathbf{E}_{1t} = \mathbf{E}_{2t}\,, \quad \mathbf{D}_{1n} - \mathbf{D}_{2n} = \sigma\mathbf{n}\,,$$
$$\hat{\mathbf{B}}_{1t} = \hat{\mathbf{B}}_{2t}\,, \quad \hat{\mathbf{H}}_{1n} - \hat{\mathbf{H}}_{2n} = \mathbf{i}\mathbf{n}\,.$$

(The normal vector $\mathbf{n}$ here has orientation from medium 2 to 1.) We see that the Lorentz-Abraham analogy is plainly brought out in these equations. We may summarize the conditions as follows: The intensity quantities $\mathbf{E}$ and $\hat{\mathbf{B}}$ have continuous components tangent to the interface, whereas the magnitude quantities $\mathbf{D}$ and $\hat{\mathbf{H}}$ have leaps of normal components when surface charges and currents occur, or have normal components continuous when surface charges and currents are absent. We call them *boundary conditions at the interface between two media.*

In the traditional language, using pseudovectors for the magnetic quantities, the conditions are written as

$$\mathbf{E}_{1t} = \mathbf{E}_{2t}\,, \quad \mathbf{D}_{1n} - \mathbf{D}_{2n} = \sigma\mathbf{n}\,,$$
$$\mathbf{B}_{1n} = \mathbf{B}_{2n}\,, \quad \mathbf{H}_{1t} - \mathbf{H}_{2t} = \mathbf{i} \times \mathbf{n}\,,$$

since the normal component of a pseudovector is replaced by the tangent component of its dual bivector and vice versa. In the case of lacking surface charges and currents, the conditions are composed differently,

$$\mathbf{E}_{1t} = \mathbf{E}_{2t}\,, \quad \mathbf{D}_{1n} = \mathbf{D}_{2n}\,,$$
$$\mathbf{H}_{1t} = \mathbf{H}_{2t}\,, \quad \mathbf{B}_{1n} = \mathbf{B}_{2n}\,,$$

which suggest the similarity of **E** to **H** and of **D** to **B**. This is called the *Hertz-Heaviside analogy* and is based mainly on the boundary conditions at the interface. We see, however, that this analogy depends essentially on the mathematical language (pseudovectors instead of bivectors) used for the description, and not the physical essence of the quantities. This is reflected in the terminology existing up to now (the electric field, the magnetic field, the electric induction, the magnetic induction), and in the relations between them:

$$\mathbf{D} = \varepsilon \mathbf{E}, \quad \mathbf{B} = \mu \mathbf{H}.$$

After replacing pseudovectors by bivectors, the boundary conditions at the interface also become the mainstay of the Lorentz-Abraham analogy. The relations between intensity and magnitude quantities should be written as

$$\mathbf{D} = \varepsilon \mathbf{E}, \quad \widehat{\mathbf{H}} = \frac{1}{\mu} \widehat{\mathbf{B}},$$

stressing in this way the correspondence between ε and $\frac{1}{\mu}$. This new correspondence is also visible in our Maxwell equation (22) where we introduced the cliffor $\tilde{j} = \frac{1}{\sqrt{\varepsilon}} \rho - \sqrt{\mu}\mathbf{j}$.

§5. Finding the Static Fields

Now we pass to the problem of solving Eq. (23) with the assumption that sources and fields are static, that is $\tilde{j}(\mathbf{r}, t) = \tilde{j}(\mathbf{r})$ and $\partial f/\partial t = 0$. In this case, Eq. (23) transforms into

$$\nabla f(\mathbf{r}) = \tilde{j}(\mathbf{r}),$$

i.e., into Eq. (0.154). Therefore, after using Lemmas 1 and 2 of Sec. 0.12, we may express its solution as

$$f(\mathbf{r}) = \frac{1}{4\pi} \int dv' \frac{(\mathbf{r} - \mathbf{r}')\, \tilde{j}\,(\mathbf{r}')}{|r - r'|^3} + f_0\,(\mathbf{r})\,. \tag{33}$$

where f_0 satisfies the homogeneous equation $\nabla f_0 = 0$. The term f_0 describes fields produced by some other sources not contained in the function $\tilde{j}$ and not included by integration — they are called *external sources*. From now on, we shall omit this term.

Let us display respective multivector parts of the integrand in (33). After substituting $\tilde{j} = \frac{1}{\sqrt{\varepsilon}}\rho - \sqrt{\mu}\mathbf{j}$, the numerator has the decomposition

$$-\sqrt{\mu}\,(\mathbf{r} - \mathbf{r}') \cdot \mathbf{j} + (\mathbf{r} - \mathbf{r}')\, \frac{1}{\sqrt{\varepsilon}} \rho - \sqrt{\mu}(\mathbf{r} - \mathbf{r}') \wedge \mathbf{j}\,.$$

We have here successively: scalar, vector and bivector parts. The scalar part should be absent at the electromagnetic cliffor f, hence the integral of the first term should vanish. This fact can be checked but for the sake of brevity, we omit it at this place and delegate it to Appendix I.

We equate separately the vector and bivector parts of (33) using the assumption that ε and μ do not depend on position:

$$\mathbf{e}(\mathbf{r}) = \frac{1}{4\pi\sqrt{\varepsilon}} \int dv' \frac{(\mathbf{r}-\mathbf{r}')\rho(\mathbf{r}')}{|\mathbf{r}-\mathbf{r}'|^3} ,$$
$$\widehat{\mathbf{b}}(\mathbf{r}) = -\frac{\sqrt{\mu}}{4\pi} \int dv' \frac{(\mathbf{r}-\mathbf{r}') \wedge \mathbf{j}(\mathbf{r}')}{|\mathbf{r}-\mathbf{r}'|^3} .$$

This implies the following expressions for the magnitude quantities

$$\mathbf{D}(\mathbf{r}) = \frac{1}{4\pi} \int dv' \frac{\rho(\mathbf{r}') \wedge (\mathbf{r}-\mathbf{r}')}{|\mathbf{r}-\mathbf{r}'|^3} , \tag{34}$$

$$\widehat{\mathbf{H}}(\mathbf{r}) = \frac{1}{4\pi} \int dv' \frac{\mathbf{j}(\mathbf{r}') \wedge (\mathbf{r}-\mathbf{r}')}{|\mathbf{r}-\mathbf{r}'|^3} . \tag{35}$$

Notice that the medium characteristics ε and μ are absent in them. The quantities $\mathbf{D}$ and $\widehat{\mathbf{H}}$ do not depend on intermediary medium — they are more directly related to the sources ρ and $\mathbf{j}$.

We apply Eq. (34) to the particular situation when only one point charge Q exists located in the point $\mathbf{r}_0$. Then, by virtue of Sec. 0.12, we substitute $\rho(\mathbf{r}) = Q\delta^3(\mathbf{r}-\mathbf{r}_0)$ and obtain from (34) the expression

$$\mathbf{D}(\mathbf{r}) = \frac{Q}{4\pi} \frac{\mathbf{r}-\mathbf{r}_0}{|\mathbf{r}-\mathbf{r}_0|^3} .$$

The well known *Coulomb field* expression

$$\mathbf{E}(\mathbf{r}) = \frac{Q}{4\pi\varepsilon} \frac{\mathbf{r}-\mathbf{r}_0}{|\mathbf{r}-\mathbf{r}_0|^3} , \tag{36}$$

follows from this. In this way, our interpretation of the Dirac delta is confirmed as the charge density of a single point charge. Equation (34), instead, may be seen as the superposition of the Coulomb fields produced by the charge elements $\rho(\mathbf{r}')dv'$ located in various points $\mathbf{r}'$.

When the electric current arises from a system of charges with the density $\rho(\mathbf{r})$ moving with velocities $\mathbf{v}(\mathbf{r})$, then one expresses the current density as $\mathbf{j}(\mathbf{r}) = \rho(\mathbf{r})\mathbf{v}(\mathbf{r})$. If it is a single point charge, the current cannot be stationary, but the current density has a similar form

$$\mathbf{j}(\mathbf{r},t) = Q\mathbf{v}(t)\delta^3(\mathbf{r}-\mathbf{r}_Q(t)) \,, \tag{37}$$

where the function $t \to \mathbf{r}_Q(t)$ is the motion of charge and $\mathbf{v}(t)$ is its velocity.

Now we transform the expression (35) to the form known in general physics. We substitute for the volume element, $dv = ds dl$, and introduce vector $d\mathbf{l} \parallel \mathbf{j}$:

$$\widehat{\mathbf{H}}(\mathbf{r}) = \frac{1}{4\pi}\iint ds' j(\mathbf{r}')\,\frac{d\mathbf{l}' \wedge (\mathbf{r}-\mathbf{r}')}{|\mathbf{r}-\mathbf{r}|^3} \,.$$

Now we assume a thin conductor, which allows us to perform the integration $\int ds' j = J$ before the integration over $d\mathbf{l}'$:

$$\widehat{\mathbf{H}}(\mathbf{r}) = \frac{J}{4\pi}\int \frac{d\mathbf{l}' \wedge (\mathbf{r}-\mathbf{r}')}{|\mathbf{r}-\mathbf{r}'|^3} \,.$$

This may be rewritten with the use of the pseudovector $\mathbf{H}$ in place of $\widehat{\mathbf{H}}$:

$$\mathbf{H}(\mathbf{r}) = \frac{J}{4\pi}\int \frac{d\mathbf{l} \times (\mathbf{r}-\mathbf{r}')}{|\mathbf{r}-\mathbf{r}'|^3} \,.$$

This is the traditional integral form of the *Biot-Savart law.*

§6. Fields Symmetric Under Translations in One Direction

Now we shall look for several static fields in situations where the sources have translational symmetry in one direction. It will turn out that the fields have the same symmetry.

6.1. Electric Field of a Charged Rectilinear Conductor

Let an infinite straight line conductor have a negligibly small cross section, let it be uniformly charged, such that it has the constant linear density of charge τ. We choose a Z-axis coinciding with the conductor. Then the volume density of charge may be taken as $\rho(x,y,z) = \tau\delta(x)\delta(y)$. After substituting this in Eq. (34), we obtain

$$\mathbf{D}(\mathbf{r}) = \frac{\tau}{4\pi}\int_{-\infty}^{\infty} dz' \frac{x\,\mathbf{e}_1 + y\,\mathbf{e}_2 + (z-z')\,\mathbf{e}_3}{[x^2+y^2+(z-z')^2]^{\frac{3}{2}}} \,.$$

Introducing the vector $\mathbf{r}_\perp = x\,\mathbf{e}_1 + y\,\mathbf{e}_2$ — this is the directed distance from the conductor to the point $\mathbf{r}$, or the component of $\mathbf{r}-\mathbf{r}'$ perpendicular to the conductor, we may write

$$\mathbf{D}\,(\mathbf{r}) = \frac{\tau}{4\pi}\left\{\mathbf{r}_\perp \int_{-\infty}^{\infty} \frac{du}{(r_\perp^2 + u^2)^{\frac{3}{2}}} - \mathbf{e}_3 \int_{-\infty}^{\infty} \frac{u\,du}{(r_\perp^2 + u^2)^{\frac{3}{2}}}\right\}\,,$$

where $u = z' - z$. The second integral is zero because the integrand is antisymmetric and the set of integration is symmetric. In this way, we obtain

$$\mathbf{D}\,(\mathbf{r}) = \frac{\tau}{2\pi r_\perp^2}\,\mathbf{r}_\perp = \frac{\tau}{2\pi}\,\mathbf{r}_\perp^{-1}\,. \tag{38}$$

This result shows that the vectors $\mathbf{D}$ and, of course, $\mathbf{E}$ are directed radially from or to the conductor (depending on whether the charge density τ is positive or negative) and always perpendicular to the conductor, and that the magnitude of the field decreases as $1/r_\perp$ as the point $\mathbf{r}$ moves away (see Fig. 32). The force lines of the electric field are radii perpendicular to the conductor, with the orientation to infinity for $\tau > 0$ or opposite for $\tau < 0$. The whole field has translational symmetry parallel to the conductor and axial symmetry around it.

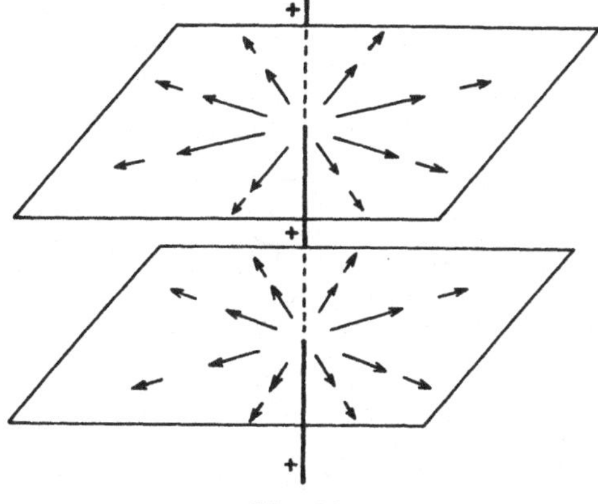

Fig. 32.

6.2. Magnetic Field of a Rectilinear Current

Let the electric current J flow in an infinite straight line conductor with a negligibly small cross-section. We choose the coordinate frame as previously

— then the current density $\mathbf{j}(x, y, z) = J\delta(x)\delta(y)\,\mathbf{e}_3$ may be substituted in (35)

$$\widehat{\mathbf{H}}(\mathbf{r}) = \frac{J}{4\pi}\int_{-\infty}^{\infty} dz' \frac{\mathbf{e}_3 \wedge (x\,\mathbf{e}_1 + y\,\mathbf{e}_2)}{[x^2 + y^2 + (z - z')^2]^{\frac{3}{2}}} = \frac{J}{4\pi}\mathbf{e}_3 \wedge \mathbf{r}_\perp \int_{-\infty}^{\infty} \frac{du}{(u^2 + r_\perp^2)^{\frac{3}{2}}} ,$$

which gives

$$\widehat{\mathbf{H}}(\mathbf{r}) = \frac{J}{2\pi}\mathbf{e}_3 \wedge \mathbf{r}_\perp^{-1} . \tag{39}$$

We see that the volutors $\widehat{\mathbf{H}}$ and, of course, $\widehat{\mathbf{B}}$ are parallel to the conductor and to $\mathbf{r}_\perp$ and the magnitude decreases as $1/r_\perp$ (see Fig. 33). The force surfaces of this field are half-planes spreading out from the conductor to infinity (see Fig. 34). Their orientation is compatible with the orientation of the current if the conductor is treated as the boundary of the surfaces. The whole magnetic field has translational symmetry parallel to the conductor and axial symmetry around it. It has also reflexion symmetry with respect to any plane containing the conductor — this symmetry is not so easily visible when the vectorial lines of the magnetic field are used. The pseudovectors merely follow a different transformation law — their reflexion must be accompanied by a multiplication of minus one.

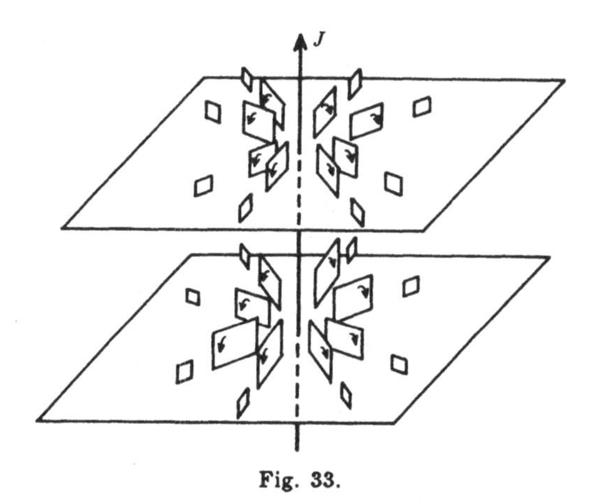

Fig. 33.

6.3. *Electromagnetic Field for Sources with Translational Symmetry in One Direction*

Notice the interesting relation between fields (38) and (39)

$$\widehat{\mathbf{H}}(\mathbf{r}) = \alpha\,\mathbf{e}_3 \wedge \mathbf{D}(\mathbf{r}) , \tag{40}$$

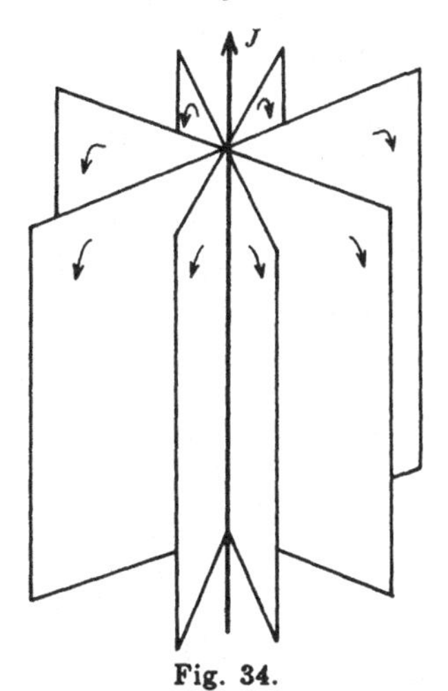

Fig. 34.

with the scalar coefficient $\alpha = \frac{J}{\tau}$. All other possible fields produced by sources invariant under translations parallel to an axis (let it be the Z-axis) can be considered as superpositions of fields (38) and (39). If one assumes proportionality of the charge and current densities

$$\mathbf{j}(\mathbf{r}) = \alpha\, \mathbf{e}_3 \wedge \rho(\mathbf{r}) = \mathbf{k} \wedge \rho(\mathbf{r}) \,, \tag{41}$$

for $\mathbf{k} = \alpha\, \mathbf{e}_3$, relation (40) remains satisfied even after the superposition. Moreover, inserting (41) directly to (35) gives

$$\hat{\mathbf{H}}(\mathbf{r}) = \frac{\mathbf{k}}{4\pi} \wedge \int dv' \frac{\rho(\mathbf{r}') \wedge (\mathbf{r} - \mathbf{r}')}{|\mathbf{r} - \mathbf{r}'|^3} = \mathbf{k} \wedge \mathbf{D}(\mathbf{r}) \,,$$

that is, the relation (40). We may also write $\hat{\mathbf{H}}(\mathbf{r}) = \mathbf{k}\,\mathbf{D}(\mathbf{r})$ with the Clifford product, because $\mathbf{k} \perp \mathbf{D}$. This result may be summarized as follows: If the proportional charge and current densities have symmetry under translation parallel to an axis, and if the current is parallel to the axis, the magnetic field is proportional to the electric field and the force surfaces of the magnetic field are formed by connecting the force lines of the electric field parallelly to the distinguished axis.

Such a situation occurs for a pair of parallel conductors charged with the same charge density and for two parallel equal currents flowing through the same conductors. We illustrate this in Figs. 35 a and b. Force lines and surfaces lying in the plane connecting both conductors have a singularity halfway between the conductors, since they have opposite orientations on both sides. This is possible because the fields are zero exactly at mid-distance.

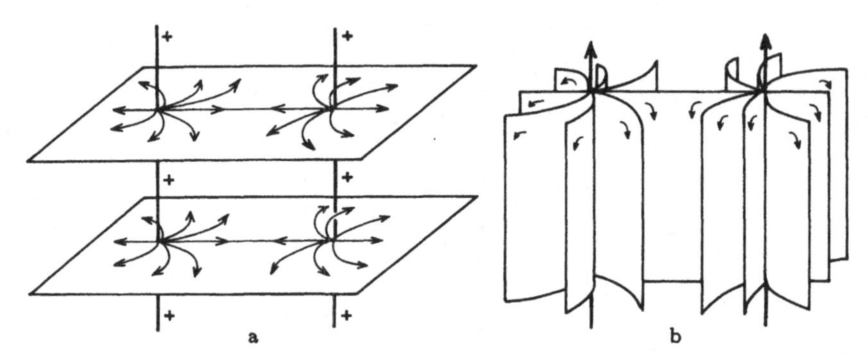

Fig. 35.

Another situation of this type occurs for two conductors with opposite charge densities and for two antiparallel opposite currents flowing in the same pair of conductors. Force lines for this situation are shown in Figs. 36 a and b.

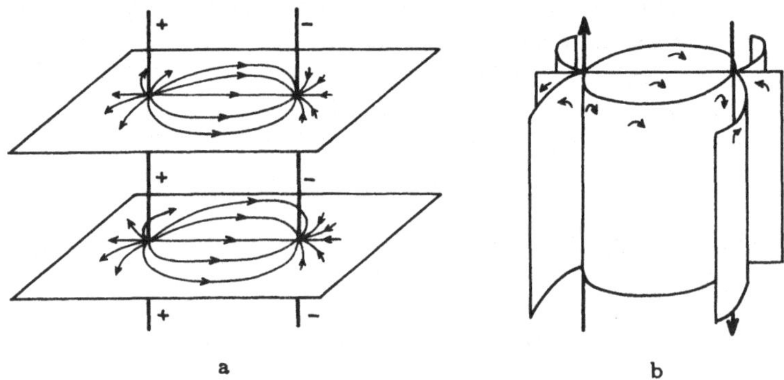

Fig. 36.

6.4. Electric and Magnetic Fields of an Infinite Plate

Let us consider an infinite plane plate with negligible thickness, uniformly charged, that is, with constant surface density charge σ. We choose the coordinate plane $(\mathbf{e}_2, \mathbf{e}_3)$ parallel to the plate and let the intersection point of the plate with X-axis have the coordinate 0. Then the volume density charge is $\rho(x, y, z) = \sigma\delta(x)$. We insert this into Eq. (34):

$$\mathbf{D}(\mathbf{r}) = \frac{\sigma}{4\pi} \iint dy'\, dz' \, \frac{\mathbf{e}_1\, x + \mathbf{e}_2\,(y - y') + \mathbf{e}_3\,(z - z')}{[x^2 + (y - y')^2 + (z - z')^2]^{\frac{3}{2}}} \,.$$

Introducing the vector $\mathbf{r}_\perp = x\,\mathbf{e}_1$ — this is the directed distance from the plate to the point $\mathbf{r}$ or the component of $\mathbf{r} - \mathbf{r}'$ perpendicular to the plate —

we write

$$\mathbf{D}(\mathbf{r}) = \frac{\sigma}{4\pi}\left[\mathbf{r}_\perp \iint \frac{du\,dv}{(r_\perp^2+u^2+v^2)^{\frac{3}{2}}} - \mathbf{e}_2 \iint \frac{u\,du\,dv}{(r_\perp^2+u^2+v^2)^{\frac{3}{2}}} - \mathbf{e}_3 \iint \frac{v\,du\,dv}{(r_\perp^2+u^2+v^2)^{\frac{3}{2}}}\right],$$

where $u = y' - y, v = z' - z$. The last two integrals are zero since the integrand is antisymmetric. In this way, we obtain

$$\mathbf{D}(\mathbf{r}) = \frac{\sigma}{2}\frac{\mathbf{r}_\perp}{r_\perp}\,. \tag{42}$$

We see that the electric field vectors are perpendicular to the plate, opposite to each other, on both sides of the plate. The orientation is away from the plate for $\sigma > 0$ and towards the plate for $\sigma < 0$. The magnitude $|\mathbf{D}|$ is constant, since $\mathbf{r}_\perp/r_\perp = \pm\,\mathbf{e}_1$ which means that the field is uniform on each side of the plate. The force lines for this field are shown in Fig. 37 for $\sigma > 0$.

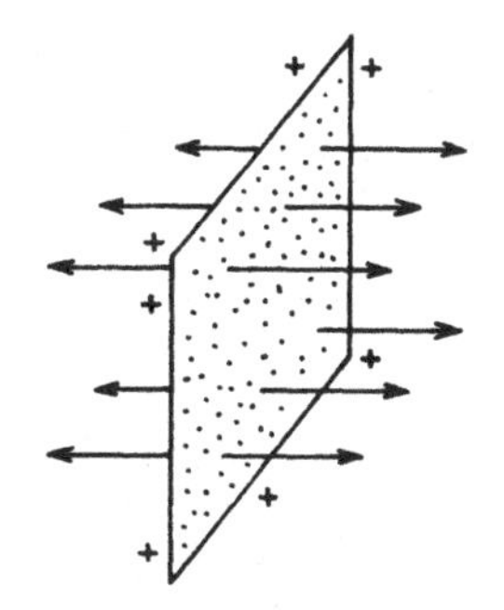

Fig. 37.

Let a uniform surface current with the linear density $\mathbf{i} = i\,\mathbf{e}_3$ flow on the same plate with the same coordinate frame. Then the surface density of the volume current is $\mathbf{j}(x, y, z) = \mathbf{i}\,\delta(x)$. This implies the following relation to the charge density of the previous example:

$$\mathbf{j} = \mathbf{k} \wedge \rho \quad \text{for } \mathbf{k} = \mathbf{i}/\sigma\,.$$

Therefore, using the considerations of Sec. 1.6.3, we may write immediately

$$\widehat{\mathbf{H}}(\mathbf{r}) = \mathbf{k} \wedge \mathbf{D}(\mathbf{r}) = \frac{\mathbf{i}}{2} \wedge \frac{\mathbf{r}_\perp}{r_\perp}\,. \tag{43}$$

In this way, the uniform magnetic field is on both sides of the plate: $\frac{1}{2}\,\mathbf{i}\wedge\mathbf{e}_1$ for $x > 0$ and $-\frac{1}{2}\,\mathbf{i}\wedge\mathbf{e}_1$ for $x < 0$. This field has the symmetry of the sources, that is, under all translations parallel to the plate, under reflexions in the plate, and under reflexions in planes perpendicular to the plate and parallel to the current. We illustrate the force surfaces for this field in Fig. 38. Of course, these force surfaces are connected with the force lines of Fig. 37 exactly as we observed in Sec. 1.6.3.

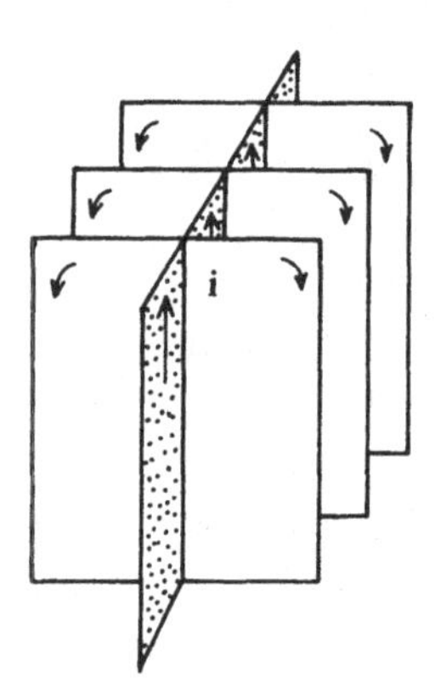

Fig. 38.

6.5. Electric and Magnetic Fields of Two Parallel Plates

We consider two parallel infinite plates, uniformly charged with exactly opposite charge densities σ and $-\sigma$. In order to find their electric field, we superpose two fields of the type (42). Both fields are equal in the space between the plates, so the net field is twice the expression (42), i.e. $\mathbf{D} = \sigma\,\mathbf{n}$, where the unit vector $\mathbf{n}$ is perpendicular to the plates and directed to the negatively-charged one.

On the other hand, both constituent fields are opposite in the space outside the plates, so that net field is zero there. In this way, we obtain the field of an ideal plane capacitor, for which the force lines are shown in Fig. 39.

Now we consider the magnetic counterpart of the plane capacitor, that is, the pair of plates with uniform surface currents and opposite current densities $\mathbf{i}$ and $-\mathbf{i}$. The magnetic field is $\widehat{\mathbf{H}} = \mathbf{i}\wedge\mathbf{n}$ between the plates, and zero outside. We show the force surfaces for this field in Fig. 40.

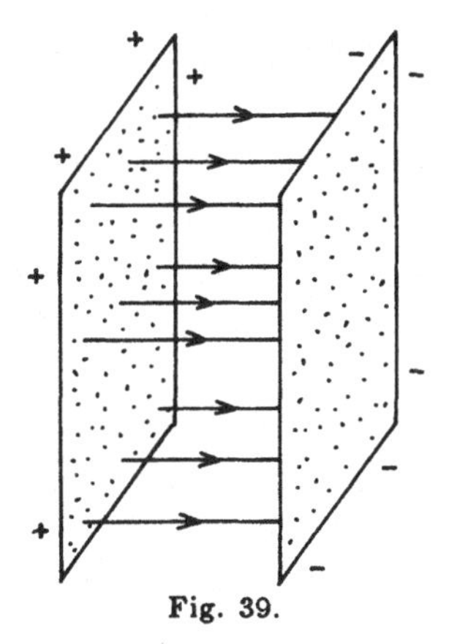

Fig. 39.

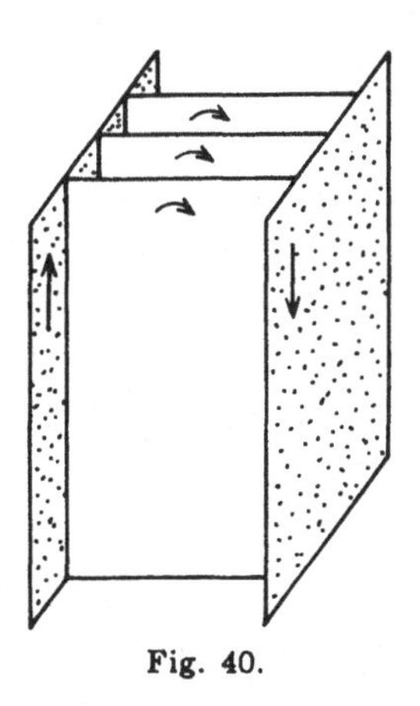
Fig. 40.

6.6. *Electric and Magnetic Fields of a Cylindrical Conductor*

Let an infinite cylindrical conductor with radius R be uniformly charged with volume density ρ. We intend to find the electric induction field by another method, directly from the differential Maxwell equation (17d) under certain simplifying assumptions. We choose the Z-axis as the cylinder axis. Because of the source symmetry under translations parallel to the Z-axis, we assume that $\mathbf{D}$ does not depend on z, i.e. $\mathbf{D}(\mathbf{r}) = \mathbf{D}(\mathbf{r}_\perp)$ where $\mathbf{r}_\perp$ is the component of $\mathbf{r}$ perpendicular to the Z-axis. The axial symmetry allows us to take $\mathbf{D}(\mathbf{r}) = \mathbf{r}_\perp \varphi(r_\perp)$ where φ is some scalar function depending on the scalar variable $r_\perp$. By virtue of Problem 0.30, we have

$$\nabla \cdot \mathbf{D} = (\nabla \cdot \mathbf{r}_\perp)\varphi + \mathbf{r}_\perp \cdot \nabla \varphi \,.$$

Since $\mathbf{r}_\perp = x\,\mathbf{e}_1 + y\,\mathbf{e}_2$, we get $\nabla \cdot \mathbf{r}_\perp = 2$ and hence

$$\nabla \cdot \mathbf{D} = 2\varphi + \mathbf{r}_\perp \cdot \frac{\mathbf{r}_\perp}{r_\perp} \frac{d\varphi}{dr_\perp} = 2\varphi + r_\perp \varphi' \ .$$

Thus Maxwell equation (17d) leads to the following differential equations for φ:

$$\begin{aligned} r_\perp \varphi' + 2\varphi &= \rho \qquad \text{for } r_\perp < R \ , \\ r_\perp \varphi' + 2\varphi &= 0 \qquad \text{for } r_\perp > R \ , \end{aligned}$$

which has the solution

$$\begin{aligned} \varphi(r_\perp) &= \frac{1}{2}\rho + C_1/r_\perp^2 \qquad \text{for } r_\perp < R \ , \\ \varphi(r_\perp) &= C_2/r_\perp^2 \qquad \text{for } r_\perp > R \ , \end{aligned}$$

where C_1, C_2 are integration constants. The function $C_1/r_\perp^2$ is singular on the axis of the conductor, which is not justified physically (the charge density is finite everywhere), so we assume $C_1 = 0$. Surface charges are absent on the boundary of the conductor, the $\mathbf{D}$ vector is perpendicular to the boundary, so, by virtue of Eq. (30), the $\mathbf{D}$ function should be continuous there. Thus we obtain the following condition $\frac{1}{2}\rho = C_2/R^2$, that is, $C_2 = \frac{1}{2}\rho R^2$. In this manner, we get the solution

$$\mathbf{D}(\mathbf{r}) = \begin{cases} \frac{1}{2}\rho\,\mathbf{r}_\perp & \text{for } x^2 + y^2 \le R^2 \ , \\ \frac{1}{2}\rho R^2\,\mathbf{r}_\perp^{-1} & \text{for } x^2 + y^2 \ge R^2 \ . \end{cases}$$

Notice that the field is of the form (38) outside the conductor, so the field is the same as if all the charge was concentrated in the cylinder axis.

Now we consider the same infinite cylindrical conductor with the uniform current J flowing through it. This means that the current density is $\mathbf{j} = (J/\pi R^2)\,\mathbf{e}_3$ in the conductor and zero outside. Thus we have the situation considered in Sec. 1.6.3. We may thus put $\mathbf{j} = \mathbf{k} \wedge \rho$ for $\mathbf{k} = \mathbf{e}_3\, J/\pi\rho R^2$ and obtain immediately the equation for the magnetic field

$$\hat{\mathbf{H}}(\mathbf{r}) = \begin{cases} \frac{J}{2\pi R^2}\,\mathbf{e}_3 \wedge \mathbf{r}_\perp & \text{for } x^2 + y^2 \le R^2 \ , \\ \frac{J}{2\pi}\,\mathbf{e}_3 \wedge \mathbf{r}_\perp^{-1} & \text{for } x^2 + y^2 \ge R^2 \ . \end{cases}$$

We see that the field is of the form (39) outside the conductor, as if all the current was concentrated in the conductor axis.

§7. Other Examples of Static Magnetic Fields

We consider an infinite surface S with translational symmetry along a straight line — let it be the Z-axis. Let surface currents of the form $\mathbf{i}(\mathbf{r}) = i_1(x,y)\,\mathbf{e}_1 + i_2(x,y)\,\mathbf{e}_2$ flow, which means that the currents also have translational symmetry parallel to the Z-axis, but — in contrast with the situation considered in the previous Sec. 1.6 — they are perpendicular to that axis. Moreover, let the magnitude $i = |\mathbf{i}|$ be constant. We know already that in the presence of surface current the volume integral (35) should be replaced by the surface integral

$$\mathbf{H}(\mathbf{r}) = \frac{1}{4\pi}\int_S ds'\,\frac{\mathbf{i}(\mathbf{r}') \wedge (\mathbf{r}-\mathbf{r}')}{|\mathbf{r}-\mathbf{r}'|^3}\ .$$

Denoting $\mathbf{r}_\perp = x\,\mathbf{e}_1 + y\,\mathbf{e}_2$, the component of the radius vector perpendicular to the Z-axis, then $\mathbf{r} = \mathbf{r}_\perp + z\,\mathbf{e}_3$ and $\mathbf{i}(\mathbf{r}) = \mathbf{i}(\mathbf{r}_\perp)$. With this notation,

$$\hat{\mathbf{H}}(\mathbf{r}) = \frac{1}{4\pi}\int_S ds'\,\frac{\mathbf{i}(\mathbf{r}'_\perp) \wedge (\mathbf{r}_\perp - \mathbf{r}'_\perp)}{|\mathbf{r}-\mathbf{r}'|^3} - \frac{\mathbf{e}_3}{4\pi} \wedge \int_S ds'\,\frac{\mathbf{i}(\mathbf{r}'_\perp)(z-z')}{|\mathbf{r}-\mathbf{r}'|^3}\ .$$

The surface S intersects the plane $(\mathbf{e}_1, \mathbf{e}_2)$ along the curve L called the *directrix* of S. We represent the surface element ds' in the form $dl'dz'$ where dl' is the length element of the directrix:

$$\hat{\mathbf{H}}(\mathbf{r}) = \frac{1}{4\pi}\int_L dl' \int_{-\infty}^{\infty} dz'\,\frac{\mathbf{i} \wedge (\mathbf{r}_\perp - \mathbf{r}'_\perp)}{|\mathbf{r}-\mathbf{r}'|^3} + \frac{\mathbf{e}_3}{4\pi} \wedge \int_L dl' \int_{-\infty}^{\infty} dz'\,\frac{\mathbf{i}(\mathbf{r}'_\perp)(z'-z)}{|\mathbf{r}-\mathbf{r}'|^3}\ .$$

The integration over z' in the second term gives zero since the integrand is antisymmetric with respect to $z'-z$. The integration over z' is easy to perform in the first term, which gives

$$\hat{\mathbf{H}}(\mathbf{r}) = \frac{1}{2\pi}\int_L \imath l'\,\frac{\mathbf{i}(\mathbf{r}'_\perp) \wedge (\mathbf{r}_\perp - \mathbf{r}'_\perp)}{|\mathbf{r}_\perp - \mathbf{r}'_\perp|^2}\ .$$

This is the curvilinear integral in the plane $(\mathbf{e}_1, \mathbf{e}_2)$ with all vectors lying in the plane, therefore we shall omit the subscript $\perp$ at $\mathbf{r}'$ from now on. We introduce the directed line element $d\mathbf{l}$, for which $\mathbf{i}\, dl = i\, d\mathbf{l}$ occurs. We assume that the magnitude i is constant, hence it may be written in front of the integral:

$$\hat{\mathbf{H}}(\mathbf{r}) = \frac{i}{2\pi}\int_L \frac{d\mathbf{l}' \wedge (\mathbf{r}_\perp - \mathbf{r}')}{|\mathbf{r}_\perp - \mathbf{r}'|^2} = \frac{i}{2\pi}\int_L (\mathbf{r}' - \mathbf{r}_\perp)^{-1} \wedge d\mathbf{l}'\ .$$

Notice that the integrand is the directed angle element $d\hat{\alpha}$, at which the line element $d\mathbf{l}'$ is seen from the point $\mathbf{r}_\perp$ (see Fig. 41). (Its magnitude is $dl' \sin\varphi / |\mathbf{r}' - \mathbf{r}_\perp| = dl'_\perp / |\mathbf{r}' - \mathbf{r}_\perp|$ where φ is the angle between $d\mathbf{l}'$ and $\mathbf{r}' - \mathbf{r}_\perp$, and $d\mathbf{l}'_\perp$ is the component of $d\mathbf{l}'$ perpendicular to $\mathbf{r}' - \mathbf{r}_\perp$.) This observation allows us to give the value of this integral for any specific surface S satisfying our assumptions and — since S is uniquely determined by its directrix L on the plane $(\mathbf{e}_1, \mathbf{e}_2)$ — also for any specific curve L. The integral

$$\int_L (\mathbf{r}' - \mathbf{r}_\perp)^{-1} \wedge d\mathbf{l}' \tag{44}$$

can be interpreted as the *visual angle* of the curve L from the point $\mathbf{r}_\perp$ on the plane $(\mathbf{e}_1, \mathbf{e}_2)$. In each case, we obtain the bivector $\hat{\mathbf{H}}$ parallel to $\mathbf{e}_1 \wedge \mathbf{e}_2$ and independent of z, which means that the field $\hat{\mathbf{H}}$ has translational symmetry along the Z-axis, like the sources.

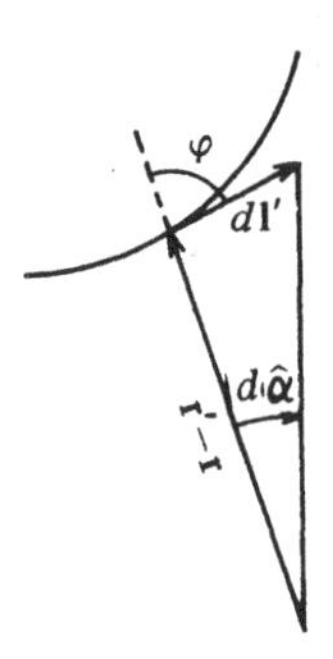

Fig. 41.

Let us apply this result to the simplest example when S is a plane, say $(\mathbf{e}_2, \mathbf{e}_3)$, that is when the directrix L is a straight line, in this case, the Y-axis. In this example, Eq. (44) gives the straight angle with opposite orientations on both sides of the plane S, namely $-\pi\, \mathbf{e}_1 \wedge \mathbf{e}_2$ for $x > 0$ and $\pi\, \mathbf{e}_1 \wedge \mathbf{e}_2$ for $x < 0$. In this way, the magnetic field is

$$\hat{\mathbf{H}}(\mathbf{r}) = \begin{cases} -\frac{i}{2}\mathbf{e}_1 \wedge \mathbf{e}_2 & \text{for } x > 0\,, \\ \frac{i}{2}\mathbf{e}_1 \wedge \mathbf{e}_2 & \text{for } x < 0\,, \end{cases}$$

which coincides with the result (43).

7.1. Magnetic Field of Two Adjoining Half-Planes

For a surface S built by two half-planes touching along the Z-axis at the angle α, as in Fig. 42 (let $0 < \alpha \leq \pi$), it is worth introducing the directed angle $\hat{\boldsymbol{\alpha}}$ as the volutor with the magnitude α and the orientation from the currents outgoing from the Z-axis to the currents incoming to it (see Fig. 42). The assumption $\alpha \leq \pi$ means that we have chosen the smaller angle of the two possible angles. The visual angle (44) of the curve L is $2\pi\frac{\hat{\boldsymbol{\alpha}}}{\alpha} - \hat{\boldsymbol{\alpha}}$ in the space enfolded by the adjoining half-planes and $-\hat{\boldsymbol{\alpha}}$ on the other side. Note that this angle is equal to the directed angle between the half-planes, but taken from the opposite side of the surface. In this way, we obtain the $\hat{\mathbf{H}}$ field equal to $i\frac{\hat{\boldsymbol{\alpha}}}{\alpha} - \frac{i}{2\pi}\hat{\boldsymbol{\alpha}}$ in the fold between the planes and $-\frac{i}{2\pi}\hat{\boldsymbol{\alpha}}$ outside. Hence the field $\hat{\mathbf{H}}$ is greater in the fold and uniform on both sides of the surface S.

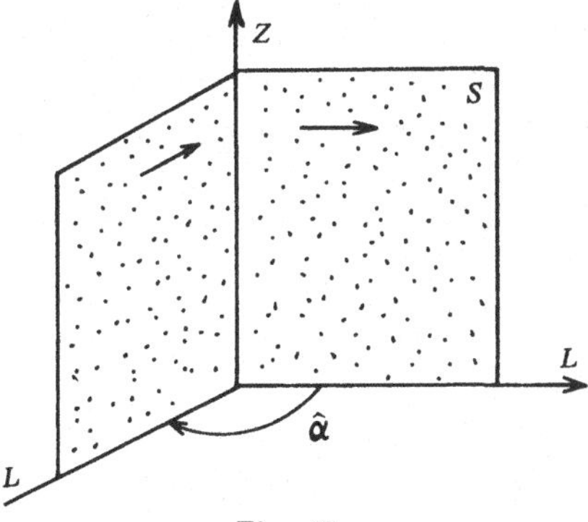

Fig. 42.

It is interesting that the result remains the same if the surface S has the shape given by the directrix L of the types shown in Fig. 43 when L does not consist of two half-straight lines but has them as its asymptotes.

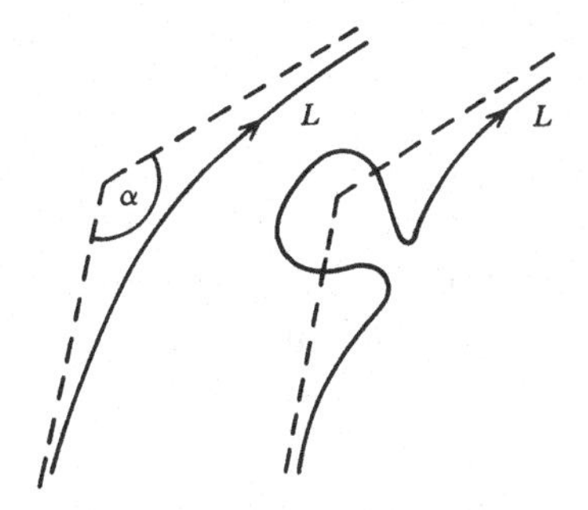

Fig. 43.

7.2. Magnetic Field of Tubular Surfaces

Another type of surface fulfilling the assumptions of the first paragraph in this section is obtained when one takes the directrix L as a closed curve with one loop. Then the surface S is a tube with the cross-section L. The integral (44) gives only two possible values in this case. If $\mathbf{r}_\perp$ lies inside the loop L (which means that the point $\mathbf{r}$ lies inside the tube S) the visual angle of L has the value 2π, whereas if $\mathbf{r}_\perp$ lies outside of L (that is, $\mathbf{r}$ lies outside S) the directed angle (44) is zero. Therefore, having introduced a unit bivector $\hat{\mathbf{n}}$ representing the direction of the directrix L, we may write the magnetic field as

$$\hat{\mathbf{H}}(\mathbf{r}) = \begin{cases} i\hat{\mathbf{n}} & \text{for } \mathbf{r} \text{ inside } S \\ 0 & \text{for } \mathbf{r} \text{ outside } S\,. \end{cases} \tag{45}$$

We illustrate the force surfaces of this field in Fig. 44 for the tube S with a rectangular section, and in Fig. 45 for S with an elliptic section. The last example corresponds to the ideal elliptic solenoid, that is, to the elliptical cylinder on which a wire is coiled so tightly that its turns can be considered as ellipses perpendicular to the cylinder axis.

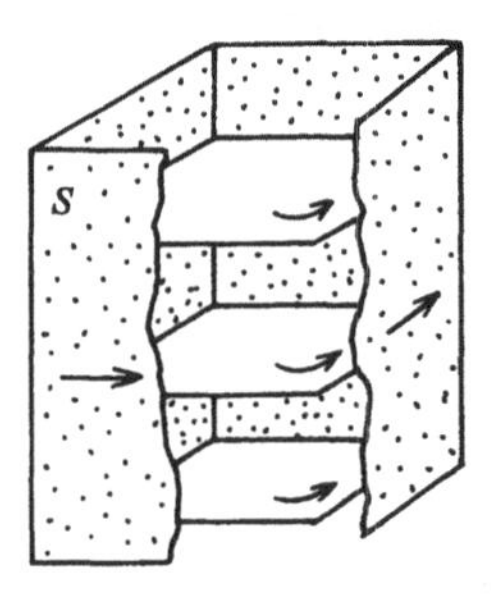

Fig. 44.

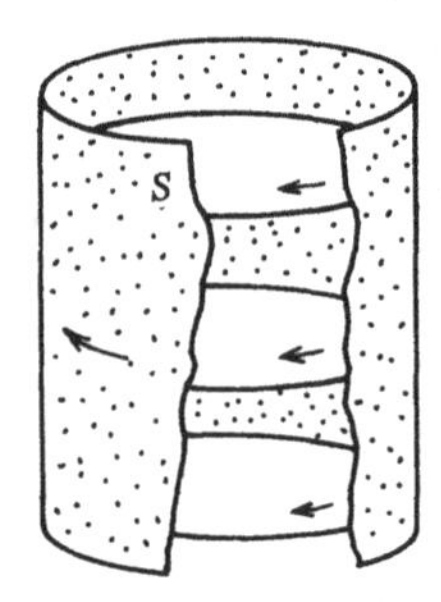

Fig. 45.

7.3. Helical Magnetic Field

We consider the magnetic field around a straight line conductor with the current J put in the external uniform field perpendicular (as the volutor) to the conductor. In this case, we look at the sum of fields corresponding to Figs. 34 and 45, that is,

$$\hat{\mathbf{H}} = H_0\mathbf{e}_2 \wedge \mathbf{e}_1 + \frac{J}{2\pi}\mathbf{e}_3 \wedge \mathbf{r}_\perp^{-1}\,. \tag{46}$$

For simplicity, we consider this expression in the plane $(\mathbf{e}_2, \mathbf{e}_3)$ when $\mathbf{r}_\perp = y\,\mathbf{e}_2$

$$\widehat{\mathbf{H}} = H_0\mathbf{e}_2 \wedge \mathbf{e}_1 + \frac{J}{2\pi y}\mathbf{e}_3 \wedge \mathbf{e}_2 = (-H_0\mathbf{e}_1 + \frac{J}{2\pi y}\mathbf{e}_3) \wedge \mathbf{e}_2 \ .$$

This volutor is parallel to the Y-axis with various inclinations to the plane $(\mathbf{e}_1, \mathbf{e}_2)$ in various points y. As the point y approaches the Z-axis, $\widehat{\mathbf{H}}$ approaches the plane $(\mathbf{e}_2, \mathbf{e}_3)$ — the straight conductor field prevails. On the other hand, as y moves away from the Z-axis, $\widehat{\mathbf{H}}$ approaches the plane $(\mathbf{e}_1, \mathbf{e}_2)$ — the uniform field prevails. We illustrate several such directions of $\mathbf{H}$ in Fig. 46. Such a volutor field is tangent to helical surfaces. Thus the force surfaces for the field (46) are helical surfaces, one of them is shown in Fig. 47.

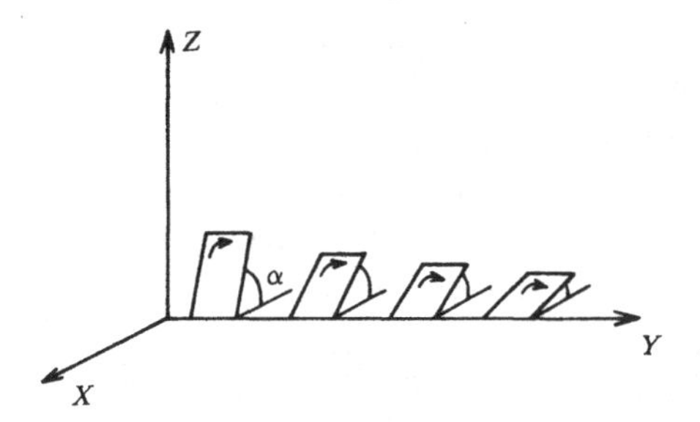

Fig. 46.

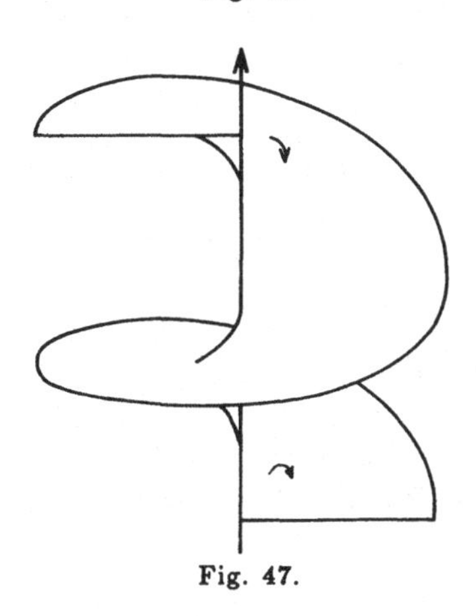

Fig. 47.

Let us find the pitch h of the helix. The intersection of the helical surface with a cylinder of the radius $r_\perp$ gives a helical line. After developing and flattening out the cylinder, the helical line becomes straight and the tangent of its inclination angle is $\tan\alpha = h/2\pi r_\perp$. On the other hand, the angle α shown in Fig. 46, due to (46), is $\tan\alpha = J/2\pi r_\perp H_0$. From a comparison of these two expressions, we obtain

$$h = \frac{J}{H_0} \,. \tag{47}$$

We now specify the external uniform field. Let $\hat{\mathbf{H}}_0$ be the magnetic field of a solenoid with the circular section. Then, by virtue of Eq. (45), $H_0 = i$. This linear density of the surface current may be expressed as $i = J_1 N$ where N is the number of turns on the unit length parallel to the axis, i.e. $N = 1/h_1$ where h_1 is the pitch of helical turns of the solenoid. We now insert $H_0 = J_1/h_1$ into Eq. (47):

$$h = \frac{Jh_1}{J_1} \,.$$

If the straight conductor closes the circuit of the solenoid, then $J = J_1$ and hence we obtain $h = h_1$. This is an interesting result: the helical force surfaces fit the helical turns of the solenoid if the current in the straight conductor is the same as in the turns of the solenoid.

7.4. Magnetic Field of Circular Circuit

We consider a circular circuit of the radius R with the current J. We introduce the cylindrical coordinates with the Z-axis coinciding with the symmetry axis of the circuit and the plane $z = 0$ coinciding with the plane of the circuit. The Cartesian coordinates (x, y, z) are related to the cylindrical ones (r, α, z) as follows

$$\begin{aligned} x &= r\cos\alpha \qquad & r &= \sqrt{x^2 + y^2} \\ y &= r\sin\alpha \qquad & \alpha &= \arctan\frac{y}{x} \\ z &= z \qquad & z &= z \,. \end{aligned}$$

The volume element is $dv = dx\;dy\;dz = r\;dr\;d\alpha\;dz$. The unit vectors tangent to the coordinate lines (see Problem 0.41) are given by the equations

$$\begin{aligned} \mathbf{e}_r &= \mathbf{e}_1\cos\alpha + \mathbf{e}_2\sin\alpha \qquad & \mathbf{e}_1 &= \mathbf{e}_r\cos\alpha - \mathbf{e}_\alpha\sin\alpha \\ \mathbf{e}_\alpha &= -\mathbf{e}_1\sin\alpha + \mathbf{e}_2\cos\alpha \qquad & \mathbf{e}_2 &= \mathbf{e}_r\sin\alpha + \mathbf{e}_\alpha\cos\alpha \\ \mathbf{e}_z &= \mathbf{e}_3 \qquad & \mathbf{e}_3 &= \mathbf{e}_z \,. \end{aligned}$$

Notice that the vectors $\mathbf{e}_r$ and $\mathbf{e}_\alpha$ depend on position, but the bivector $\mathbf{e}_r \wedge \mathbf{e}_\alpha = \mathbf{e}_1 \wedge \mathbf{e}_2$ is constant.

We may express the current density as $\mathbf{j}(\mathbf{r}) = \mathbf{e}_\alpha J\delta(r - R)\delta(z)$. We plug this into Eq. (35):

$$\hat{\mathbf{H}}(\mathbf{r}) = \frac{J}{4\pi}\int_0^{2\pi} d\alpha' \, R \frac{\mathbf{e}_{\alpha'} \wedge (\mathbf{r} - \mathbf{r}')}{|\mathbf{r} - \mathbf{r}'|^3},$$

where $\mathbf{r}'$ satisfies $|\mathbf{r}'| = R, z' = 0$. This integral is not expressible by elementary functions for arbitrary points $\mathbf{r}$. But, in the particular case, when $\mathbf{r}$ lies on the Z-axis, that is $\mathbf{r} = z\,\mathbf{e}_3$, the difference $\mathbf{r} - \mathbf{r}'$ has constant magnitude $\sqrt{z^2 + R^2}$ in the integration set. Moreover, $\mathbf{e}_{\alpha'} \wedge (\mathbf{r} - \mathbf{r}') = \mathbf{e}_{\alpha'} \wedge z\,\mathbf{e}_3 - \mathbf{e}_{\alpha'} \wedge R\,\mathbf{e}_{r'} = z\,\mathbf{e}_{\alpha'} \wedge \mathbf{e}_3 + R\,\mathbf{e}_1 \wedge \mathbf{e}_2$ and hence

$$\hat{\mathbf{H}}(\mathbf{r}) = \frac{J}{4\pi}\frac{R^2 \mathbf{e}_1 \wedge \mathbf{e}_2}{(z^2 + R^2)^{\frac{3}{2}}}\int_0^{2\pi} d\alpha' - \frac{z\,\mathbf{e}_3}{(z^2 + R^2)^{\frac{3}{2}}} \wedge \int_0^{2\pi} R\,\mathbf{e}_{\alpha'} d\alpha' .$$

The expression $R\,\mathbf{e}_{\alpha'} d\alpha' = d\mathbf{l}'$ is the length element of the circuit, hence the second integral is $\int d\mathbf{l}'$ and is zero since it is the integral over the closed curve. Therefore the expression

$$\hat{\mathbf{H}}(\mathbf{r}) = \frac{JR^2}{2(z^2 + R^2)^{\frac{3}{2}}}\,\mathbf{e}_1 \wedge \mathbf{e}_2$$

remains for the magnetic field on the symmetry axis of the circuit.

Nevertheless, in order to illustrate the force surfaces of the whole magnetic field, we use the well known images of vectorial lines for this field. We display them in Fig. 48 in a plane perpendicular to the circuit and passing through its center. Since the force surfaces are to be orthogonal to these lines, their intersection with the plane of drawing gives the curves shown in Fig. 49. The required surfaces can be obtained by rotation of the curves around the symmetry axis of the circuit (see Fig. 50). One of the force surfaces is the plane, in which the circuit lies, with opposite orientations inside the circuit and outside it.

* * *

We conclude with several remarks summarizing the last two sections. In all the examples considered, we succeeded in finding the force surfaces of the magnetic fields. This does not, however, imply that such surfaces exist for any magnetic fields possible in nature. At first sight, it would seem that force surfaces always exist when vectorial lines exist for the magnetic field. But

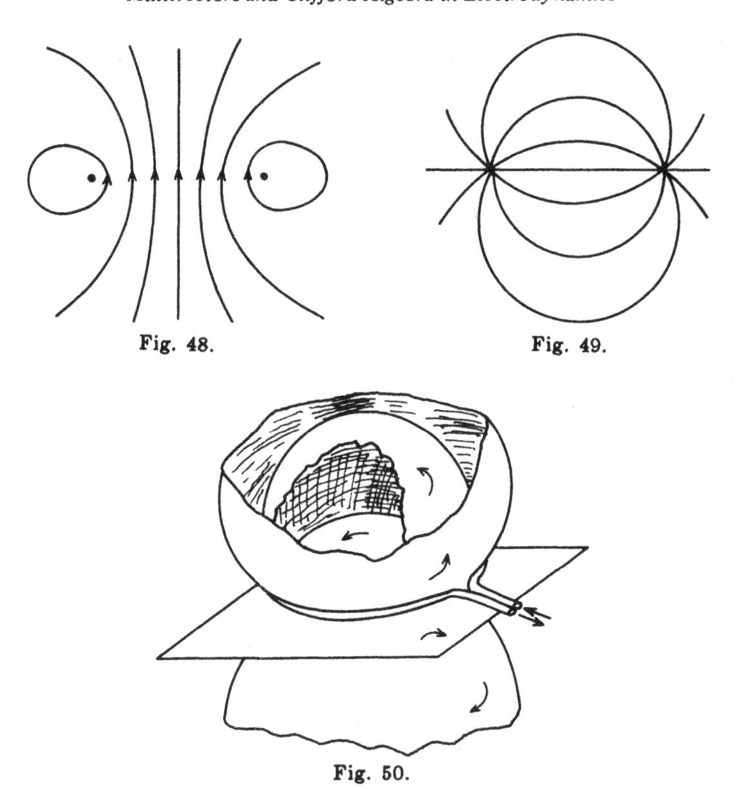

Fig. 48.

Fig. 49.

Fig. 50.

this is not the case — one may only locally at each point of the vectorial line assign a surface element perpendicular to it. The question of joining them into surfaces is called the *integrability problem* and does not always have a solution. We consider this point in Appendix II, where the following answer is found: The force surfaces of the magnetic field exist in isotropic media when the scalar product of $\mathbf{j} + \frac{\partial \mathbf{D}}{\partial t}$ with $\mathbf{H}$ is zero. Therefore we may state that in magnetostatics (that is when $\partial \mathbf{D}/\partial t = 0$), for the isotropic media in the regions devoid of currents (that is $\mathbf{j} = 0$), the force surfaces exist. All our examples refer to such situations.

The examples discussed display basic features of the magnetic field. The force lines of the electrostatic field begin and end on the charges (if they end at all). Similarly, the force surfaces of the magnetic field have their boundaries on the currents which are sources of the field. Moreover, the orientation of the force surfaces is always compatible with that of the currents at their boundaries. In the case of a uniform field, this is excellently visible for the solenoid, where the force surfaces are orthogonal to the axis of the solenoid end on the currents, and have orientations conforming to the current.

One more feature of the magnetic force surfaces similar to a property of electric force lines is worth pointing out, namely the *density of force surfaces* i.e. the adequately defined "number of surfaces" intersecting a line of unit length perpendicular to them is proportional to the magnitude of the magnetic induction in regions free of currents. In the case of the uniform field, this density is constant since force surfaces have constant distance between each other. In the case of the linear conductor, the density of force surfaces decreases as $1/r_\perp$ when the distance $r_\perp$ from the conductor grows. This is not so perceptible when one draws the vectorial lines of the magnetic field which, in this example, are concentric circles and must be artificially drawn further from each other as the distance from the conductor grows.

§8. Simplest Non-Static Electromagnetic Fields

We now consider the simplest non-static solutions to the homogeneous Maxwell equation (24) for a medium homogeneous in space and constant in time. By "the simplest" we mean here that the fields are linearly dependent on time and space coordinates.

If one looks carefully at the Maxwell equation

$$\nabla f + \frac{1}{u}\frac{\partial f}{\partial t} = 0 \,, \tag{24}$$

one notices that the first term contains the rank changing operator nabla. Therefore, if the time dependent term consists only of the electric or magnetic part, the space dependent term must contain the other part. Thus we wish to consider several examples in which the one part (electric or magnetic) of the electromagnetic cliffor f has definite linear space dependence, and appropriately fit the time dependence of the other part.

8.1. Fields with Plane Symmetry

We assume that the electromagnetic field has two-dimensional translational symmetry parallel to some distinguished plane S — therefore we call it *plane symmetry.* Let the plane pass through the origin and let $\mathbf{n}$ be a unit vector perpendicular to it. The simplest linear function of the position fulfilling such a symmetry is $\mathbf{r} \to (\mathbf{n}\cdot\mathbf{r})$. We introduce another unit vector $\mathbf{m}$ parallel to the plane S and build the following bivector-valued function

$$\widehat{\mathbf{b}}(\mathbf{r}) = -k\,(\mathbf{n}\cdot\mathbf{r})\,\mathbf{n}\wedge\mathbf{m} = -k\,(\mathbf{n}\cdot\mathbf{r})\,\mathbf{n}\mathbf{m} \,,$$

with a dimensional coefficient k. The nabla acting on it gives

$$\nabla \widehat{\mathbf{b}} = -k\,\mathbf{n}\,\mathbf{n}\,\mathbf{m} = -k\,\mathbf{m}\,.$$

If $\widehat{\mathbf{b}}$ is the only position-dependent part in $f = \mathbf{e} + \widehat{\mathbf{b}}$, then $\nabla f = \nabla \widehat{\mathbf{b}}$ and the Maxwell equation (24) implies $\frac{1}{u}\frac{\partial f}{\partial t} = k\,\mathbf{m}$ and

$$\mathbf{e}\,(t) = ukt\,\mathbf{m} = \omega t\,\mathbf{m}\,,$$

where $\omega = uk$. In this way, we obtain the following solution

$$f(\mathbf{r}, t) = \omega t\,\mathbf{m} - k\,(\mathbf{n} \cdot \mathbf{r})\,\mathbf{n}\,\mathbf{m} \tag{48}$$

to the Maxwell equation. Its magnetic part is constant in time and grows linearly as the distance from S grows, its electric part is uniform in space, parallel to S and increases linearly with time. We illustrate this field in Fig. 51 in the plane $(\mathbf{n}, \mathbf{m})$. Notice that the field has additionally reflexion symmetry with respect to S.

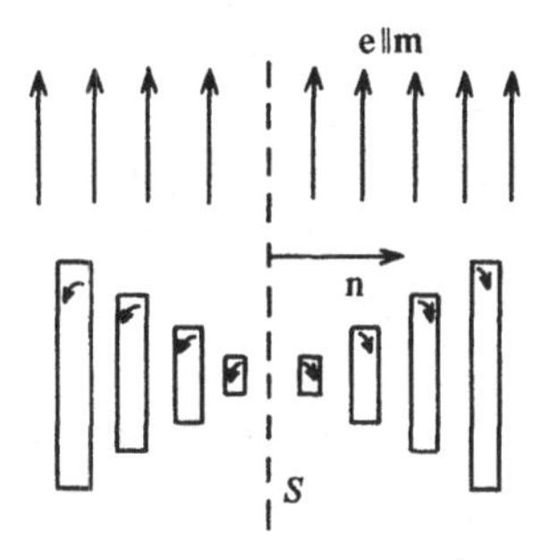

Fig. 51.

We may find *a posteriori* sources producing such a field. Since the electric field is of the shape produced by a capacitor, let the sources lie on two infinite plates A and B perpendicular to the electric field (see Fig. 52). Surface charges of densities σ_A and σ_B must be spread over the plates with the values

$$\sigma_A = -\sqrt{\varepsilon}\omega t\,, \qquad \sigma_B = \sqrt{\varepsilon}\omega t \tag{49}$$

in order to satisfy the boundary conditions (30) on the plates such that the field is zero below B and above A. Moreover, surface currents $\mathbf{i}_A$ and $\mathbf{i}_B$ must flow on the plates, with the values

$$\mathbf{i}_A = \frac{k}{\sqrt{\mu}}(\mathbf{n} \cdot \mathbf{r})\,\mathbf{n}\,, \qquad \mathbf{i}_B = -\frac{k}{\sqrt{\mu}}(\mathbf{n} \cdot \mathbf{r})\,\mathbf{n}\,, \tag{50}$$

which satisfy the boundary conditions (32) such that the magnetic field is zero below B and above A. The source densities (49) and (50) fulfil the continuity equation

$$\frac{\partial \sigma}{\partial t} + \nabla \cdot \mathbf{i} = 0\,,$$

which is a test of compatibility for expressions (49) and (50).

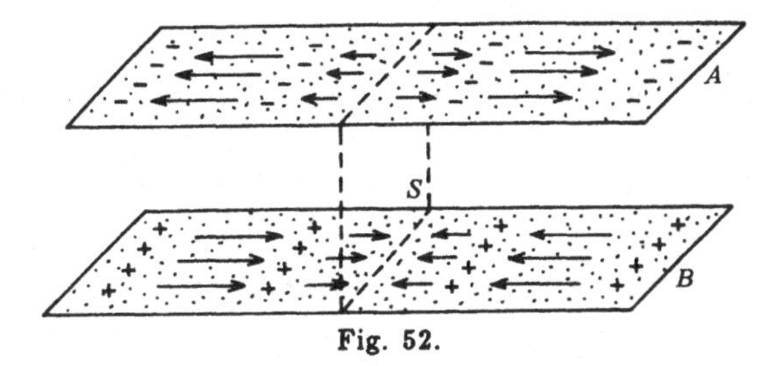

Fig. 52.

In this manner, we see that the electromagnetic field (48) is confined to the space between the plates A and B. This is still an idealization since the plates are to be infinite.

The same space dependence $\mathbf{r} \to (\mathbf{n}\cdot\mathbf{r})$ may be applied to the electric part. Therefore we build the following vector-valued function for the electric part: $\mathbf{e}(\mathbf{r}) = k(\mathbf{n}\cdot\mathbf{r})\,\mathbf{m}$. As previously, we find the corresponding magnetic part $\widehat{\mathbf{b}}(t) = \omega t\,\mathbf{m}\,\mathbf{n}$. In this way, we obtain the next solution

$$f(\mathbf{r}, t) = k(\mathbf{n}\cdot\mathbf{r})\,\mathbf{m} - \omega t\,\mathbf{n}\,\mathbf{m} \tag{51}$$

to the Maxwell equation. Its electric part is constant in time and grows linearly with the distance from S, its magnetic part is uniform in space and increases with time. We illustrate this field in Fig. 53. Notice that the field is additionally antisymmetric under reflexion in S ("antisymmetric" means that after reflexion the opposite field to the one on the other side of S is obtained).

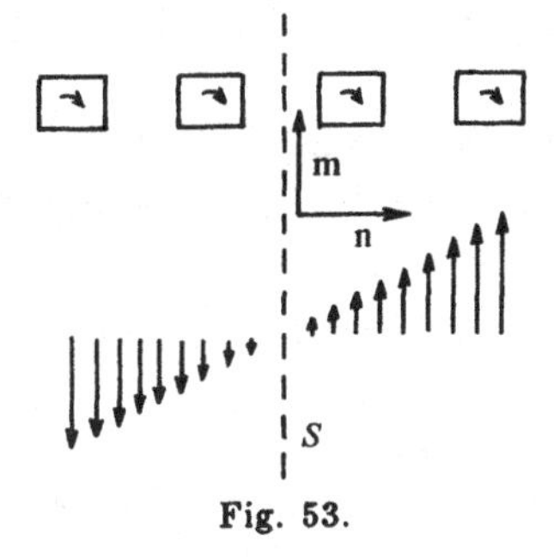

Fig. 53.

The field (51) may be obtained by multiplying (48) by $\mathbf{n}$ from the right. Therefore a similar transformation applied to sources (49) and (50) yields

$$\mathbf{i}_A = \frac{kt}{\sqrt{\mu}}\mathbf{n}\,, \qquad \mathbf{i}_B = -\frac{kt}{\sqrt{\mu}}\mathbf{n}\,,$$
$$\sigma_A = -\sqrt{\varepsilon}\omega(\mathbf{n}\cdot\mathbf{r})\,, \qquad \sigma_B = \sqrt{\varepsilon}\omega(\mathbf{n}\cdot\mathbf{r})\,.$$

These source densities (which are illustrated in Fig. 54) satisfy the continuity equation trivially — both terms are zero. They guarantee that the electromagnetic field (51) is confined to the space between plates A and B.

Of course, the symmetry described is not a necessary feature of the fields. One may add a uniform and constant electric field or magnetic field, or both, and by the superposition principle, the Maxwell equation would still be satisfied.

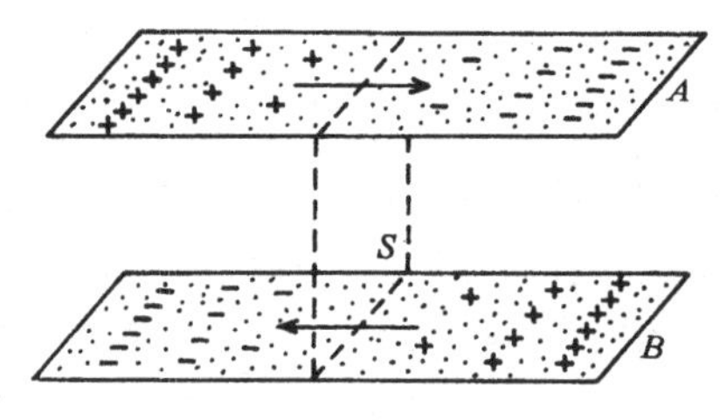

Fig. 54.

8.2. Fields with Axial Symmetry

We start with the bivector expression $\widehat{\mathbf{b}}(\mathbf{r}) = \frac{k}{2}\mathbf{n}\wedge\mathbf{r}$ (here $\mathbf{n}$ is a unit vector, k is a dimensional scalar coefficient) which is a linear function of the radius vector $\mathbf{r}$ and has axial symmetry around the axis of direction $\mathbf{n}$ passing through the origin. Its Clifford derivative is

$$\nabla\widehat{\mathbf{b}} = \frac{k}{2}\nabla(-\mathbf{r}_\perp\mathbf{n}) = -\frac{k}{2}2\mathbf{n} = -k\mathbf{n}\,.$$

In order to satisfy the Maxwell equation (24), we find $e(t) = \omega t\,\mathbf{n}$, where $\omega = uk$. In this manner, we arrive at the following solution

$$f(\mathbf{r},t) = \omega t\,\mathbf{n} + \frac{k}{2}\mathbf{n}\wedge\mathbf{r}\,, \tag{52}$$

to the Maxwell equation. We illustrate this field in Fig. 55. Such a field may be produced by two infinite plates A and B perpendicular to the electric field with the surface charge densities $\sigma_A = -\sqrt{\varepsilon}\,\omega t$, $\sigma_B = \sqrt{\varepsilon}\,\omega t$ and with the surface current densities $\mathbf{i}_A = \frac{k}{2\sqrt{\mu}}\mathbf{r}_\perp$, $\mathbf{i}_B = -\frac{k}{2\sqrt{\mu}}\mathbf{r}_\perp$. For these sources illustrated in Fig. 56, the electromagnetic field is confined to the space between the plates.

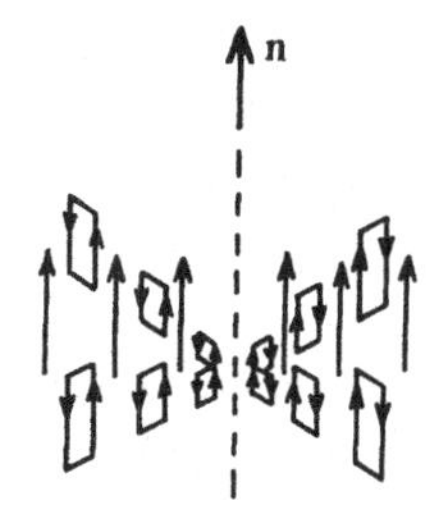

Fig. 55.

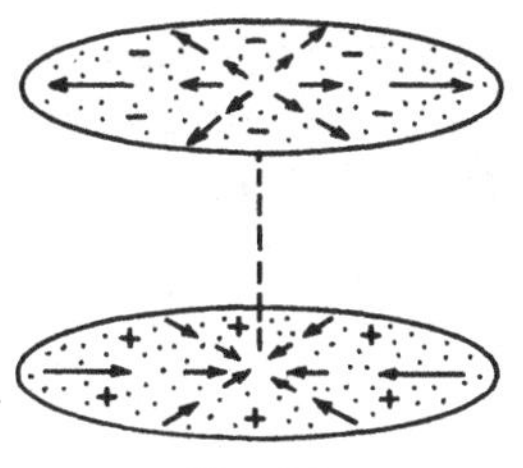

Fig. 56.

Another possibility is the vector expression $\mathbf{e}\,(\mathbf{r}) = \frac{k}{2}\,\widehat{\mathbf{n}}\cdot\mathbf{r}$ (here $\widehat{\mathbf{n}}$ is a unit volutor) which also has axial symmetry around the axis perpendicular to $\mathbf{n}$ and passing through the origin. Its Clifford derivative is

$$\nabla\mathbf{e} = \frac{k}{2}\nabla(-\mathbf{r}_{\|}\widehat{\mathbf{n}}) = -\frac{k}{2}2\,\widehat{\mathbf{n}} = -k\,\widehat{\mathbf{n}}\;.$$

This implies time dependence of the form $\widehat{\mathbf{b}}(t) = \omega t\,\widehat{\mathbf{n}}$ and the solution

$$f(\mathbf{r},t) = \frac{k}{2}\,\widehat{\mathbf{n}}\cdot\mathbf{r} + \omega t\,\widehat{\mathbf{n}} \tag{53}$$

to the Maxwell equation. We show this field in Fig. 57. This is good illustration of the statement: "A changing magnetic field is accompanied by a vortical electric field".

The field (53) can be produced by idealized circular solenoid of radius R with the axis coinciding with the symmetry axis of the field. A surface current should flow on it of the form

$$\mathbf{i} = \omega t\,\mathbf{m}\cdot\widehat{\mathbf{n}}\;, \tag{54}$$

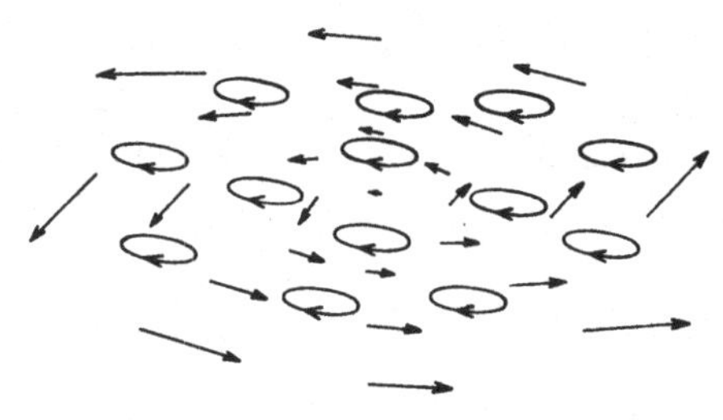

Fig. 57.

where $\mathbf{m} = \mathbf{r}/R$ for $|\mathbf{r}| = R$. A surface charge need not be present in this case (compare with Problem 12). Field (53) cannot be confined to the interior of the solenoid. The static electric field $\mathbf{e}(\mathbf{r}) = (kR^2/2)\,(\widehat{\mathbf{n}} \cdot \mathbf{r})/r^2$ for $|\mathbf{r}| > R$ must be admitted in order to fulfil the boundary condition $\mathbf{E}_{1t} = \mathbf{E}_{2t}$ on the surface of the solenoid. This field satisfies the electrostatic Maxwell equation $\nabla \mathbf{e} = 0$ and has an interesting property: the circulation $\int_C d\mathbf{l} \cdot \mathbf{e}$ over a loop C is non-zero when C encircles the solenoid. This is a perspicuous illustration of the Faraday law: $\int_C d\mathbf{l} \cdot \mathbf{E} = -\frac{d\Phi}{dt}$ where Φ is the magnetic flux confined in the solenoid.

Problems

1. Show on drawings the following relation: If the angle between $\mathbf{e}$ and $\mathbf{b}$ is acute, the trivector $\mathbf{e} \wedge \widehat{\mathbf{b}}$ is right-handed, if the angle is obtuse, $\mathbf{e} \wedge \widehat{\mathbf{b}}$ is left-handed.
2. Demonstrate that the Maxwell equations (17) may be obtained directly from the traditional differential Maxwell equations

$$\operatorname{curl}\mathbf{E} = -\frac{\partial \mathbf{B}}{\partial t}\,, \quad \operatorname{div}\mathbf{B} = 0\,, \quad \operatorname{curl}\mathbf{H} = \frac{\partial \mathbf{D}}{\partial t} + \mathbf{j}\,, \operatorname{div}\mathbf{D} = \rho\,.$$

   Hint: Use the Hodge map and the identities (0.75), (0.121 – 123).
3. Show that the integral $\int_\Omega dv\,\mathbf{j}$ is zero when the stationary currents flow in a bounded region, and the integration set Ω includes that region.
4. Find the deflection law for the electric force lines on the interface between two dielectric media. When is a "turning back" of the lines possible?
5. Find the deflection law for the magnetic force surfaces on the interface between two magnetic media. Show analogies with electrostatics.
6. Calculate the integral $\int (d\mathbf{l} \cdot \mathbf{r})/r^3$ over an infinite straight line conductor.

7. Check for Examples 6.4, 7.1 and 7.2 whether the boundary conditions are satisfied when the surfaces with charges or currents are treated as interface between two media.
8. Find the magnetic field produced by two infinite surfaces parallel to the Z-axis which have intersections with the plane $(\mathbf{e}_1, \mathbf{e}_2)$ shown in Fig. 58. Equal surface currents perpendicular to the Z-axis flow in these surfaces. Consider the cases of concordant and opposite currents.

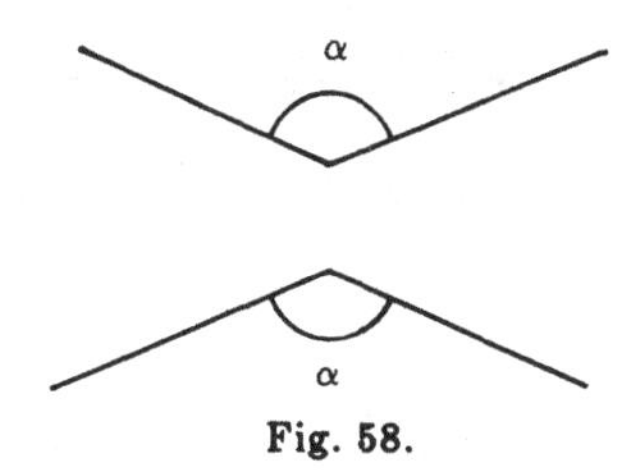

Fig. 58.

9. Find the electric field inside and outside a uniformly spatially charged sphere.
10. Find the magnetic field of an infinite circular cylinder on which a uniform surface current flows parallel to the axis. Compare the result with the electric field of the same cylinder with uniform surface charge.
11. Find the magnetic field of a nonideal solenoid, that is one for which currents flowing on the cylindrical surfaces are helical. Hint: Make superposition of the field found in Problem 10 with the field of the ideal solenoid.
12. Show that the volume current density $\mathbf{j} = \frac{\omega t}{R}(\mathbf{r} \cdot \hat{\mathbf{n}})\,\delta(\sqrt{x^2+y^2} - R)$ corresponds to the surface current density (54), when the Z-axis is chosen perpendicular to $\hat{\mathbf{n}}$. Check that $\mathbf{j}$ satisfies the equation $\nabla \cdot \mathbf{j} = 0$.

Chapter 2

ELECTROMAGNETIC POTENTIALS

§1. Introducing Electromagnetic Potentials

The homogeneous Maxwell equations (1.17a,b) allow us to introduce the so-called *electromagnetic potentials.* With the aid of the Poincaré Lemma of Sec. 0.10, Eq. (1.17b) implies that, in a region contractible to a point, a vector field $\mathbf{A}(\mathbf{r}, t)$ exists such that the magnetic induction may be expressed as

$$\hat{\mathbf{B}} = \boldsymbol{\nabla} \wedge \mathbf{A} . \tag{1}$$

Field $\mathbf{A}$ is called the *vector potential of the electromagnetic field.* We insert this in Eq. (1.17a):

$$\mathbf{E} + \frac{\partial \mathbf{A}}{\partial t} = 0 .$$

The Poincaré Lemma now states that in a region contractible to a point, a scalar field $\varphi(\mathbf{r}, t)$ exists such that $\mathbf{E} + \frac{\partial \mathbf{A}}{\partial t} = -\boldsymbol{\nabla} \wedge \varphi$. The field φ is called the *scalar potential of the electromagnetic field.* Hence the following expression for the electric field is obtained

$$\mathbf{E} = -\boldsymbol{\nabla} \wedge \varphi - \frac{\partial \mathbf{A}}{\partial t} . \tag{2}$$

For stationary systems, when φ and $\mathbf{A}$ do not depend on time, we have

$$\mathbf{E} = -\text{grad}\ \varphi . \tag{3}$$

We know that the potential φ is defined in electrostatics up to an additive constant, since such a constant does not affect the field $\mathbf{E}$ obtained according to (3). A similar situation occurs with the vector potential, because the derivatives of an arbitrary constant vector-valued function are also zero. However, one more arbitrariness takes place here. If the gradient of a scalar-valued function χ is added to $\mathbf{A}$, $\widehat{\mathbf{B}}$ given by (1) remains unchanged, since $\nabla \wedge (\mathbf{A} + \nabla \wedge \chi) = \nabla \wedge \mathbf{A} = \widehat{\mathbf{B}}$ by dint of Problem 0.15. Therefore the vector potential is defined up to the gradient of a scalar function. Of course, an appropriate change of the scalar potential must accompany the change $\mathbf{A} \to \mathbf{A} + \nabla\chi$ in order that relation (2) is preserved. This may be found by inserting $\mathbf{A} + \nabla\chi$ into the right-hand side of (2) with an addition of ψ to the scalar potential:

$$-\nabla(\phi + \psi) - \frac{\partial}{\partial t}(\mathbf{A} + \nabla\chi) = -\nabla\varphi - \frac{\partial \mathbf{A}}{\partial t} - \nabla\left(\psi + \frac{\partial\chi}{\partial t}\right) .$$

If all this is to give $\mathbf{E}$, the equality $\psi + \frac{\partial\chi}{\partial t} = f(t)$ must be satisfied. In this way, we obtain

$$\psi = -\frac{\partial\chi}{\partial t} + f .$$

The function f does not depend on position, so it does not influence the field $\mathbf{E}$ obtained through (2), therefore f will henceforth be omitted. In this manner, we have shown that the following change of potentials

$$\mathbf{A} \to \mathbf{A} + \nabla\chi , \qquad \varphi \to \varphi - \frac{\partial\chi}{\partial t} , \tag{4}$$

is possible without affecting the intensity quantities $\mathbf{E}$ and $\widehat{\mathbf{B}}$. We call this change *gauge transformation*, χ — *gauge function*, and the fact that $\mathbf{E}$ and $\widehat{\mathbf{B}}$ remain unchanged — *gauge invariance*.

The two homogeneous Maxwell equations allowed us to state that vector and scalar potentials of the electromagnetic field exist. Now we shall find out what conditions are imposed on the potentials by the other Maxwell equations. We have from (1) and (2)

$$\widehat{\mathbf{H}} = \frac{1}{\mu}\nabla \wedge \mathbf{A} , \qquad \mathbf{D} = -\varepsilon\left(\nabla\varphi + \frac{\partial \mathbf{A}}{\partial t}\right) . \tag{5}$$

Having assumed that the medium is uniform in space and stationary in time (this means that the derivatives do not affect ε and μ), after the substitution of (5) into (1.17c) we obtain

$$\frac{1}{\mu}\nabla \cdot (\nabla \wedge \mathbf{A}) - \varepsilon\frac{\partial}{\partial t}\nabla\varphi - \varepsilon\frac{\partial^2 \mathbf{A}}{\partial t^2} = -\mathbf{j} .$$

By virtue of the identity $\nabla \cdot (\nabla \wedge \mathbf{A}) = -\nabla \wedge (\nabla \cdot \mathbf{A}) + \Delta\, \mathbf{A}$, following from (0.126), we obtain

$$-\nabla \wedge \left(\nabla \cdot \mathbf{A} + \varepsilon\mu \frac{\partial \varphi}{\partial t}\right) + \Delta\, \mathbf{A} - \varepsilon\mu \frac{\partial^2 \mathbf{A}}{\partial t^2} = -\mu \mathbf{j} \,. \tag{6}$$

We now use the above-mentioned arbitrariness in introducing the potentials $\mathbf{A}, \varphi$ and postulate that the expression in parenthesis vanishes:

$$\nabla \cdot \mathbf{A} + \varepsilon\mu \frac{\partial \varphi}{\partial t} = 0 \,.$$

This equality is called the *Lorentz condition*. It can be achieved through an appropriate gauge transformation with a gauge function satisfying an appropriate condition (see Problem 1). In this case, Eq. (6) takes the form

$$\Box \mathbf{A} = -\mu \mathbf{j} \,, \tag{7}$$

where $\Box = \Delta - \varepsilon\mu \frac{\partial^2}{\partial t^2}$ is the Dalambertian.

In order to find a similar equation for the scalar potential, we insert (5) into the last Maxwell equation (1.17d):

$$\nabla^2 \varphi + \frac{\partial}{\partial t} \nabla \cdot \mathbf{A} = -\frac{1}{\varepsilon} \rho \,.$$

We substitute $\nabla \cdot \mathbf{A} = -\varepsilon\mu \frac{\partial \varphi}{\partial t}$ from the Lorentz condition and obtain

$$\Box \varphi = -\frac{1}{\varepsilon} \rho \,. \tag{8}$$

Equations (7) and (8) are called *d'Alembert equations*. In the stationary case, when $\mathbf{A}$ and φ do not depend on time, they are known as the *Poisson equations*

$$\Delta\, \mathbf{A} = -\mu \mathbf{j} \,, \qquad \Delta \varphi = -\frac{1}{\varepsilon} \rho \,.$$

The d'Alembert and Poisson equations bring into relief the analogy between μ and $\frac{1}{\varepsilon}$ which was mentioned in Sec. 1.4.

Notice that, for the static case, the roles of potentials are separated: φ describes the electric field, $\mathbf{A}$ describes the magnetic field only (compare Problem 6). The Lorentz condition reduces then to $\nabla \cdot \mathbf{A} = 0$.

Because of the arbitrariness in the choice of vector potential, it is not considered as a directly measurable physical quantity. Another situation occurs with the curvilinear integral of this potential over a directed loop ∂S, that is — according to the terminology introduced in Sec. 0.11 — with the adherence of $\mathbf{A}$ to the loop ∂S. Now, by dint of Stokes theorem (0.139)

$$\int_{\partial S} d\mathbf{l} \cdot \mathbf{A} = -\int_S d\hat{\mathbf{s}} \cdot (\boldsymbol{\nabla} \wedge \mathbf{A}) = -\int_S d\hat{\mathbf{s}} \cdot \hat{\mathbf{B}} \,. \tag{9}$$

We noticed in Sec. 1.2 that the right-hand side represents the magnetic flux through the surface S. In this way, we see that the adherence of the vector potential to a closed directed curve does not depend on gauge — therefore, it is a measurable quantity. Expression (9) corresponds to the potential difference occurring in electrostatics — it also does not depend on the choice of a potential describing the same field.

It is worth noticing that the magnetic flux (9) does not depend on the surface S itself, but only on its boundary ∂S. Indeed, for two directed surfaces S_1 and S_2 with the same boundary (see Fig. 59) we may consider the closed surface $S_1 \cup (-S_2)$ where the minus sign denotes the opposite orientation. Due to the integral Maxwell equation (1.10), we have

$$\int_{S_1 \cup (-S_2)} d\hat{\mathbf{s}} \cdot \hat{\mathbf{B}} = 0 \,.$$

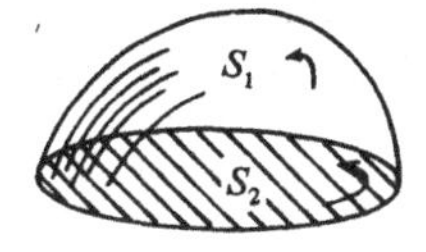

Fig. 59.

Hence, after replacing the left-hand side by the difference of the two integrals, we get

$$\int_{S_1} d\hat{\mathbf{s}} \cdot \hat{\mathbf{B}} = \int_{S_2} d\hat{\mathbf{s}} \cdot \hat{\mathbf{B}} \,,$$

which is what we wanted to prove.

The relations between the potentials and the intensity quantities may also be written in terms of the Clifford algebra. Let us form the *cliffor of the electromagnetic potentials*:

$$\widetilde{A} = \frac{1}{\sqrt{\mu}} \mathbf{A} - \sqrt{\varepsilon}\, \varphi \,.$$

The coefficients are chosen so as to allow us to add quantities of the same dimension. We have picked up the dimension $\sqrt{A\,\mathrm{Wb/m}} = \sqrt{\mathrm{J/m}}$. Let the operator $\mathcal{D}^* = \boldsymbol{\nabla} - \sqrt{\varepsilon\mu}\frac{\partial}{\partial t}$ act on this cliffor (remember that ε and μ are constants):

$$\left(\boldsymbol{\nabla} - \sqrt{\varepsilon\mu}\,\frac{\partial}{\partial t}\right)\left(\frac{1}{\sqrt{\mu}}\mathbf{A} - \sqrt{\varepsilon}\,\varphi\right) = \varepsilon\sqrt{\mu}\,\frac{\partial\varphi}{\partial t} - \sqrt{\varepsilon}\frac{\partial\,\mathbf{A}}{\partial t} - \sqrt{\varepsilon}\boldsymbol{\nabla}\varphi + \frac{1}{\sqrt{\mu}}\boldsymbol{\nabla}\mathbf{A}\,.$$

The last term is the Clifford derivative, so

$$\begin{aligned}\mathcal{D}^*\widetilde{A} &= \varepsilon\sqrt{\mu}\,\frac{\partial\varphi}{\partial t} - \sqrt{\varepsilon}\left(\frac{\partial\,\mathbf{A}}{\partial t} + \boldsymbol{\nabla}\varphi\right) + \frac{1}{\sqrt{\mu}}\boldsymbol{\nabla}\cdot\mathbf{A} + \frac{1}{\sqrt{\mu}}\boldsymbol{\nabla}\wedge\mathbf{A}\\ &= \sqrt{\varepsilon}\,\mathbf{E} + \frac{1}{\sqrt{\mu}}\widehat{\mathbf{B}} + \frac{1}{\sqrt{\mu}}\left(\varepsilon\mu\,\frac{\partial\varphi}{\partial t} + \boldsymbol{\nabla}\cdot\mathbf{A}\right)\,.\end{aligned}$$

In accordance with the Lorentz condition, the last term vanishes and the first two terms constitute the electromagnetic cliffor. In this manner, we obtain

$$\mathcal{D}^*\widetilde{A} = f^{\cdot}\,, \tag{10}$$

as the relation between the cliffor of electromagnetic potentials and the electromagnetic cliffor.

The operators $\mathcal{D}$ and $\mathcal{D}^*$ commute (as was shown in Sec. 0.10), and their product is equal to the Dalambertian. Thus $\mathcal{D}$ acting on (10) gives $\Box\,\widetilde{A} = \mathcal{D}f$ and, by virtue of the Maxwell equation (1.23),

$$\Box\,\widetilde{A} = \widetilde{j}\,. \tag{11}$$

The same equation can be obtained as a linear combination of (7) and (8) with the appropriate coefficients.

Another potential may be introduced in a special case. The Maxwell equation (1.14) reduces to $\boldsymbol{\nabla}\cdot\widehat{\mathbf{H}} = 0$ in the stationary case (that is, when $\dot{\mathbf{D}} = 0$) for regions devoid of currents (that is $\mathbf{j} = 0$). By means of (0.121), this can be rewritten as $\boldsymbol{\nabla}\times\mathbf{H} = 0$ and, by virtue of (0.120), as $\boldsymbol{\nabla}\wedge\mathbf{H} = 0$. Now we would like to use the Poincaré Lemma to ascertain the existence of a pseudoscalar potential ψ such that

$$\mathbf{H} = \boldsymbol{\nabla}\wedge\psi\,, \tag{12}$$

in regions contractible to a point. The whole region outside the currents can never be contractible to a point (the lines of the stationary currents must be infinite or closed), therefore we can admit the existence of ψ only locally, not in the whole space devoid of currents.

Let us multiply both sides of (12) by I:

$$I\mathbf{H} = I(\nabla \wedge \psi) \ ,$$

and use the identities (0.76) and (0.123)

$$\widehat{\mathbf{H}} = \nabla \cdot (I\psi) \ .$$

We now introduce the trivector $\widehat{\Psi} = I\,\psi$ and obtain

$$\widehat{\mathbf{H}} = \nabla \cdot \widehat{\Psi} \ . \tag{13}$$

As a result of the Poincaré Lemma, this equality states that, in magnetostatics, in contractible parts of the region devoid of currents, a *trivector magnetic potential* $\widehat{\Psi}$ exists such that the magnetic field $\widehat{\mathbf{H}}$ is its inner derivative.

The equipotential surfaces of $\widehat{\Psi}$ are simultaneously equipotential surfaces of ψ, hence due to (12) and to the wellknown properties of the gradient, they are perpendicular to the vector field $\mathbf{H}$, and consequently, they are tangent to the volutor field $\widehat{\mathbf{H}}$. Thus for the isotropic medium, that is, when $\widehat{\mathbf{B}} = \mu\,\widehat{\mathbf{H}}$, the equipotential surfaces of $\widehat{\Psi}$ are good representatives for the force surfaces of the magnetic field if the orientation is added to them.

§2. Finding Potentials from Given Sources

In order to find solutions to Eq. (11), we use Lemmas 1 and 3 from Sec. 0.12, which state that two types of solutions:

$$\widetilde{A}^{(\pm)}(\mathbf{r},t) = -\frac{1}{4\pi}\int dt' \int dv' \frac{1}{|\mathbf{r}-\mathbf{r}'|}\,\delta\Big(t - t' \pm \frac{|\mathbf{r}-\mathbf{r}'|}{u}\Big)\widetilde{j}(\mathbf{r}',t') + \widetilde{A}_0(\mathbf{r},t)$$

exist for Eq. (11), where $u = 1/\sqrt{\varepsilon\mu}$ is the speed of electromagnetic waves in the medium, as will be shown in Sec. 4. The integration over t' may be performed due to the presence of the one-dimensional Dirac delta

$$\widetilde{A}^{(\pm)}(\mathbf{r},t) = -\frac{1}{4\pi}\int dv' \frac{\widetilde{j}(\mathbf{r}', t \pm \frac{|\mathbf{r}-\mathbf{r}'|}{u})}{|\mathbf{r}-\mathbf{r}'|} + \widetilde{A}_0(\mathbf{r},t) \ . \tag{14}$$

In the expression for $\widetilde{A}^{(-)}$, the sources are taken *earlier in time* by the amount $|\mathbf{r}-\mathbf{r}'|/u$ needed for the electromagnetic signal to travel from point $\mathbf{r}'$, in which the source is located, to point $\mathbf{r}$, in which the potential is calculated.

In this manner, the changes of the potential are delayed with respect to that of the sources, therefore $\tilde{A}^{(-)}$ is called the *retarded potential*. On the other hand, in the expression for $\tilde{A}^{(+)}$, the sources appear later in time by the amount $|\mathbf{r}-\mathbf{r}'|/u$ needed for the electromagnetic signal to cover the distance from $\mathbf{r}$ to $\mathbf{r}'$. Thus the changes of this potential outpace the corresponding changes of the sources, therefore $\tilde{A}^{(+)}$ is called the *advanced potential*. By virtue of the *causality principle* (the cause $\tilde{j}$ should be earlier than the effect $\tilde{A}$), we usually discard the second type of solution.

In the case of stationary sources, when $\tilde{j}$ does not depend on time, we obtain

$$\tilde{A}^{(\pm)}(\mathbf{r},t) = -\frac{1}{4\pi}\int dv' \frac{\tilde{j}(\mathbf{r}')}{|\mathbf{r}-\mathbf{r}'|} + \tilde{A}_0(\mathbf{r},t) \ .$$

We see that the difference between the two types of solution disappears. For this reason, only one type of solution to the Maxwell equation was sufficient in Sec. 1.5.

Equation (14) concern the general solutions when, in addition to the fields immediately produced by the sources $\tilde{j}$, some other fields $\tilde{A}_0$ are present which originate from other sources not included in the function $\tilde{j}$ or arrive from infinity in the form of electromagnetic waves. From now on, we shall omit the term $\tilde{A}_0$, since we are interested only in potentials arising directly from given sources. In this manner, the formula

$$\tilde{A}(\mathbf{r},t) = -\frac{1}{4\pi}\int dv' \frac{\tilde{j}(\mathbf{r}', t - \frac{|\mathbf{r}-\mathbf{r}'|}{u})}{|\mathbf{r}-\mathbf{r}'|} \tag{15}$$

remains. The scalar and vector parts of it, taken separately, give

$$\varphi(\mathbf{r},t) = \frac{1}{4\pi\varepsilon}\int dv' \frac{\rho(\mathbf{r}', t - \frac{|\mathbf{r}-\mathbf{r}'|}{u})}{|\mathbf{r}-\mathbf{r}'|} , \tag{16}$$

$$\mathbf{A}(\mathbf{r},t) = \frac{\mu}{4\pi}\int dv' \frac{\mathbf{j}(\mathbf{r}', t - \frac{|\mathbf{r}-\mathbf{r}'|}{u})}{|\mathbf{r}-\mathbf{r}'|} . \tag{17}$$

In the case of stationary charges and currents, the following expressions for static potentials

$$\varphi(\mathbf{r}) = \frac{1}{4\pi\varepsilon}\int dv' \frac{\rho(\mathbf{r}')}{|\mathbf{r}-\mathbf{r}'|} , \tag{18}$$

$$\mathbf{A}(\mathbf{r}) = \frac{\mu}{4\pi}\int dv' \frac{\mathbf{j}(\mathbf{r}')}{|\mathbf{r}-\mathbf{r}'|} \tag{19}$$

are obtained. By means of the identity $\nabla_r \frac{1}{|\mathbf{r}-\mathbf{r}'|} = -\frac{\mathbf{r}-\mathbf{r}'}{|\mathbf{r}-\mathbf{r}'|^3}$, the following formulae for the intensity quantities

$$\mathbf{E}(\mathbf{r}) = -\nabla\varphi(\mathbf{r}) = \frac{1}{4\pi\varepsilon}\int dv' \frac{(\mathbf{r}-\mathbf{r}')\,\rho(\mathbf{r}')}{|\mathbf{r}-\mathbf{r}'|^3}, \tag{20}$$

$$\hat{\mathbf{B}}(\mathbf{r}) = \nabla\wedge\mathbf{A}(\mathbf{r}) = -\frac{\mu}{4\pi}\int dv' \frac{(\mathbf{r}-\mathbf{r}')\wedge\mathbf{j}(\mathbf{r}')}{|\mathbf{r}-\mathbf{r}'|^3} \tag{21}$$

are obtained, which are compatible with Eqs. (1.34) and (1.35) for the magnitude quantities.

Equations (18) and (19) are less universal than (20) and (21) because there are situations (for instance, the infinite uniformly charged plate, the infinite straight-line conductor) in which the integrals (20,21) give finite results for points outside the sources, whereas (18,19) give infinity. Thus Eqs. (18) and (19) cannot be applied in the case of sources extending to infinity, which do not fall fast enough.

Now we must check whether the potentials obtained satisfy the Lorentz condition. As we shall see, the assumption that the currents vanish outside some bounded region is sufficient for this. We have from (17)

$$\nabla\cdot\mathbf{A}(\mathbf{r},t) = \frac{\mu}{4\pi}\int_V dv'\,\nabla\cdot\frac{\mathbf{j}\left(\mathbf{r}', t-\frac{|\mathbf{r}-\mathbf{r}'|}{u}\right)}{|\mathbf{r}-\mathbf{r}'|}. \tag{22}$$

The variable $\mathbf{r}$ occurs under the integral in two places: in the denominator and in the time argument of the function $\mathbf{j}$, which we denote by $t' = t - |\mathbf{r}-\mathbf{r}'|/u$. Thus

$$\nabla\cdot\frac{\mathbf{j}(\mathbf{r}',t')}{|\mathbf{r}-\mathbf{r}'|} = \left(\nabla\frac{1}{|\mathbf{r}-\mathbf{r}'|}\right)\cdot\mathbf{j}(\mathbf{r}',t') + \frac{1}{|\mathbf{r}-\mathbf{r}'|}\nabla\cdot\mathbf{j}(\mathbf{r}',t').$$

The operator ∇ may be replaced by $-\nabla'$ when acting on a function of the difference $\mathbf{r}-\mathbf{r}'$, but in the second term at the right-hand side, $\mathbf{r}'$ also appears independently of t', therefore we have

$$\nabla\cdot\frac{\mathbf{j}(\mathbf{r}',t')}{|\mathbf{r}-\mathbf{r}'|} = -\left(\nabla'\frac{1}{|\mathbf{r}-\mathbf{r}'|}\right)\cdot\mathbf{j}(\mathbf{r}',t') - \frac{1}{|\mathbf{r}-\mathbf{r}'|}\nabla'\cdot\mathbf{j}(\mathbf{r}',t')$$
$$+\frac{1}{|\mathbf{r}-\mathbf{r}'|}\nabla'\cdot\mathbf{j}(\mathbf{r}',t')\Big|_{t'=\mathrm{const}}.$$

The last term contains differentiation only with respect to $\mathbf{r}'$ standing independently of t'. Therefore, uniting the first two terms we obtain

$$\nabla\cdot\frac{\mathbf{j}(\mathbf{r}',t')}{|\mathbf{r}-\mathbf{r}'|} = -\nabla'\cdot\frac{\mathbf{j}(\mathbf{r}',t')}{|\mathbf{r}-\mathbf{r}'|} + \frac{1}{|\mathbf{r}-\mathbf{r}'|}\nabla'\cdot\mathbf{j}(\mathbf{r}',t')\Big|_{t'=\mathrm{const}}.$$

We insert this into (22)

$$\nabla\cdot\mathbf{A}(\mathbf{r},t)=-\frac{\mu}{4\pi}\int_V dv'\nabla'\cdot\frac{\mathbf{j}(\mathbf{r}',t')}{|\mathbf{r}-\mathbf{r}'|}+\frac{\mu}{4\pi}\int_V\frac{dv'}{|\mathbf{r}-\mathbf{r}'|}\nabla'\cdot\mathbf{j}(\mathbf{r}',t')\Big|_{t'=\text{const}}\,.$$

Due to the Gauss theorem, we change the first term into the surface integral:

$$\int_V dv'\nabla'\cdot\frac{\mathbf{j}(\mathbf{r}',t')}{|\mathbf{r}-\mathbf{r}'|}=\int_{\partial V}d\mathbf{s}'\cdot\frac{\mathbf{j}(\mathbf{r}',t')}{|\mathbf{r}-\mathbf{r}'|}\,.$$

We choose the integration set V larger than the region containing the sources. This means that the current density $\mathbf{j}$ vanishes on the boundary ∂V, hence this integral is zero. Therefore we obtain

$$\nabla\cdot\mathbf{A}(\mathbf{r},t)=\frac{\mu}{4\pi}\int_V\frac{dv'}{|\mathbf{r}-\mathbf{r}'|}\nabla'\cdot\mathbf{j}(\mathbf{r}',t)\Big|_{t'=\text{const}}\,.$$

On the other hand, we have from (16):

$$\frac{\partial\varphi}{\partial t}=\frac{1}{4\pi\varepsilon}\int_V\frac{dv'}{|\mathbf{r}-\mathbf{r}'|}\frac{\partial\rho(\mathbf{r}',t')}{\partial t}=\frac{1}{4\pi\varepsilon}\int_V\frac{dv'}{|\mathbf{r}-\mathbf{r}'|}\frac{\partial\rho(\mathbf{r}',t')}{\partial t'}\,.$$

We put the last two expressions in the Lorentz condition:

$$\nabla\cdot\mathbf{A}+\varepsilon\mu\frac{\partial\varphi}{\partial t}=\frac{\mu}{4\pi}\int_V\frac{dv'}{|\mathbf{r}-\mathbf{r}'|}\left[\nabla'\cdot\mathbf{j}(\mathbf{r}',t')\Big|_{t'=\text{const}}+\frac{\partial}{\partial t'}\rho(\mathbf{r}',t')\right]\,.$$

The expression in square brackets vanishes due to the continuity equation (1.28), hence the Lorentz condition is satisfied.

§3. Examples of Static Potentials

3.1. Potentials for the Fields of a Rectilinear Conductor

We consider an infinite straight-line conductor of a negligibly small cross-section, with constant linear density of charge τ. We choose the same coordinate system as in Example 6.1 of Chap. 1. Then integral (18) reduces to the linear integral

$$\phi(\mathbf{r})=\frac{\tau}{4\pi\varepsilon}\int\frac{dl'}{|\mathbf{r}-\mathbf{r}'|}=\frac{\tau}{4\pi\varepsilon}\int_{-\infty}^{\infty}\frac{dz'}{\sqrt{x^2+y^2+(z'-z)^2}}\,.$$

This integral is divergent — we warned earlier against such a situation — Eq. (18) does not apply in this case.

We thus employ another method. We make use of the field obtained in (1.38):

$$\mathbf{E}(\mathbf{r}) = \frac{\tau}{2\pi\varepsilon}\mathbf{r}_\perp^{-1} . \tag{23}$$

This field has axial symmetry, hence we seek function φ of the same symmetry, that is, $\varphi(\mathbf{r}) = \varphi(\mathbf{r}_\perp) = \varphi(r_\perp)$. In such a case, $\nabla\varphi(r_\perp) = \frac{\mathbf{r}_\perp}{r_\perp}\frac{d\varphi}{dr_\perp}$ and $\mathbf{E} = -\nabla\varphi = -\frac{\mathbf{r}_\perp}{r_\perp}\varphi'$. We insert this into Eq. (23):

$$-\frac{\mathbf{r}_\perp}{r_\perp}\varphi' = \frac{\tau}{2\pi\varepsilon}\frac{\mathbf{r}_\perp}{r_\perp^2} ,$$

and obtain the differential equation

$$\varphi' = -\frac{\tau}{2\pi\varepsilon}\frac{1}{r_\perp} ,$$

which has the solution

$$\varphi(r_\perp) = -\frac{\tau}{2\pi\varepsilon}\log\frac{r_\perp}{r_0} , \tag{24}$$

where r_0 is the integration constant. It defines the distance from the conductor at which the potential is zero.

Now let the current J flow in the straight-line conductor in the positive direction of $\mathbf{e}_3$. Since the current ranges up to infinity, we cannot apply formula (19). Therefore, we start from expression (1.39) which yields

$$\widehat{\mathbf{B}}(\mathbf{r}) = \frac{\mu J}{2\pi}\mathbf{e}_3 \wedge \mathbf{r}_\perp^{-1} .$$

This is an example of the volutor field of the form $\widehat{\mathbf{B}}(\mathbf{r}) = \mathbf{k}\wedge\mathbf{E}(\mathbf{r})$ with the constant vector $\mathbf{k} = (\mu\varepsilon J/\tau)\,\mathbf{e}_3$ and $\mathbf{E}$ given by (23). In such a case, by virtue of Problem 0.26, we have

$$\nabla\wedge(\mathbf{k}\wedge\varphi) = -\mathbf{k}\wedge(\nabla\wedge\varphi) = \mathbf{k}\wedge\mathbf{E} ,$$

and we may assume $\mathbf{A} = \mathbf{k}\wedge\varphi$ for φ given by (24):

$$\mathbf{A}(\mathbf{r}) = -\frac{\mu J}{2\pi}\mathbf{e}_3\log\frac{r_\perp}{r_0} . \tag{25}$$

The logarithm is negative for $r_\perp < r_0$, so $\mathbf{A}$ has the direction $\mathbf{e}_3$ for $r_\perp < r_0$, that is, the direction of the current. On the other hand, $\mathbf{A}$ is opposite to the current for $r_\perp > r_0$. Field (25) is illustrated in Fig. 60 on a straight line intersecting the conductor perpendicularly. This field has the full symmetry of the sources — it is invariant under rotations around the conductor and under translations parallel to it. The directed squares in Fig. 60 represent the volutors of magnetic induction.

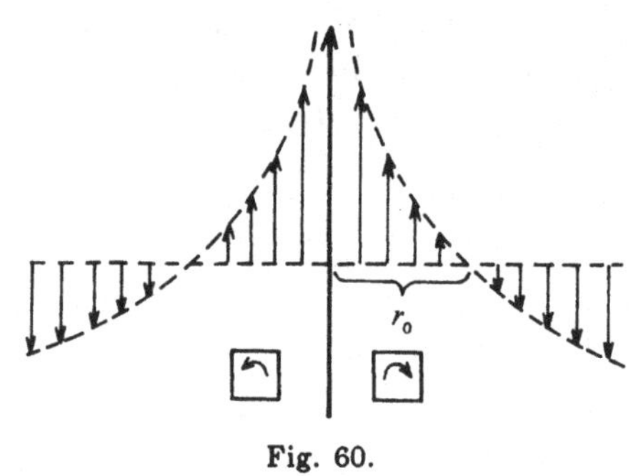

Fig. 60.

The potential (25) satisfies, moreover, the Lorentz condition (see Problem 3).

3.2. The Vector Potential of Two Parallel Conductors with Currents

We consider two parallel rectilinear conductors with currents of equal values, at first with the same orientations and then with opposite orientations. Making use of the superposition principle, we may represent the potential for this situation as the sum of two expressions of the form (25). Let one of the conductors coincide with the Z-axis, let the other one intersect the plane $(\mathbf{e}_1, \mathbf{e}_2)$ at the point $\mathbf{a}$. We choose $r_0 = a$ in order that the potential produced by one of the conductors vanishes on the other. In this way,

$$\mathbf{A}(\mathbf{r}) = -\frac{\mu J}{2\pi}\mathbf{e}_3\left(\log\frac{r_\perp}{a} \pm \log\frac{|\mathbf{r}_\perp - \mathbf{a}|}{a}\right).$$

We thus obtain

$$\mathbf{A}(\mathbf{r}) = -\frac{\mu J}{2\pi}\mathbf{e}_3 \log\frac{r_\perp|\mathbf{r}_\perp - \mathbf{a}|}{a^2} \tag{26}$$

for the compatible currents, and

$$\mathbf{A}(\mathbf{r}) = -\frac{\mu J}{2\pi}\mathbf{e}_3 \log\frac{r_\perp}{|\mathbf{r}_\perp - \mathbf{a}|}, \tag{27}$$

for the opposite currents.

We depict these fields in Fig. 61 on a straight line intersecting the two conductors. For compatible currents, (a) the potential is zero on a cylindrical surface encircling the two conductors and given by the equation $r_\perp |\mathbf{r}_\perp - \mathbf{a}| = a^2$, whereas for this opposite currents, (b) the potential vanishes in the plane of equal distances from the conductors described by the equation $r_\perp = |\mathbf{r}_\perp - \mathbf{a}|$. Both vector potentials are parallel to the currents in their vicinities and have translational symmetry parallel to the conductors. Moreover, potential (26) is symmetric under reflexion in the plane of equal distances, whilst (27) is antisymmetric under this operation.

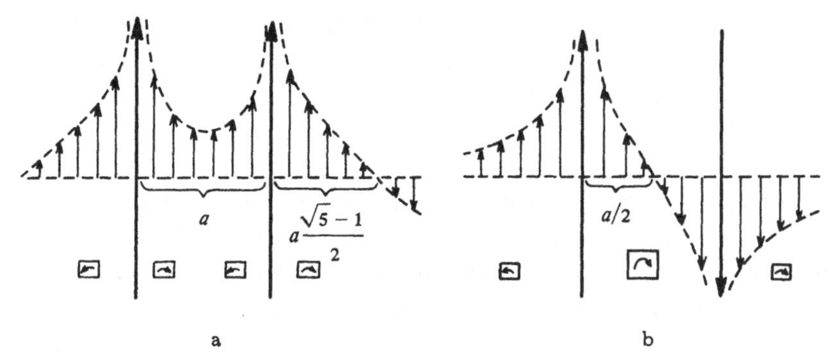

Fig. 61.

Both potentials satisfy the Lorentz condition as the combinations of the potentials of the form (25).

3.3. Potentials for the Fields of an Infinite Plate

Let an infinite plate with uniform surface charge of density σ coincide with the plane $(\mathbf{e}_2, \mathbf{e}_3)$. We know from Sec. 1.6 that the electric field on both sides of the plate is uniform

$$\mathbf{E}(\mathbf{r}) = \begin{cases} \frac{\sigma}{2\epsilon}\mathbf{e}_1 & \text{for } x_1 > 0\,, \\ -\frac{\sigma}{2\epsilon}\mathbf{e}_1 & \text{for } x_1 < 0\,. \end{cases} \tag{28}$$

Formula (18) is not applicable here. But we know (from Problem 0.34, for instance) that the gradient of the linear function $\varphi(\mathbf{r}) = \mathbf{b}\cdot\mathbf{r}$ is constant and equal to $\mathbf{b}$, so we may assume

$$\varphi(\mathbf{r}) = \begin{cases} -\frac{\sigma}{2\epsilon}\mathbf{e}_1\cdot\mathbf{r} & \text{for } x_1 > 0\,, \\ \frac{\sigma}{2\epsilon}\mathbf{e}_1\cdot\mathbf{r} & \text{for } x_1 < 0\,. \end{cases}$$

These expressions can be contained in a single formula:

$$\varphi(\mathbf{r}) = -\frac{\sigma}{2\varepsilon}|\mathbf{e}_1\cdot\mathbf{r}| = -\frac{\sigma}{2\varepsilon}|x_1|\,. \tag{29}$$

If the charge of the plate is positive, the potential falls linearly as the distance from the plate grows.

We now consider the same plate with a surface current of linear density $\mathbf{i} = \imath\, \mathbf{e}_3$. In this case, due to the result of Sec. 1.7, the magnetic induction is

$$\widehat{\mathbf{B}}(\mathbf{r}) = \begin{cases} \frac{\mu\imath}{2}\, \mathbf{e}_3 \wedge \mathbf{e}_1 & \text{for } x_1 > 0\,, \\ -\frac{\mu\imath}{2}\, \mathbf{e}_3 \wedge \mathbf{e}_1 & \text{for } x_1 < 0\,. \end{cases} \tag{30}$$

This field is related to (29) through $\widehat{\mathbf{B}}(\mathbf{r}) = \mathbf{k} \wedge \mathbf{E}(\mathbf{r})$ with $\mathbf{k} = (\mu\varepsilon\imath/\sigma)\, \mathbf{e}_3$, so we may, in analogy with 3.1, write $\mathbf{A} = \mathbf{k} \wedge \varphi$ for φ is given by (29):

$$\mathbf{A}(\mathbf{r}) = -\frac{\mu\imath}{2}\, \mathbf{e}_3 |\mathbf{e}_1 \cdot \mathbf{r}_1| = -\frac{\mu}{2}\, \mathbf{i}\, |x_1|\,.$$

This potential is everywhere parallel to the surface current, but has opposite orientation and is zero on the plate itself. In order to ensure an orientation coinciding with that of the currents in their vicinity, it is enough to add a constant vector with an appropriate direction, for instance

$$\mathbf{A}(\mathbf{r}) = -\frac{\mu}{2}\, \mathbf{i}\, |x_1| + \frac{\mu a}{2}\, \mathbf{i} = \frac{\mu}{2}\, \mathbf{i}\, (a - |x_1|)\,, \tag{31}$$

for the constant a with the dimension of length. We illustrate this field in Fig. 62 in a plane perpendicular to the plate and parallel to the current. It has full symmetry of the sources, that is, invariance under reflexions in the plate and under translations parallel to the plate. Potential (31) satisfies the Lorentz condition (see Problem 3). The obtained function $\mathbf{A}(\mathbf{r})$ is everywhere finite, which was not the case in previous examples, and is continuous, though the magnetic induction is discontinuous on the plate.

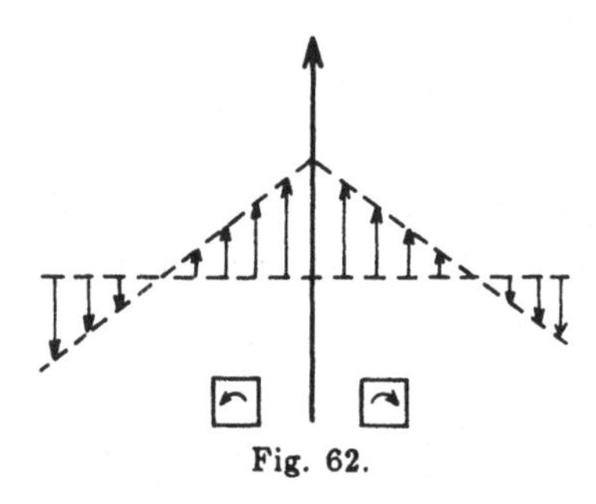

Fig. 62.

The same magnetic field can also be described by another vector potential:

$$\mathbf{A}'(\mathbf{r}) = \alpha\, \mathbf{m}\, |\mathbf{e}_1 \cdot \mathbf{r}| = \alpha\, \mathbf{m}\, |x_1|\,,$$

where α is a constant and $\mathbf{m}$ is a unit vector lying in the plane $(\mathbf{e}_1, \mathbf{e}_3)$, that is, in the plane of Fig. 62. Let β be the angle between $\mathbf{m}$ and $\mathbf{e}_3$: $\mathbf{m} = \mathbf{e}_3 \cos\beta + \mathbf{e}_1 \sin\beta$. We find α from the condition $\nabla \wedge \mathbf{A}' = \widehat{\mathbf{B}}$ for $\widehat{\mathbf{B}}$ expressed by (30), and obtain

$$\mathbf{A}'(\mathbf{r}) = -\frac{i}{2\cos\beta}\mathbf{m}\,|\mathbf{e}_1 \cdot \mathbf{r}|\,. \tag{32}$$

We see that the angle β cannot be arbitrary, β must be different from $\pi/2$.

We illustrate this field in Fig. 63. It is zero on the plate and has smaller symmetry than that of the sources. The potential (32) does not satisfy the Lorentz condition (see Problem 3).

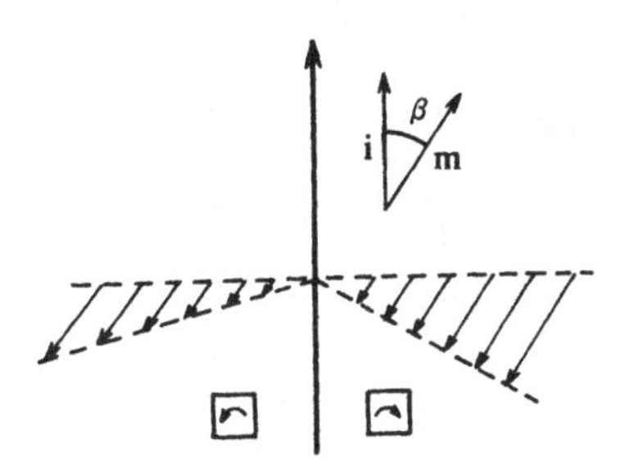

Fig. 63.

3.4. Potentials for the Fields of Two Parallel Plates

Consider two parallel infinite plates charged oppositely with the same modulus of surface charge density σ. Let the distance between the plates be a. As we know from Sec. 1.6.5, the electric field is nonzero only in the space between the plates and is $\mathbf{E} = \frac{\sigma}{\epsilon}\mathbf{n}$, where $\mathbf{n}$ is the unit vector perpendicular to the plates and oriented from the positively to the negatively charged plate. We choose the origin of coordinates exactly half-way between the plates with the X-axis parallel to $\mathbf{n}$. In this case, we may write:

$$\varphi(\mathbf{r}) = \begin{cases} \text{const} & \text{for } \mathbf{n}\cdot\mathbf{r} < -a/2\,, \\ -\frac{\sigma}{\epsilon}\mathbf{n}\cdot\mathbf{r} & \text{for } -a/2 < \mathbf{n}\cdot\mathbf{r} < a/2\,, \\ \text{const} & \text{for } a/2 < \mathbf{n}\cdot\mathbf{r}\,. \end{cases}$$

In order to make this function continuous, we adjust the proper constant values:

$$\varphi(\mathbf{r}) = \begin{cases} \frac{\sigma}{\epsilon}\frac{a}{2} & \text{for } \mathbf{n}\cdot\mathbf{r} \le -\frac{a}{2} \\ -\frac{\sigma}{\epsilon}\mathbf{n}\cdot\mathbf{r} & \text{for } -\frac{a}{2} \le \mathbf{n}\cdot\mathbf{r} \le \frac{a}{2} \\ -\frac{\sigma}{\epsilon}\frac{a}{2} & \text{for } \frac{a}{2} \le \mathbf{n}\cdot\mathbf{r}\,. \end{cases}$$

In the case of two uniform surface currents with densities $\mathbf{i}$ and $-\mathbf{i}$, respectively, we substitute $\mathbf{A} = \mathbf{k} \wedge \varphi$ with an appropriate $\mathbf{k}$, and obtain

$$\mathbf{A}(\mathbf{r}) = \begin{cases} \mu\,\mathbf{i}\,\frac{a}{2} & \text{for } \mathbf{n}\cdot\mathbf{r} \leq -\frac{a}{2}\,, \\ -\mu\,\mathbf{i}\,(\mathbf{n}\cdot\mathbf{r}) & \text{for } -\frac{a}{2} \leq \mathbf{n}\cdot\mathbf{r} \leq \frac{a}{2}\,, \\ -\mu\,\mathbf{i}\,\frac{a}{2} & \text{for } \frac{a}{2} \leq \mathbf{n}\cdot\mathbf{r}\,. \end{cases}$$

We illustrate this potential in Fig. 64. The obtained field $\mathbf{A}$ is continuous, satisfies the Lorentz condition, is parallel to the currents in their vicinity and has the full symmetry of the sources.

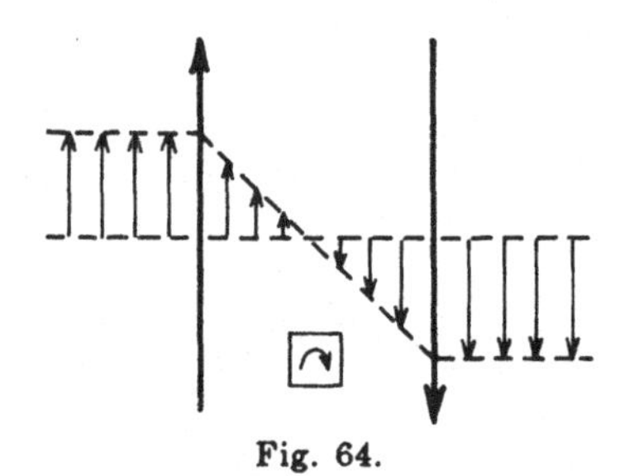

Fig. 64.

3.5. *Vector Potential of Two Intersecting Plates*

We consider two infinite plates with uniform surface currents of densities $\mathbf{i}_1, \mathbf{i}_2$ perpendicular to the edge of intersection of the plates. The angle β between the vectors $\mathbf{i}_1, \mathbf{i}_2$ is also the angle between the plates. The space is divided by the plates into four parts; in each of them, the uniform magnetic field exists as a superposition of the fields of the form (1.43). Introduce unit vectors $\mathbf{n}_1, \mathbf{n}_2$ perpendicular to the respective plates so that the volutors $\mathbf{i}_1 \wedge \mathbf{n}_1$ and $\mathbf{i}_2 \wedge \mathbf{n}_2$ have the same orientation (see Fig. 65). We regard the vector potential arising from such a system of currents as the sum of two potentials of the form (32):

$$\mathbf{A}(\mathbf{r}) = -\frac{\mu i_1}{2\cos\,\beta_1}\,\mathbf{m}_1|\,\mathbf{n}_1\cdot\mathbf{r}\,| - \frac{\mu i_2}{2\cos\,\beta_2}\,\mathbf{m}_2|\,\mathbf{n}_2\cdot\mathbf{r}\,|\,.$$

We now choose the vectors $\mathbf{m}_1$ and $\mathbf{m}_2$ such that the field $\mathbf{A}$ is parallel to the currents in their vicinity, that is, $\mathbf{m}_1 = \mathbf{i}_2/i_2, \mathbf{m}_2 = \mathbf{i}_1/i_1$. Then $\beta_1 = \beta_2 = \beta$ and we obtain

$$\mathbf{A}(\mathbf{r}) = -\frac{\mu i_1}{2\cos\,\beta}\,\frac{\mathbf{i}_2}{i_2}|\,\mathbf{n}_1\cdot\mathbf{r}\,| - \frac{\mu i_2}{2\cos\,\beta}\,\frac{\mathbf{i}_1}{i_1}|\,\mathbf{n}_2\cdot\mathbf{r}\,|\,.$$

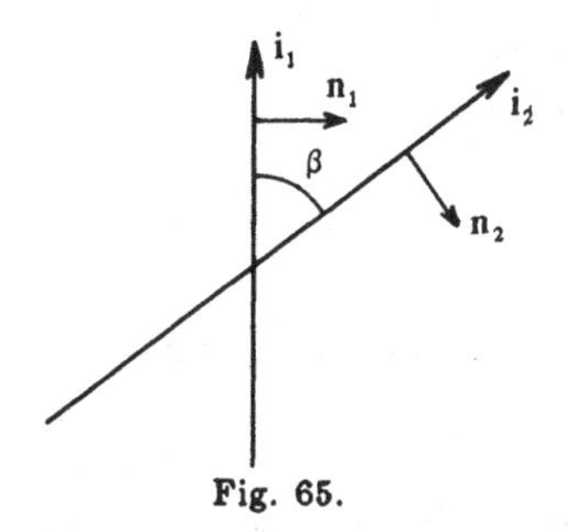

Fig. 65.

This field is continuous but does not fulfil the Lorentz condition. The first term is zero on the first plate, hence only the second term remains at its points — this is the vector parallel to $\mathbf{i}_1$ (with the opposite orientation for $\beta < \frac{\pi}{2}$ and with the same orientation for $\beta > \frac{\pi}{2}$). An analogous situation occurs at the second plate. In this way, for $\beta \neq \frac{\pi}{2}$, we ensure that potential A is parallel to the currents in the close vicinity, but not always with the same orientation. We depict this field in Fig. 66 for $\beta < \frac{\pi}{2}$ and in Fig. 67 for $\beta > \frac{\pi}{2}$. We have chosen $i_1 > i_2$ for the two figures. The field $\widehat{\mathbf{B}}$ is less in the region between the arms of the angle β than in the two neighbouring regions. We picture this by directed squares of various areas in Figs. 66 and 67.

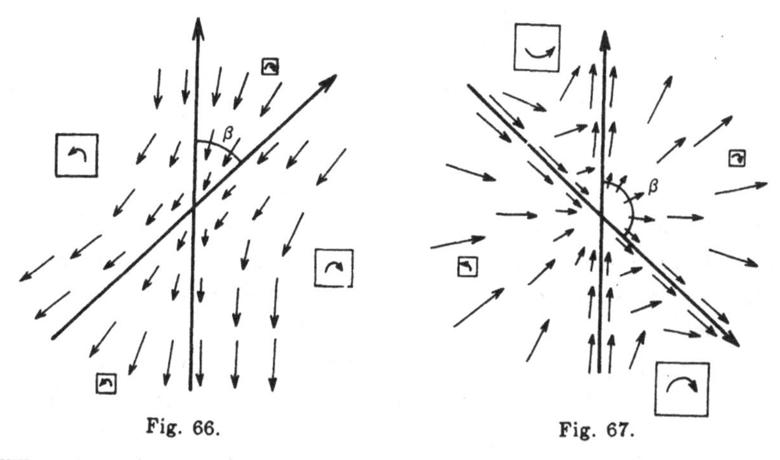

Fig. 66. Fig. 67.

When $i_1 = i_2 = i$, the magnetic field vanishes in two of the four regions separated by the plates and the vector potential takes the form

$$\mathbf{A}(\mathbf{r}) = -\frac{\mu}{2\cos\beta}\left(\mathbf{i}_2|\mathbf{n}_1\cdot\mathbf{r}| + \mathbf{i}_1|\mathbf{n}_2\cdot\mathbf{r}|\right).$$

The formula reduces to

$$\mathbf{A}(\mathbf{r}) = -\frac{\mu i\tan\beta}{2}\mathbf{r}_\perp$$

in the region between the arms of β; here $\mathbf{r}_\perp$ is the component of $\mathbf{r}$ orthogonal to the intersection edge of the plates. On the other hand,

$$\mathbf{A}(\mathbf{r}) = \frac{\mu i \tan\beta}{2} \mathbf{r}_\perp$$

on the opposite side of the edge. It is easy to check that the last two fields have zero outer derivatives. In this way, we convince ourselves that the magnetic field vanishes in the two regions under consideration.

If the densities of currents have equal magnitudes, the plates may be disrupted on the intersection edge without violation of the current conservation, and then separated, without any rotation, to form two inclined plates as shown in Fig. 68. The potential $\mathbf{A}$ is easy to guess for such a situation, we show it graphically in Fig. 68.

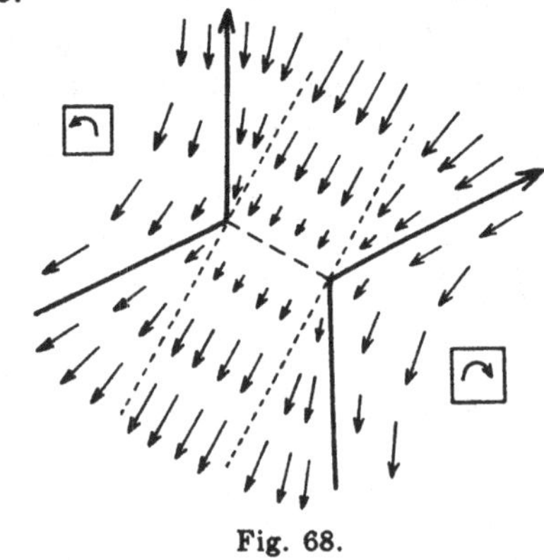

Fig. 68.

The potentials illustrated in Figs. 66 – 68 are continuous and have full symmetry of the sources, namely invariance under translations parallel to the intersection edge, and, for $i_1 = i_2$, under reflexions transforming one plate into the other.

We have not found potentials compatible with the directions of currents in their vicinities in all the examples. Perhaps the reader will succeed in this task!

3.6. Vector Potential of the Solenoid

We consider an ideal circular solenoid with radius R. We know from Sec. 1.7.2 that its magnetic induction may be expressed as

$$\hat{\mathbf{B}}(\mathbf{r}) = \begin{cases} \mu i \hat{\mathbf{n}} & \text{for } \mathbf{r} \text{ inside the solenoid ,} \\ 0 & \text{for } \mathbf{r} \text{ outside the solenoid ,} \end{cases}$$

where $\widehat{\mathbf{n}}$ is the unit volutor perpendicular to the axis of the cylinder and with orientation compatible with the surface current of the solenoid.

We still cannot use formula (19) since the currents extend to infinity. Therefore we use another method, that is, the explicit formula for $\mathbf{A}$ satisfying $\nabla \wedge \mathbf{A} = \widehat{\mathbf{B}}$ for a given field $\widehat{\mathbf{B}}$:

$$A_i(\mathbf{r}) = \int_0^1 ds\, s\, x_j\, B_{ji}(s\,\mathbf{r}) \ .$$

Here $\mathbf{r}_\perp$ is the component of $\mathbf{r}$ perpendicular to the solenoid axis. We know that $\widehat{\mathbf{B}}$ is constant for $r_\perp < R$, hence we obtain $A_i(\mathbf{r}) = \frac{1}{2} x_j B_{ji}$ for $r_\perp \leq R$, that is

$$\mathbf{A}(\mathbf{r}) = \frac{1}{2}\mathbf{r}\cdot\widehat{\mathbf{B}} = \frac{\mu i}{2}\mathbf{r}\cdot\widehat{\mathbf{n}} \qquad \text{for } r_\perp \leq R \ . \tag{33}$$

On the other hand, for $r_\perp > R$, only a part of the integration interval gives the nonzero contribution, namely for $s < s_0 = R/r_\perp$. Thus we obtain

$$\mathbf{A}(\mathbf{r}) = \frac{1}{2}s_0^2\mathbf{r}\cdot\widehat{\mathbf{B}} = \frac{1}{2}\frac{R^2}{r_\perp^2}\mathbf{r}\cdot\widehat{\mathbf{B}} = \frac{\mu i}{2}\frac{R^2}{r_\perp^2}\mathbf{r}\cdot\widehat{\mathbf{n}} \qquad \text{for } r_\perp \geq R \ .$$

The fourth property of the inner product (0.17) allows us to write $\mathbf{r}\cdot\widehat{\mathbf{n}} = \mathbf{r}_\perp\cdot\widehat{\mathbf{n}}$, hence

$$\mathbf{A}(\mathbf{r}) = \begin{cases} \frac{\mu i}{2}\mathbf{r}_\perp\cdot\widehat{\mathbf{n}} & \text{for } r_\perp \leq R \ , \\ \frac{\mu i}{2}\frac{R^2}{r_\perp^2}\mathbf{r}_\perp\cdot\widehat{\mathbf{n}} & \text{for } r_\perp \geq R \ . \end{cases} \tag{34}$$

We denote by $\mathbf{i}(\mathbf{r}_\perp)$ the current density at the intersection point of the direction line of $\mathbf{r}_\perp$ with the surface of the solenoid. The vector $\mathbf{r}_\perp \cdot \widehat{\mathbf{n}}$ is orthgonal to $\mathbf{r}_\perp$, hence it is parallel to $\mathbf{i}(\mathbf{r}_\perp)$. Moreover, its value is $|\mathbf{r}_\perp\cdot\widehat{\mathbf{n}}| = r_\perp$, hence we may write $i\,\mathbf{r}_\perp \cdot \widehat{\mathbf{n}} = r_\perp \mathbf{i}(\mathbf{r}_\perp)$ and consequently,

$$\mathbf{A}(\mathbf{r}) = \begin{cases} \frac{\mu r_\perp}{2}\,\mathbf{i}(\mathbf{r}_\perp) & \text{for } r_\perp \leq R \ , \\ \frac{\mu R^2}{2r_\perp}\,\mathbf{i}(\mathbf{r}_\perp) & \text{for } r_\perp \geq R \ . \end{cases}$$

In this manner, we see explicitly that the vector potential obtained is parallel to the currents in their vicinity and has the same orientation. We show this in Fig. 69. It is continuous and has full symmetry of the sources. Moreover, it satisfies the Lorentz condition (see Problem 3).

Notice that for $r_\perp > R$ this potential is proportional to $1/r_\perp$, which guarantees the constant value of the adherence

$$\int_C d\mathbf{l}\cdot\mathbf{A} = \pi\mu i R^2$$

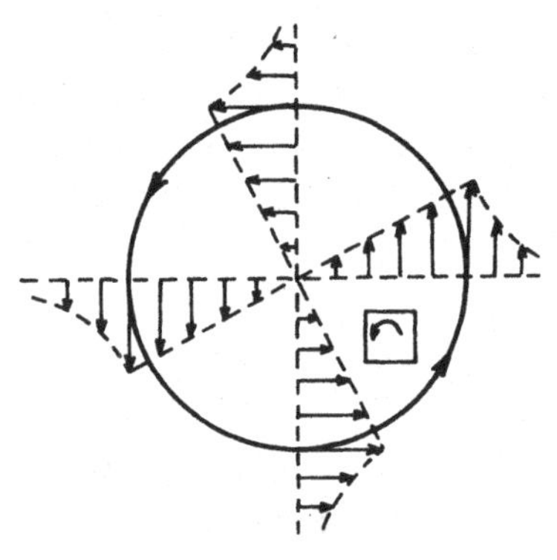

Fig. 69.

to circumference C of constant $r_\perp$. Recalling the observation of Sec. 1 about the interpretation of this integral, we find that this corresponds to the constant magnetic flux through surfaces enclosed by C. This is clear because the whole magnetic flux is confined to the circle of radius R, therefore changing $r_\perp$ for $r_\perp > R$ does not influence its value.

It is worth realizing that in the Examples 3.3 – 3.6, the vector potentials describe in principle the same uniform magnetic field. We write "in principle" because the "uniform" magnetic fields are not the same in the whole space. The potentials under consideration are different, since we tried to make them parallel to the currents in their vicinities. All the potentials are, of course, connected to each other by gauge transformations (see Problem 4).

§4. Multipole Expansion

Let us consider an electric field produced by a system of point charges contained in a bounded region, at far distances in comparison with the dimensions of that region. We choose the origin of coordinates somewhere between the charges. We denote the radii-vectors of separate charges as $\mathbf{r}_{(s)}$, where $s \in \{1, 2, 3, ...\}$. The scalar potential produced by all the charges is

$$\varphi(\mathbf{r}) = \frac{1}{4\pi\varepsilon} \sum_s \frac{q_s}{|\mathbf{r} - \mathbf{r}_{(s)}|} \ .$$

We wish to consider this expression for large $\mathbf{r}$, that is, for $r >> r_{(s)}$.

We develop a scalar function f of difference $\mathbf{r} - \mathbf{r}_{(s)}$ into the Taylor series:

$$f(\mathbf{r} - \mathbf{r}_{(s)}) = f(\mathbf{r}) - x_{(s)i} \frac{\partial f}{\partial x_i} + \frac{1}{2} x_{(s)i} x_{(s)j} \frac{\partial^2 f}{\partial x_{(s)i} \partial x_{(s)j}} + \dots \ . \tag{35}$$

If we restrict ourselves to terms of the first order with respect to $\mathbf{r}_{(s)}$, we obtain

$$\varphi(\mathbf{r}) = \frac{1}{4\pi\varepsilon}\sum_s q_s\left(\frac{1}{r} - \mathbf{r}_{(s)}\cdot\nabla\frac{1}{r}\right) = \frac{1}{4\pi\varepsilon r}\sum_s q_s - \frac{1}{4\pi\varepsilon}\sum_s q_s\mathbf{r}(s)\cdot\nabla\frac{1}{r}\,.$$

The sum $\mathbf{d} = \Sigma_s q_s \mathbf{r}_{(s)}$ is called the *electric dipole moment* with respect to the origin of coordinates. The expression $Q = \Sigma_s q_s$ is obviously the total charge of the system. If $Q = 0$, the dipole moment is independent of the choice of origin. (To see this, it is sufficient to substitute $\mathbf{r}'_{(s)} = \mathbf{r}_{(s)} + \mathbf{a}$ and calculate the moment $\mathbf{d}' = \Sigma_s q_s \mathbf{r}'_{(s)}$ which turns out to be equal to $\mathbf{d}$.) In this manner, we arrive at the approximate formula for the electric potential

$$\varphi(\mathbf{r}) = \frac{1}{4\pi\varepsilon}\left(\frac{Q}{r} - \mathbf{d}\cdot\nabla\frac{1}{r}\right)\,.$$

When $Q = 0$ and further terms are absent in expansion (35), such a system of charges is called the *electric dipole*. The simplest example of the dipole is the system of two opposite charges $\pm q$ with directed distance $\mathbf{l}$ from the negative to the positive charge. Then, independently of the choice of the origin, $\mathbf{d} = |q|\,\mathbf{l}$.

Thus the electric potential of the electric dipole has the following approximate form

$$\varphi^{(1)}(r) = -\frac{1}{4\pi\varepsilon}\mathbf{d}\cdot\nabla\frac{1}{r} = \frac{1}{4\pi\varepsilon}\frac{\mathbf{d}\cdot\mathbf{r}}{r^3}\,. \tag{36}$$

It becomes more and more accurate as $\mathbf{r}$ tends to infinity. The same formula can also be obtained from the sum of Coulomb potentials of two opposite charges $\pm q$ when the vector $\mathbf{l}$ joining them tends to zero and the charge q tends to infinity so that $|q|\,\mathbf{l} = \mathbf{d} = \text{const}$ (see Problem 10). We call the limit of such a system of charges the *ideal dipole*.

The same potential $\varphi^{(1)}$ may be obtained also from formula (18) with the substitution of the following generalized function

$$\rho(\mathbf{r}) = -\mathbf{d}\cdot\nabla\delta^3(\mathbf{r})\,. \tag{37}$$

Therefore we admit (37) as the charge density corresponding to the ideal electric dipole.

When such a dipole has the moment varying with time, $\mathbf{d} = \mathbf{d}(t)$, then the density (37) is also a function of time, and — by virtue of the continuity equation (1.28) — a nonzero current density $\mathbf{j}$ must appear. It is possible to verify that (1.28) is satisfied by the generalized function:

$$\mathbf{j}(\mathbf{r},t) = \frac{d\,\mathbf{d}(t)}{dt}\,\delta^3(\mathbf{r})\,. \tag{38}$$

The potential $\varphi^{(1)}$ yields the following electric field

$$\mathbf{E}(\mathbf{r}) = -\nabla\varphi^{(1)}(\mathbf{r}) = \frac{3(\mathbf{n}\cdot\mathbf{d})\,\mathbf{n} - \mathbf{d}}{4\pi\varepsilon r^3}\,, \tag{39}$$

where $\mathbf{n} = \mathbf{r}/r$ (see Problem 0.34). We remind the reader that this expression is obtained from the expansion of type (35) in which the terms of order greater than one are omitted and Q is set equal to zero. The terms omitted contain $\mathbf{r}$ in powers less then minus two in the potential function and less than minus three in the electric field function. Therefore we may claim that the electric potential produced by a system of charges with total charge zero is inversely proportional to the square of the distance for large distances, and the electric field is inversely proportional to the cube of the distance. The field (39) has axial symmetry around the direction of $\mathbf{d}$.

Now we pass to the next term in the expression (35):

$$\varphi^{(2)}(\mathbf{r}) = \frac{1}{8\pi\varepsilon}\sum_s q_s\, x_{(s)i}\, x_{(s)j}\frac{\partial^2}{\partial x_i \partial x_j}\frac{1}{r}\,.$$

It is called the *quadrupole potential*, since it acquires importance when at least four charges of various signs are present. This happens when $Q = 0$ and $\mathbf{d} = 0$. If, under these assumptions, the next terms in (35) are absent, such a system of charges is called the *electric quadrupole*.

Nine quantities $\Sigma_s q_s x_{(s)i} x_{(s)j}$ for $i, j \in \{1,2,3\}$ enter the potential $\varphi^{(2)}$ and form a second rank tensor. We notice that it is a symmetric tensor, so only six coordinates may be independent in it. Yet one of these coordinates must depend on the others, since the function $1/r$ satisfies the *Laplace equation:*

$$\delta_{ij}\frac{\partial^2}{\partial x_i \partial x_j}\frac{1}{r} = 0\,,$$

for $\mathbf{r} \neq 0$ (see Problem 0.39). This zero expression may be added to $\varphi^{(2)}$ with an arbitrary coefficient. Let us do this in the following manner

$$\varphi^{(2)}(\mathbf{r}) = \frac{1}{8\pi\varepsilon}\sum_s q_s\left(x_{(s)i}\, x_{(s)j} - \frac{1}{3}r_{(s)}^2\,\delta_{ij}\right)\frac{\partial^2}{\partial x_i \partial x_j}\frac{1}{r}\,.$$

The tensor $D_{ij} = \Sigma_s q_s\,(3x_{(s)i}\, x_{(s)j} - r_{(s)}^2\,\delta_{ij})$ is called the *electric quadrupole moment.* We see that it is a symmetric tensor. Its *trace* (i.e., the sum of the coordinates with equal indices) is zero: $D_{jj} = 0$. Therefore D_{ij} has only five independent coordinates. We may now write

$$\varphi^{(2)}(\mathbf{r}) = \frac{D_{ij}}{24\pi\varepsilon}\frac{\partial^2}{\partial x_i\,\partial x_j}\frac{1}{r}\,.$$

The differentiation yields $\frac{\partial^2}{\partial x_i \partial x_j}\frac{1}{r} = \frac{3x_i x_j}{r^5} - \frac{\delta_{ij}}{r^3}$, so

$$\varphi^{(2)}(\mathbf{r}) = \frac{1}{8\pi\varepsilon}\left(\frac{D_{ij}\, x_i\, x_j}{r^5} - \frac{D_{jj}}{3r^3}\right) = \frac{D_{ij}\, n_i\, n_j}{8\pi\varepsilon r^3} \,.$$

We may be convinced that the quadrupole moment is independent of the choice of origin for $Q = 0, \mathbf{d} = 0$. In such a case, the electric potential is inversely proportional to the cube of the distance and the electric field to the fourth power of the distance for large distances.

The next terms of the expansion (35) may be written using the same method. Then it turns out that the lth term is defined by the lth rank tensor called the *electric 2^lth-pole moment.* It is a symmetric tensor under all indices and is zero after summing over each pair of equal indices. This implies that it has $2l + 1$ independent coordinates.

We now consider the magnetic field produced by a system of stationary currents contained in a bounded region. We choose the origin of the coordinates somewhere among the currents. We start with formula (19) for the vector potential. We assume $r >> r'$, expand the function $1/|\mathbf{r} - \mathbf{r}'|$ in the powers of $\mathbf{r}'$ and restrict ourselves to the first power:

$$\mathbf{A}(\mathbf{r}) = \frac{\mu}{4\pi r}\int dv'\, \mathbf{j}(\mathbf{r}') - \frac{\mu}{4\pi}\int dv'\, \mathbf{j}(\mathbf{r}')\left(\mathbf{r}' \cdot \nabla \frac{1}{r}\right) .$$

The first term gives zero when the integration set includes all the currents (see Problem 1.3). Hence the expression

$$\mathbf{A}(\mathbf{r}) = -\frac{\mu}{4\pi}\int dv'\, \mathbf{j}(\mathbf{r}')\left(-\frac{\mathbf{r}' \cdot \mathbf{r}}{r^3}\right) = \frac{\mu}{4\pi r^3}\int dv'\, \mathbf{j}(\mathbf{r}')(\mathbf{r}' \cdot \mathbf{r})$$

remains. We obtain from $(\mathbf{j} \wedge \mathbf{r}') \cdot \mathbf{r} = \mathbf{j}(\mathbf{r}' \cdot \mathbf{r}) - \mathbf{r}'(\mathbf{j} \cdot \mathbf{r})$ (see (0.21)) the identity

$$\mathbf{j}(\mathbf{r}' \cdot \mathbf{r}) = \frac{1}{2}[\mathbf{j}(\mathbf{r}' \cdot \mathbf{r}) + \mathbf{r}'(\mathbf{j} \cdot \mathbf{r})] + \frac{1}{2}(\mathbf{j} \wedge \mathbf{r}') \cdot \mathbf{r} \,. \qquad (40)$$

We first consider the expression in the square brackets. We multiply it by an arbitrary constant vector $\mathbf{a}$ and notice the following equality

$$T = (\mathbf{a} \cdot \mathbf{j})(\mathbf{r}' \cdot \mathbf{r}) + (\mathbf{a} \cdot \mathbf{r}')(\mathbf{j} \cdot \mathbf{r}) = \mathbf{j} \cdot \nabla'[(\mathbf{a} \cdot \mathbf{r}')(\mathbf{r} \cdot \mathbf{r}')] \,.$$

Moreover, because of $\nabla \cdot (\mathbf{j}\, f) = f\, \nabla \cdot \mathbf{j} + \mathbf{j} \cdot \nabla f$, we may write

$$T = \nabla' \cdot [\mathbf{j}\,(\mathbf{a} \cdot \mathbf{r}')(\mathbf{r} \cdot \mathbf{r}')] - (\mathbf{a} \cdot \mathbf{r}')(\mathbf{r} \cdot \mathbf{r}')\nabla' \cdot \mathbf{j} \,.$$

We consider stationary systems, that is, $\frac{\partial \rho}{\partial t} = 0$, hence $\nabla' \cdot \mathbf{j} = 0$ due to the continuity equation, and we get

$$T = \nabla' \cdot [\mathbf{j}\,(\mathbf{a}\cdot\mathbf{r}')(\mathbf{r}\cdot\mathbf{r}')] \,.$$

Thus we may apply the Gauss theorem to the integral of the square bracket in (40) multiplied by $\mathbf{a}$:

$$\int_V dv'\,\nabla' \cdot [\mathbf{j}\,(\mathbf{a}\cdot\mathbf{r}')(\mathbf{r}\cdot\mathbf{r}')] = \int_{\partial V} d\mathbf{s}' \cdot \mathbf{j}\,(\mathbf{r}')(\mathbf{r}\cdot\mathbf{r}')(\mathbf{a}\cdot\mathbf{r}') \,.$$

Since $\mathbf{a}$ is arbitrary, we obtain

$$\int_V dv' [\mathbf{j}\,(\mathbf{r}\cdot\mathbf{r}') + \mathbf{r}'\,(\mathbf{j}\cdot\mathbf{r})] = \int_{\partial V} [d\mathbf{s}' \cdot \mathbf{j}\,(\mathbf{r}')](\mathbf{r}\cdot\mathbf{r}')\,\mathbf{r}' \,.$$

We choose the set V such that it includes all the currents, so $\mathbf{j}$ vanishes on its boundary ∂V and the surface integral also vanishes. In this case, only the second term in (40) contributes to the potential integral:

$$\mathbf{A}\,(\mathbf{r}) = \frac{\mu}{8\pi r^3} \int dv' [\mathbf{j}\,(\mathbf{r}') \wedge \mathbf{r}'] \cdot \mathbf{r} \,.$$

The expression $\widehat{\mathbf{M}} = \frac{1}{2}\int dv'\,\mathbf{r}' \wedge \mathbf{j}(\mathbf{r}')$ is called the *magnetic (dipole) moment* of the system of currents. One writes the vector potential with its use as

$$\mathbf{A}\,(\mathbf{r}) = \frac{\mu}{4\pi}\,\frac{\mathbf{r}\cdot\widehat{\mathbf{M}}}{r^3} \,, \tag{41}$$

in this approximation. Notice its similarity to the electric dipole potential $\varphi^{(1)}$.

Let us calculate the magnetic moment of an electric circuit. For this purpose, we represent the volume element as $dv' = ds'\,dl'$ for $d\mathbf{l}' \| \mathbf{j}\,(\mathbf{r}')$, which means that $\mathbf{j}\,dl' = j\,d\mathbf{l}'$, and we obtain

$$\widehat{\mathbf{M}} = \frac{1}{2}\iint ds' j\,\mathbf{r}' \wedge d\mathbf{l}' = \frac{J}{2}\int \mathbf{r}' \wedge d\mathbf{l}' \,, \tag{42}$$

where $J = \int ds' j$ is the electric current in the circuit. The expression

$$\widehat{\mathbf{S}} = \frac{1}{2}\int_C \mathbf{r}' \wedge d\mathbf{l}' \,, \tag{43}$$

for — not necessarily flat — closed curve C is independent of the choice of origin (see Problem 11). Therefore it depends only on C and is a characteristic of a given circuit. That is why we accept (43) as the definition of the *directed area* of an arbitrary circuit C. It is possible to check that the volutor (43) has magnitude equal to the area of the figure encircled by C when C is a flat curve. (Compare remarks on Pg. 2 about the physical model of a volutor.)

Now the formula for the magnetic moment of the electric circuit

$$\widehat{\mathbf{M}} = J\hat{\mathbf{S}}\,, \tag{44}$$

follows from (42). Notice its similarity to the formula $\mathbf{d} = q\,\mathbf{l}$ for the electric moment of two opposite charges. Expression (44) is merely higher by one dimension.

Vector potential (41) may be obtained also from (19) if one substitutes the generalized function

$$\mathbf{j}\,(\mathbf{r}) = (\widehat{\mathbf{M}} \cdot \nabla)\delta^3(\mathbf{r}) \tag{45}$$

for the current density. Therefore we admit (45) as the current density corresponding to the *ideal magnetic dipole*. Its divergence $\nabla \cdot \mathbf{j}\,(\mathbf{r}) = (\frac{\partial}{\partial x_j} M_{jk} \frac{\partial}{\partial x_k})\delta^3\,(\mathbf{r})$ is zero because of the antisymmetry of the coordinates M_{jk} under interchange of indices. Thus

$$\nabla \cdot \mathbf{j}\,(\mathbf{r}) = 0 \tag{46}$$

for the current density (45). When the magnetic dipole has the moment varying with time, then the density (45) still fulfils (46), hence in order to satisfy the continuity equation (1.28) the charge density must be constant in time. In particular, it may be $\rho(\mathbf{r}, t) = 0$.

By virtue of Problem 0.34, the outer derivative of field (41) yields the following magnetic induction

$$\hat{\mathbf{B}}\,(\mathbf{r}) = \frac{\mu}{4\pi}\,\frac{2\widehat{\mathbf{M}} - 3\,\mathbf{n} \wedge (\mathbf{n} \cdot \widehat{\mathbf{M}})}{r^3}\,.$$

where $\mathbf{n} = \mathbf{r}/r$. This formula hardly resembles the electric field (39) of the electric dipole, but by means of the identity $\mathbf{n} \wedge (\mathbf{n} \cdot \widehat{\mathbf{M}}) + \mathbf{n} \cdot (\mathbf{n} \wedge \widehat{\mathbf{M}}) = \widehat{\mathbf{M}}$ (compare with Problem 0.22), we obtain

$$\hat{\mathbf{B}}\,(\mathbf{r}) = \frac{\mu}{4\pi}\,\frac{3\,\mathbf{n} \cdot (\mathbf{n} \wedge \widehat{\mathbf{M}}) - \widehat{\mathbf{M}}}{r^3}\,. \tag{47}$$

After replacement of $\widehat{\mathbf{M}}$ and $\widehat{\mathbf{B}}$ by $\mathbf{M}$ and $\mathbf{B}$, we get the formula

$$\mathbf{B}(\mathbf{r}) = \frac{\mu}{4\pi}\frac{3\mathbf{n}(\mathbf{n}\cdot\mathbf{M}) - \mathbf{M}}{r^3}, \tag{48}$$

which is more similar to (39). It states that if one uses the vectorial lines of the magnetic field, one obtains a similar picture of the force lines of the electric dipole.

Field (47) has rotational symmetry in the plane of $\widehat{\mathbf{M}}$ or around the axis of $\mathbf{M}$. In particular, for $\mathbf{r} \perp \widehat{\mathbf{M}}$, we have $\mathbf{n}\cdot(\mathbf{n}\wedge\widehat{\mathbf{M}}) = \widehat{\mathbf{M}}$, hence (47) yields $\widehat{\mathbf{B}}(\mathbf{r}) = \frac{\mu}{4\pi}\frac{2\widehat{\mathbf{M}}}{r^3}$, that is, the volutors $\widehat{\mathbf{B}}$ and $\widehat{\mathbf{M}}$ have the same orientation on the symmetry axis. A similar situation holds for the field (39) of the electric dipole. For this, one has $\mathbf{E}(\mathbf{r}) = \frac{1}{4\pi\varepsilon}\frac{2\mathbf{d}}{r^3}$ for $\mathbf{r}\parallel\mathbf{d}$, that is, vectors $\mathbf{E}$ and $\mathbf{d}$ have the same orientation on the symmetry axis.

When a stationary system of charges and currents has simultaneously electric moment $\mathbf{d}$ and magnetic moment $\widehat{\mathbf{M}}$, then its electromagnetic potential can be expressed by the single formula

$$A(\mathbf{r}) = \frac{1}{\sqrt{\mu}}\mathbf{A} - \sqrt{\varepsilon}\phi = \frac{1}{4\pi}\frac{\mathbf{r}}{r^3}\left(\sqrt{\mu}\widehat{\mathbf{M}} - \frac{1}{\sqrt{\varepsilon}}\mathbf{d}\right).$$

We see that a new quantity is worth introducing, namely the *cliffor of the electromagnetic moment* $K = \sqrt{\mu}\widehat{\mathbf{M}} - \frac{1}{\sqrt{\varepsilon}}\mathbf{d}$. Its physical dimension is $\mathrm{m}\sqrt{\mathrm{VCm}} = \sqrt{\mathrm{Jm}^3}$. Then the electromagnetic potential is expressed by the formula

$$\widetilde{A}(\mathbf{r}) = \frac{1}{4\pi}\frac{\mathbf{r}\cdot K}{r^3}.$$

§5. A Geometric Interpretation of the Force Surfaces

We obtain from (48) the following formula

$$\mathbf{H}(\mathbf{r}) = \frac{1}{4\pi}\frac{3(\mathbf{n}\cdot\mathbf{M})\mathbf{n} - \mathbf{M}}{r^3} \tag{49}$$

for the magnetic field produced by the magnetic dipole. The similarity of this expression to (39) and the relation (12) (analogous to $\mathbf{E} = -\nabla\wedge\varphi$) imply that

$$\psi(\mathbf{r}) = -\frac{1}{4\pi}\frac{\mathbf{M}\cdot\mathbf{n}}{r^2},$$

is the pseudoscalar potential for field (49). Now by (0.76) and (0.47),

$$\widehat{\Psi}(\mathbf{r}) = -\frac{1}{4\pi}I\frac{\mathbf{M}\cdot\mathbf{n}}{r^2} = -\frac{1}{4\pi}\frac{\widehat{\mathbf{M}}\wedge\mathbf{n}}{r^2} \tag{50}$$

is the trivector potential for the magnetic field of the magnetic dipole.

Let us consider a family of magnetic dipoles placed tightly close to one another on some surface S (see Fig. 70). Let their orientations be compatible, which means that the currents of neighbouring dipoles are opposite. Such a family of magnetic dipoles is called the *magnetic shell.* One may introduce the notion of the *magnetic moment density* ν for the shell through the formula $d\widehat{\mathbf{M}} = \nu\, d\hat{\mathbf{s}}$ where $d\hat{\mathbf{s}}$ is the directed surface element. Comparison with (44) helps us to notice that ν is equal to $J(\mathbf{r})$, the electric current of the magnetic dipole in a given point $\mathbf{r}$ of S. In this way, the total magnetic moment of the magnetic shell is $\widehat{\mathbf{M}} = \int_S J(\mathbf{r}) d\hat{\mathbf{s}}$.

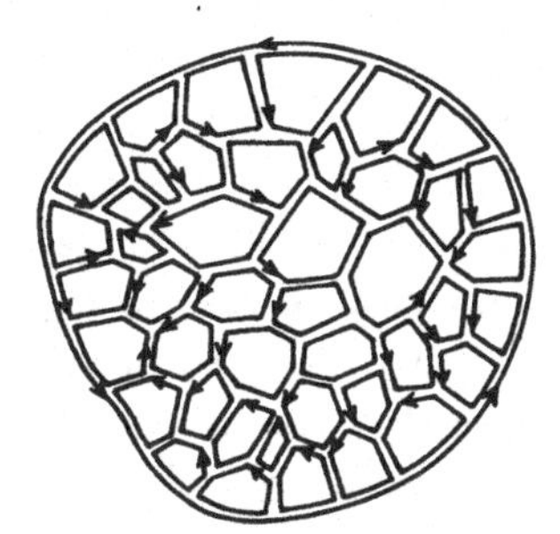

Fig. 70.

If the currents of all constituent dipoles of the shell have equal values J, the magnetic field produced by pairs of neighbouring currents is cancelled and only the segments lying on the boundary ∂S contribute to the magnetic field of the whole shell. In this manner, the magnetic shell with constant magnetic moment density, that is, of constant electric current J, is equivalent to an electric circuit with the current J flowing in its boundary ∂S.

We now calculate the trivector potential of the magnetic shell. Formula (50) gives

$$\hat{\Psi}(\mathbf{r}) = -\frac{1}{4\pi}\int_S \frac{d\widehat{\mathbf{M}} \wedge (\mathbf{r}-\mathbf{r}')}{|\,\mathbf{r}-\mathbf{r}'\,|^3} = \frac{1}{4\pi}\int_S J(\mathbf{r}')\frac{d\hat{\mathbf{s}}' \wedge \mathbf{n}}{|\,\mathbf{r}'-\mathbf{r}\,|^2}\,,$$

where $\mathbf{n} = (\mathbf{r}'-\mathbf{r})/|\,\mathbf{r}'-\mathbf{r}\,|$. For $J(\mathbf{r}) = J =$ constant, we obtain

$$\hat{\Psi}(\mathbf{r}) = \frac{J}{4\pi}\int_S \frac{d\hat{\mathbf{s}}' \wedge \mathbf{n}}{|\,\mathbf{r}'-\mathbf{r}\,|^2}\,.$$

We now recognize that the integral is the directed solid visual angle $\widehat{\Omega}(\mathbf{r})$ of the surface S from the point $\mathbf{r}$. In this manner, we obtain

$$\hat{\Psi}(\mathbf{r}) = \frac{J}{4\pi}\widehat{\Omega}(\mathbf{r})\,. \tag{51}$$

As we know from the discussion in Sec. 1, the trivector magnetic potential $\hat{\Psi}$ exists only locally, not in the whole space devoid of currents. It follows from our construction that function (51) is not defined on S. Its values in points close to S are different on opposite sides of S. For flat circuits, when the surface S can be chosen flat, these values must be $\pm\frac{J}{2}I$, since the visual solid angle of the arbitrary flat figure seen from points close to its surface is half of the full solid angle, that is 2π. In this manner, S becomes the discontinuity surface for the function $\hat{\Psi}$.

We know already from Sec. 1 that the equipotential surfaces of $\hat{\Psi}$ are simultaneously the force surfaces for the magnetic field. In this way, we obtain the geometric interpretation of force surfaces: *a magnetic force surface around a circuit is the locus of points of a constant solid visual angle of the circuit.*

This interpretation allows us to treat the force surfaces discussed in Chap. 1 in a new manner. For instance, the magnetic field of the circular current shown in Fig. 50 suggests the following geometric problem: What is the locus of points in which the circle is seen at a given constant solid angle? As far as the author is aware, no solution to this problem has been found up to now. We may only be sure that the loci are not parts of spheres, since, otherwise, the magnetic field at arbitrary points could be easily expressed by elementary functions. This, however, is possible exclusively in points on the symmetry axis of the circle, as mentioned in Sec. 1.7.4.

By merely looking at the geometrical configuration, we may conclude how the function $\hat{\Psi}(\mathbf{r})$ behaves in this case. Let us choose the symmetry axis as the Z-axis with the positive direction above the plane of the current shown in Fig. 50. If the current flows as in that Figure, $\hat{\Omega}(\mathbf{r})$ changes from zero to $-2\pi I$ for z ranging from $+\infty$ to 0, and changes from $+2\pi I$ to zero for z ranging from 0 to $-\infty$. Therefore the respective intervals for $\hat{\Psi}$ are $(0, -\frac{J}{2}I)$ and $(\frac{J}{2}I, 0)$. The part of the circuit plane outside the circuit itself is one of the equipotential surfaces, namely for $\hat{\Psi} = 0$. The exact values of $\hat{\Psi}$ on the Z-axis are calculated in Appendix III.

Similar geometric considerations may be applied to the magnetic field of the rectilinear current shown in Figs. 33 and 34. First of all, we need a closed circuit. For this purpose, we assume that the linear conductor is closed at infinity by a hemicycle. The half-plane S of this hemicycle becomes the discontinuity surface of $\hat{\Psi}$. On considering the matter further, one can conclude that the surfaces of constant visual angle of S are also half-planes passing through the straight line of the conductor. If the plane angle between a given half-plane A and the half-plane S is α, the solid visual angle of S from a point $\mathbf{r}$ on A is $2(\pi - \alpha)$ (see Fig. 71). Therefore in cylindrical coordinates (ρ, α, z) natural for this situation, we may write $\widehat{\Omega}(\mathbf{r}) = -2(\pi-\alpha)I$, and hence

$$\widehat{\Psi}(\mathbf{r}) = \frac{\alpha - \pi}{2\pi} JI \, . \tag{52}$$

Its extremal values are: $\widehat{\Psi} = -\frac{1}{2}JI$ for $\alpha = 0$ and $\widehat{\Psi} = \frac{1}{2}JI$ for $\alpha = 2\pi$.

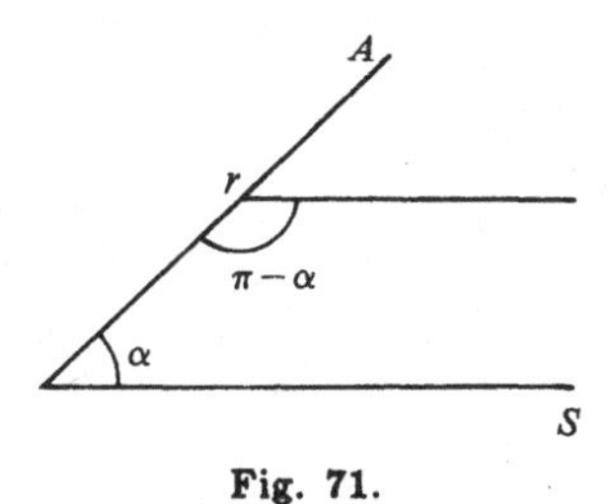

Fig. 71.

The new way of looking at two examples known from Chap. 1, which is presented here, is helpful for finding the force surfaces of the combined magnetic field. When considering the magnetic field produced by the circular conductor (with the current J_1) along with the linear conductor (with the current J_2) put in the symmetry axis of the circle, one should simply add their trivector potentials $\widehat{\Psi}_1$ and $\widehat{\Psi}_2$ and consider loci of points $\mathbf{r}$ staisfying the condition $\widehat{\Psi}_1(\mathbf{r}) + \widehat{\Psi}_2(\mathbf{r}) = \text{constant}$, which form the force surfaces of the combined field. If we are interested only qualitatively in the shape of these surfaces, we may assume that equipotential surfaces for $\widehat{\Psi}_1$ function corresponding to the circular current are parts of spheres. Under these assumptions, one may find that one of the possible force surfaces for equal currents is as shown in Fig. 72. The details are explained in Appendix III. We want only to point out here that the two currents form the boundary for this surface and a half-plane exists, passing through the linear current, which is a kind of asymptote for the surface.

§6. Energy and Momentum of Electromagnetic Field

It is known from general physics that the product of the electric charge q by the value of stationary potential φ at the position of the charge is its potential energy $W_e = q\varphi$. If only two charges exist, their potential energy may be written in two ways:

$$W_e = q_1 \varphi_{1,2} = q_2 \varphi_{2,1} \ ,$$

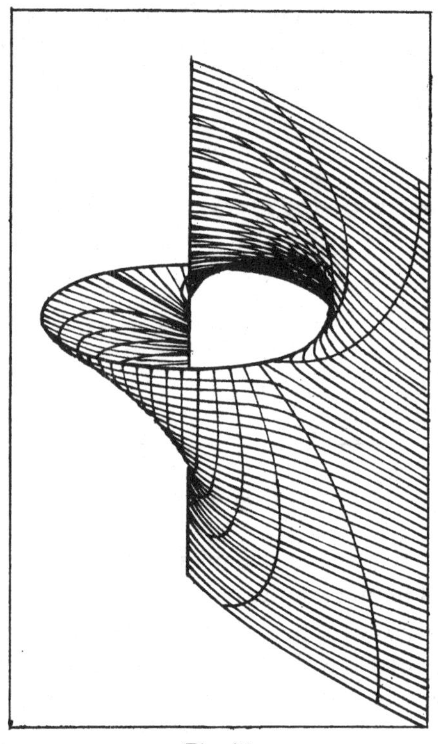

Fig. 72.

where $\varphi_{i,j} = q_j/4\pi\varepsilon|\mathbf{r}_i - \mathbf{r}_j|$ is the potential in point $\mathbf{r}_i$ produced by charge q_j. For the sake of symmetry, we may write

$$W_e = \frac{1}{2}(q_1 \varphi_{1,2} + q_2 \varphi_{2,1}) .$$

The potential energy of the three charges is

$$\begin{aligned} W_e &= \frac{1}{4\pi\varepsilon}\left(\frac{q_1 q_2}{|\mathbf{r}_1 - \mathbf{r}_2|} + \frac{q_2 q_3}{|\mathbf{r}_2 - \mathbf{r}_3|} + \frac{q_3 q_1}{|\mathbf{r}_3 - \mathbf{r}_1|}\right) \\ &= q_1 \varphi_{1,2} + q_2 \varphi_{2,3} + q_3 \varphi_{3,1} = q_2 \varphi_{2,1} + q_3 \varphi_{3,2} + q_1 \varphi_{1,3} . \end{aligned}$$

Because of the equality $q_i \varphi_{i,j} = q_j \varphi_{j,i}$, we may also write

$$\begin{aligned} W_e &= \frac{1}{2}[q_1(\varphi_{1,2} + \varphi_{1,3}) + q_2(\varphi_{2,3} + \varphi_{2,1}) + q_3(\varphi_{3,1} + \varphi_{3,2})] \\ &= \frac{1}{2}(q_1 \varphi_1 + q_2 \varphi_2 + q_3 \varphi_3) , \end{aligned}$$

where

$$\varphi_i = \sum_{j \neq i} \varphi_{i,j} \tag{53}$$

is the potential in point $\mathbf{r}_i$ produced by all other charges.

The argument for an arbitrary number N of electric charges can be used in the same way and, with notation (53), yields the result

$$W_e = \frac{1}{2} \sum_{i=1}^{N} q_i \, \varphi_i \, .$$

The coefficient $\frac{1}{2}$ stems from the fact that the potential energy of each pair of charges is taken twice in the sum. Now we generalize this into the continuous distribution of charges described by their spatial density ρ:

$$W_e = \frac{1}{2} \int dv \, \rho \, \varphi \, .$$

The integration extends over the region in which electric charges exist, that is, where the integrand is non-zero.

We know from Maxwell equation (1.17d) that $\rho = \nabla \cdot \mathbf{D}$, so

$$W_e = \frac{1}{2} \int dv \, \varphi \nabla \cdot \mathbf{D} \, . \tag{54}$$

By applying Problem 0.30 to the product $\varphi \mathbf{D}$ and using the equality $\nabla \varphi = -\mathbf{E}$, we obtain

$$\varphi \nabla \cdot \mathbf{D} = \nabla \cdot (\varphi \mathbf{D}) + \mathbf{E} \cdot \mathbf{D} \, .$$

The terms on the right-hand side are non-zero also outside the region occupied by charges (the sum, however, is still zero outside that region since such a property belongs to the left-hand side). Afetr inserting this identity to (54), we get

$$W_e = \frac{1}{2} \int_{\Omega} dv \nabla \cdot (\varphi \mathbf{D}) + \frac{1}{2} \int_{\Omega} dv \, \mathbf{E} \cdot \mathbf{D} \, .$$

Now we integrate over set Ω greater than the region occupied by charges. By virtue of the Gauss theorem (0.140), we may write the first integral differently:

$$W_e = \frac{1}{2} \int_{\partial\Omega} d\mathbf{s} \cdot (\varphi \mathbf{D}) + \int_{\Omega} dv \, \mathbf{E} \cdot \mathbf{D} \, .$$

We intend to get rid of the first term. For this purpose, we make the integration set Ω so big that the field $\mathbf{D}$ vanishes on its boundary and outside it. We may also pass to the limit $\Omega \to \mathbb{R}^3$ if the product $\varphi \dot{\mathbf{D}}$ decreases so fast that the first integral tends to zero (compare with Problem 13). After this change, the integration must extend over the whole space filled by the non-zero field $\mathbf{D}$ — it is much greater than the region occupied by the charges. Then we are left with the expression

$$W_e = \frac{1}{2} \int_{\Omega} dv \, \mathbf{E} \cdot \mathbf{D} \,. \tag{55}$$

We interpret this by stating that the energy of the electric field is distributed over the whole space occupied by the field, with the spatial density

$$w_e = \frac{1}{2} \mathbf{E} \cdot \mathbf{D} = \frac{1}{2} \varepsilon \, \mathbf{E}^2 = \frac{1}{2} \mathbf{e}^2 \,. \tag{56}$$

Non-zero contributions to (55) enter from regions devoid of charges, in which only the electric field exists. Thus formula (55) corresponds to the notion that *the electric field itself carries an energy.*

We now proceed to find a similar expression for the magnetic field. We start with the (presumably known) formula for the energy of a circuit placed in the external magnetic field: $W_m = J\Phi$, where $\Phi = \int_S d\mathbf{s} \cdot \mathbf{B} = -\int_S d\widehat{\mathbf{s}} \cdot \widehat{\mathbf{B}}$ is the magnetic flux through the surface S enchanced by the circuit with the current J. By dint of (9), after denoting $\partial S = C$, we obtain

$$W_m = J \int_C d\mathbf{l} \cdot \mathbf{A} \,.$$

So far we have treated the circuit as a curved line of negligible thickness. Now we abandon this idealization and represent the current in the form $J = \int_P ds \, j$ where P is the cross-section of the conductor and j is the current density magnitude. We introduce the unit vector $\mathbf{n}$ through the formula $d\mathbf{l} = \mathbf{n} \, dl$, then

$$W_m = \int_P ds \int_C dl \, j \, \mathbf{n} \cdot \mathbf{A} \,.$$

We now introduce $\mathbf{j} = \mathbf{n} \, j$:

$$W_m = \int_P \int_C ds dl \, \mathbf{j} \cdot \mathbf{A} = \int_V dv \, \mathbf{j} \cdot \mathbf{A} \,,$$

where V stands for the region occupied by the conductor.

This expression refers to the current of density $\mathbf{j}$ placed in the external magnetic field of the vector potential $\mathbf{A}$. If we also take into account the currents producing the field **A** and include them in the integration set, then this energy of circuits is taken twice, just as in previous calculations for the energy of the electric field. Thus we have to divide the result by two:

$$W_m = \frac{1}{2}\int dv\, \mathbf{A}\cdot\mathbf{j} \,.$$

The integration extends over all regions containing the currents.

We know from Maxwell equation (1.17c) that for stationary currents, $\mathbf{j} = -\boldsymbol{\nabla}\cdot\widehat{\mathbf{H}}$ so

$$W_m = -\frac{1}{2}\int dv\, \mathbf{A}\cdot(\boldsymbol{\nabla}\cdot\widehat{\mathbf{H}}) \,.$$

By virtue of the formula $\mathbf{A}\cdot(\boldsymbol{\nabla}\cdot\widehat{\mathbf{H}}) = \widehat{\mathbf{B}}\cdot\widehat{\mathbf{H}} - \boldsymbol{\nabla}\cdot(\mathbf{A}\cdot\widehat{\mathbf{H}})$ (Problem 0.28), and one gets

$$W_m = \frac{1}{2}\int dv\, \boldsymbol{\nabla}\cdot(\mathbf{A}\cdot\widehat{\mathbf{H}}) - \frac{1}{2}\int dv\, \widehat{\mathbf{B}}\cdot\widehat{\mathbf{H}} \,.$$

We get rid of the first integral as previously: by means of the Gauss theorem, we change it into the surface integral, then extend the integration set so that the magnetic field vanishes on its boundary. In this way, we get

$$W_m = -\frac{1}{2}\int_\Omega dv\, \widehat{\mathbf{B}}\cdot\widehat{\mathbf{H}} \,. \qquad (57)$$

This formula demonstrates that the energy of the magnetic field is distributed over the whole space occupied by the field, with the density

$$w_m = -\frac{1}{2}\widehat{\mathbf{B}}\cdot\widehat{\mathbf{H}} = -\frac{1}{2\mu}\widehat{\mathbf{B}}^2 = \frac{1}{2\mu}|\widehat{\mathbf{B}}|^2 = \frac{1}{2}|\widehat{\mathbf{b}}|^2 \,. \qquad (58)$$

This expression, analogous to (56), is nonnegative and manifests that *the magnetic field itself carries an energy.*

When we have the electric and magnetic field simultaneously, we call the *energy of the electromagnetic field* the sum of (55) and (57):

$$W = \frac{1}{2}\int_\Omega dv(\mathbf{E}\cdot\mathbf{D} - \widehat{\mathbf{B}}\cdot\widehat{\mathbf{H}}) \,,$$

assuming thereby that the electric and magnetic fields carry the energy independently of each other. Thus we interpret the expression

$$w = \frac{1}{2}(\mathbf{E}\cdot\mathbf{D} - \widehat{\mathbf{B}}\cdot\widehat{\mathbf{H}}) = \frac{1}{2}(\varepsilon|\mathbf{E}|^2 + \frac{1}{\mu}|\widehat{\mathbf{B}}|^2) = \frac{1}{2}(|\mathbf{e}|^2 + |\widehat{\mathbf{b}}|^2) \,,$$

as the ***energy density of the electromagnetic field.*** We have derived this under the assumption that all fields are independent of time. This, however, describes the energy density in all other situations as well.

This expression may also be written by means of the electromagnetic clifor. Now, using (1.4) and (0.63), we have

$$w = \frac{1}{2}|f|^2 . \tag{59}$$

In this way, the physical meaning of the cliffor f comes into sight. Its physical dimension is $\sqrt{\mathrm{J/m^3}}$, which corresponds to the square root of the energy density. The relation between f and w resembles the relation between the probability amplitude and the probability density occurring in quantum mechanics. (Bear in mind the similarity of the Maxwell equation to the Dirac equation discussed in Sec. 1.2.)

We now restrict the integration set to the arbitrarily chosen region V, being aware that this is only part W_V of the energy of the electromagnetic field, and we consider whether W_V changes with time or not. We calculate the time rate of the energy increase without changing the integration set:

$$\frac{dW_V}{dt} = \frac{1}{2}\int_V dv \frac{\partial}{\partial t}\left(\varepsilon \mathbf{E}^2 - \frac{1}{\mu}\widehat{\mathbf{B}}^2\right) .$$

We assume that ε and μ are time independent:

$$\frac{dW_V}{dt} = \frac{1}{2}\int_V dv\left(2\varepsilon \mathbf{E}\cdot\frac{\partial \mathbf{E}}{\partial t} - \frac{2}{\mu}\widehat{\mathbf{B}}\cdot\frac{\partial \widehat{\mathbf{B}}}{\partial t}\right) = \int_V dv\left(\mathbf{E}\cdot\frac{\partial \mathbf{D}}{\partial t} - \widehat{\mathbf{H}}\cdot\frac{\partial \widehat{\mathbf{B}}}{\partial t}\right) .$$

The Maxwell equations (1.17a,c) allow us to eliminate $\frac{\partial \mathbf{D}}{\partial t}$ and $\frac{\partial \widehat{\mathbf{B}}}{\partial t}$:

$$\frac{dW_V}{dt} = -\int_V dv\, \mathbf{j}\cdot\mathbf{E} + \int_V dv[\widehat{\mathbf{H}}\cdot(\boldsymbol{\nabla}\wedge\mathbf{E}) - \mathbf{E}\cdot(\boldsymbol{\nabla}\cdot\mathbf{H})] .$$

Due to Problem (0.28) and Gauss theorem (0.140),

$$\frac{dW_V}{dt} = -\int_V dv\, \mathbf{j}\cdot\mathbf{E} + \int_{\partial V} d\mathbf{s}\cdot(\mathbf{E}\cdot\widehat{\mathbf{H}}) . \tag{60}$$

The first term on the right-hand side represents the work done per unit time by the fields, known under the name *Joule heat.* The minus in front of it means that the energy converted into heat is lost from the electromagnetic field energy. The second term is to be interpreted as the time rate of energy flowing into region V through its boundary ∂V, that is, the energy flux through the surface ∂V. In the convention of formula (0.140), however, the vector $d\mathbf{s}$ is oriented outwards from ∂V, therefore we should accept

$$\mathbf{S} = -\mathbf{E} \cdot \widehat{\mathbf{H}} = \widehat{\mathbf{H}} \cdot \mathbf{E} \,, \tag{61}$$

as the *energy flux density* at a given point. It is also called the *Poynting vector*.

The relation (60), known as the *Poynting theorem*, is the *electromagnetic energy conservation law*. It states that the energy of the electromagnetic field contained in a region V increases by the energy transported through the boundary ∂V and decreases by the energy converted into heat.

The expression (61) is closely connected with another quantity, namely, the *momentum density of the electromagnetic field*:

$$\mathbf{g} = -\varepsilon\mu \mathbf{E} \cdot \widehat{\mathbf{H}} = \varepsilon\mu \mathbf{S} \,. \tag{62}$$

One may find its derivation in Ref. 2, for instance. The two expressions (59) and (62) may be included in a single quantity. It is enough to write $\mathbf{g} = -\sqrt{\varepsilon\mu}\,\mathbf{e} \cdot \widehat{\mathbf{b}} = \sqrt{\varepsilon\mu}\,\widehat{\mathbf{b}} \cdot \mathbf{e}$ and notice the identity (1.6): $f f^{\dagger} = |f|^2 + 2\widehat{\mathbf{b}} \cdot \mathbf{e}$. In that case,

$$\frac{1}{2} f f^{\dagger} = \frac{1}{2} |f|^2 + \widehat{\mathbf{b}} \cdot \mathbf{e} = w + \frac{1}{\sqrt{\varepsilon\mu}} \mathbf{g} = w + u\mathbf{g} \,,$$

where $u = 1/\sqrt{\varepsilon\mu}$ has the dimension of velocity. We see from this that the scalar part of $\frac{1}{2} f f^{\dagger}$ is the energy density whereas the vector part is proportional to the momentum density of the electromagnetic field. We therefore call $\frac{1}{2} f f^{\dagger}$ the *energy-momentum cliffor* of the electromagnetic field. It may also be written by means of the Poynting vector:

$$\frac{1}{2} f f^{\dagger} = w + \sqrt{\varepsilon\mu}\,\mathbf{S} = w + \frac{1}{u} \mathbf{S} \,. \tag{63}$$

We are going to show that its vector part is not greater than the scalar part. We start with the obvious inequality $(|\mathbf{e}| - |\widehat{\mathbf{b}}|)^2 \geq 0$ and transform it:

$$\frac{1}{2}(|\mathbf{e}|^2 + |\widehat{\mathbf{b}}|^2 - 2|\mathbf{e}||\widehat{\mathbf{b}}|) \geq 0 \,,$$
$$\frac{1}{2}(|\mathbf{e}|^2 + |\widehat{\mathbf{b}}|^2) \geq |\mathbf{e}||\widehat{\mathbf{b}}| \geq |\mathbf{e} \cdot \widehat{\mathbf{b}}| \,.$$

The last inequality has been written using (0.20). This means that $|\widehat{\mathbf{b}}\cdot\mathbf{e}|\le \frac{1}{2}|f|^2$ or $\frac{1}{u}|\mathbf{S}|\le w$, which is what we wanted to show. In this way, we obtain the following inequality for the Poynting vector

$$|\mathbf{S}|\le uw\,. \tag{64}$$

We shall now find the velocity $\mathbf{v}$ of the energy transport. Let us consider a cylinder of height $d\mathbf{h}$ parallel to $\mathbf{S}$, with the base area s perpendicular to $d\mathbf{h}$. The energy of the electromagnetic field contained in the cylinder is $dW = ws\,dh$. If we take $d\mathbf{h} = \mathbf{v}\,dt$, we admit that the energy is transported with the velocity $\mathbf{v}$. Then $dW = wsv\,dt$ and the energy flow is $dW/dt = wsv$. Its density is, therefore, $\frac{1}{s}\frac{dW}{dt} = wv$. If we treat this quantity as a vector, we can equate it to the Poynting vector: $w\mathbf{v} = \mathbf{S}$. Hence we obtain the formula that we were looking for:

$$\mathbf{v} = \frac{1}{w}\mathbf{S}\,. \tag{65}$$

Inequality (64) states that the possible values of this velocity are bounded from above: $v\le u$. Thus $u = 1/\sqrt{\varepsilon\mu}$ is the maximal possible velocity of the energy transport in a given medium. That value is achieved if, and only if, all the inequalities preceding (64) become equalities. This means, firstly, that $|\widehat{\mathbf{b}}\cdot\mathbf{e}| = |\widehat{\mathbf{b}}||\mathbf{e}|$ or $\mathbf{e}\parallel\widehat{\mathbf{b}}$, and secondly, that $|\mathbf{e}| = |\widehat{\mathbf{b}}|$. We shall see in Chap. 4 that u is the velocity of light, in particular $u = c$ for the vacuum.

We check at last the behaviour of the energy-momentum cliffor under the Larmor-Reinich transformation (1.27). Denote $f' = e^{I\alpha}f$ and calculate:

$$f'f'^{\dagger} = e^{I\alpha}f(e^{I\alpha}f)^{\dagger} = e^{I\alpha}ff^{\dagger}e^{-I\alpha}\,.$$

Due to the commutativity of I with all cliffors,

$$f'f'^{\dagger} = ff^{\dagger}e^{I\alpha}e^{-I\alpha} = ff^{\dagger}\,.$$

This means that the energy density and the energy flux do not change under (1.27). Thus the Larmor-Reinich transformations are symmetries of the energy-momentum cliffor.

Problems

1. Potentials $\mathbf{A}$ and φ do not fulfil the Lorentz condition. Show that a gauge transformation exists, after which new potentials $\mathbf{A}'$ and φ' fulfil the condition.

2. Find vector and scalar potentials for constant and uniform electric and magnetic fields existing simultaneously. Look after the Lorentz condition.
3. Verify the Lorentz condition for all potentials found in Sec. 3.
4. Several vector potentials corresponding to the uniform magnetic field have been found in Sec. 3. Find the gauge functions joining them pairwise in gauge transformations.
5. Let $\widehat{\mathbf{B}}$ be a constant volutor. Show that $\mathbf{A}(\mathbf{r}) = (\mathbf{n}\cdot\mathbf{r})(\mathbf{n}\cdot\widehat{\mathbf{B}})$ is the vector potential for the uniform magnetic field of induction $\widehat{\mathbf{B}}$, if $\mathbf{n} \parallel \widehat{\mathbf{B}}$. Is the Lorentz condition satisfied? Find the gauge function joining it with the potential found in Sec. 3.6.
6. To which electromagnetic field do the potentials $\mathbf{A}' = \frac{1}{2}\mathbf{r}\cdot\widehat{\mathbf{B}}' - \frac{1}{2}t\,\mathbf{E}', \varphi' = -\frac{1}{2}\mathbf{r}\cdot\mathbf{E}'$ correspond, where $\widehat{\mathbf{B}}', \mathbf{E}'$ are constant quantities? Find the gauge transformation joining them with the potentials found earlier for the same field.
7. Let $f(\mathbf{r})$ be a scalar function independent of time. Show that the potentials $\mathbf{A} = t \operatorname{grad} f, \varphi = 0$ describe the electrostatic field. Find the gauge function joining them with the potentials $\mathbf{A}' = 0, \varphi' = f$.
8. Find the scalar potential inside and outside the uniformly spatially charged sphere.
9. The uniformly superficially charged sphere rotates around its diameter with a constant angular velocity. Find the vector potential and the magnetic induction inside and outside it. Hint: Apply formula (19) adapted to surface currents and use Problems 0.42 and 1.9.
10. Find the scalar potential of an ideal electric dipole as a limit of the potential of the two opposite charges $\pm q$ when the directed distance $\mathbf{l}$ between them tends to zero with the condition $q\,\mathbf{l} =$ constant.
11. Show that the expression $\widehat{\mathbf{S}} = \frac{1}{2}\int \mathbf{r}\wedge d\mathbf{l}$ with the integration over a closed directed curve does not depend on the choice of the origin.
12. Find the relationship between the magnetic moment and the angular momentum of a system of moving charges of equal q/m ratios.
13. Check that the integral $\int_{\partial\Omega} d\mathbf{s}\cdot(\varphi\mathbf{D})$ tends to zero when Ω is a sphere with its radius tending to infinity, and a charge Q is located in its center. How does this integral behave for a finite system of point charges?
14. By virtue of formula $W_m = J\Phi$, calculate the energy of a rectangular circuit placed in an external magnetic field. How does W_m depend on the angle between the circuit plane and the force surfaces of the field? When is this energy minimal?
15. Determine the Poynting vector inside and outside a cylindrical conduc-

tor in which a uniform current flows. What is happening to the energy described by the flux of the Poynting vector?

16. Find the energy-momentum cliffor for the field described in Sec. 1.8.

Chapter 3

CHARGES IN THE ELECTROMAGNETIC FIELD

§1. Motion of a Charge in a Uniform Field

The motion of a charged particle in the electromagnetic field is governed by the *Lorentz equation*

$$\frac{d\mathbf{p}}{dt} = q\mathbf{E} - q\mathbf{v} \cdot \hat{\mathbf{B}} \,. \tag{1}$$

Here $\mathbf{v}$ is the velocity of the particle, $\mathbf{p}$ — its momentum and q — its electric charge. Information about the time rate of changes of kinetic energy T may be deduced from this. We take the formula known from mechanics $dT = \mathbf{v} \cdot d\mathbf{p}$ for the energy increases. (It is generally valid, both in the nonrelativistic and relativistic case.) Hence,

$$\frac{dT}{dt} = \mathbf{v} \cdot \frac{d\mathbf{p}}{dt} \,.$$

We take the scalar product of (1) with $\mathbf{v}$:

$$\mathbf{v} \cdot \frac{d\mathbf{p}}{dt} = q\mathbf{v} \cdot \mathbf{E} - q\mathbf{v} \cdot (\mathbf{v} \cdot \hat{\mathbf{B}}) \,.$$

The second term on the right-hand side vanishes (see Problem 0.16), such that

$$\frac{dT}{dt} = q\mathbf{v} \cdot \mathbf{E} \,.$$

We find that only the electric field influences the kinetic energy of the charged particle. Therefore, we conclude that the magnetic field changes only the direction of velocity, not its magnitude.

Finding solutions to the equation of motion (1) is very difficult in a general electromagnetic field. In a few cases, it is practically possible. The static uniform field is one of them. One may consider three cases: the electric field alone, the magnetic field alone, or the two fields together. The motion of a charged particle in a uniform electric field is well known — it reduces to the motion of a particle subject to a constant gravitational force. This is a superposition of the uniformly accelerated motion in nonrelativistic theory in the direction of $q\mathbf{E}$ with the uniform motion in the perpendicular direction.

Let us take up the solution to (1) in a static uniform magnetic field. In principle, we now solve it nonrelativistically, but we use the relativistic expression for the momentum: $\mathbf{p} = m\mathbf{v} = \frac{\mathcal{E}}{c^2}\mathbf{v}$. Here $\mathcal{E}$ is the total energy, that is, the sum of kinetic and rest energy, which — as we already know — is constant. In this manner, we obtain from (1)

$$\frac{d\mathbf{p}}{dt} = \frac{\mathcal{E}}{c^2}\frac{d\mathbf{v}}{dt} = -q\mathbf{v}\cdot\widehat{\mathbf{B}} ,$$

whence

$$\frac{d\mathbf{v}}{dt} = -\frac{qc^2}{\mathcal{E}}\mathbf{v}\cdot\widehat{\mathbf{B}} . \tag{2}$$

Now we refer to the considerations of Sec. 0.8. We found there that the function

$$\mathbf{v}(t) = e^{-\frac{1}{2}\hat{\omega}t}\,\mathbf{v}_0\,e^{\frac{1}{2}\hat{\omega}t} \tag{3}$$

representing a rotation of the initial vector $\mathbf{v}_0$ with constant angular velocity $\hat{\omega}$, satisfies the equation $d\mathbf{v}/dt = \mathbf{v}\cdot\hat{\omega}$. As a matter of fact, this is Eq. (2) with the substitution

$$\hat{\omega} = -\frac{qc^2}{\varepsilon}\widehat{\mathbf{B}} . \tag{4}$$

Hence we may state that the velocity of the charge changes with time according to Eq. (3), i.e., rotates uniformly with angular velocity (4).

In order to find $\mathbf{r}(t)$, that is, to integrate Eq. (3), we decompose the initial velocity $\mathbf{v}_0$ onto the component $\mathbf{v}_{0\|}$ parallel to $\widehat{\mathbf{B}}$, and $\mathbf{v}_{0\perp}$ — orthogonal to $\widehat{\mathbf{B}}$. The vector $\mathbf{v}_{0\|}$ anticommutes, and $\mathbf{v}_{0\perp}$ commutes with $\widehat{\mathbf{B}}$, hence

$$\mathbf{v}(t) = e^{-\frac{1}{2}\hat{\omega}t}\,\mathbf{v}_{0\|}\,e^{\frac{1}{2}\hat{\omega}t} + e^{-\frac{1}{2}\omega t}\,\mathbf{v}_{o\perp}\,e^{\frac{1}{2}\hat{\omega}t} = e^{-\hat{\omega}t}\,\mathbf{v}_{0\|} + \mathbf{v}_{0\perp} .$$

In this way, for the respective components of $\mathbf{v}(t)$, we obtain

$$\mathbf{v}_{\|}(t) = e^{-\hat{\omega}t}\,\mathbf{v}_{0\|}\,, \quad \mathbf{v}_{\perp}(t) = \mathbf{v}_{0\perp} = \text{const}\,.$$

These equations are easy to integrate (compare with Eq. (0.115)):

$$\mathbf{r}_{\|} = -\hat{\omega}^{-1}\,e^{-\hat{\omega}t}\,\mathbf{v}_{0\|} + \mathbf{r}_{0\|} = e^{-\hat{\omega}t}\mathbf{v}_{0\|}\hat{\omega}^{-1} + r_{0\|}\,.$$

Thus we obtain

$$\mathbf{r}_{\|}(t) = e^{-\hat{\omega}t}\,(\mathbf{v}_0 \cdot \hat{\omega}^{-1}) + \mathbf{r}_{0\|}\,. \tag{5a}$$

$$\mathbf{r}_{\perp}(t) = \mathbf{v}_{0\perp}\,t + r_{0\perp}\,. \tag{5b}$$

The first formula states that the function $t \to \mathbf{r}_{\|}(t)$ describes the rotation around the point $\mathbf{r}_{0\|}$ with the angular velocity $\hat{\omega}$ over a circumference with the radius $v_{0\|}/\omega = v_0\,\mathcal{E}/qc^2\,B$. Notice that for small velocities $\mathbf{v}$ the angular velocity does not depend on the initial conditions (because $\mathcal{E}/c^2 = m_0$, the rest mass),

$$\hat{\omega} = -\frac{q}{m}\,\hat{\mathbf{B}}\,. \tag{6}$$

Only the radius is proportional to the parallel component of the initial velocity. The independence of $\hat{\omega}$ on the initial condition is crucial for the focusing of charged particle beams in cyclotrones, therefore $\hat{\omega}$ is called the *cyclotron frequency*.

Equation (5b) states that the motion orthogonal to the magnetic field (treated as a volutor) is uniform. After superposing the two motions, we may ascertain that the solution to (2) is the helical motion with the helix axis perpendicular to the force surfaces or parallel to the vectorial lines of the magnetic field. We illustrate this in Fig. 73 on a background of vectorial lines. The trivector $\mathbf{v}_0 \wedge \hat{\omega} = \mathbf{v}_{0\perp} \wedge \hat{\omega} = \mathbf{v}_{\perp} \wedge \hat{\omega} = \mathbf{v} \wedge \hat{\omega}$ determines the three-dimensional orientation of the helix. Dus to (6), the same orientation is described by the trivector $-\mathbf{v}_{\perp} \wedge \frac{1}{q}\,\hat{\mathbf{B}} = -\mathbf{v} \wedge \frac{1}{q}\,\hat{\mathbf{B}}$. $T = 2\pi/\omega$ is the period of circular motion, thus the pitch of the helix is

$$h = v_{\perp}T = \frac{2\pi v_{\perp}}{\omega} = \frac{2\pi m v_{\perp}}{q|\hat{\mathbf{B}}|}\,.$$

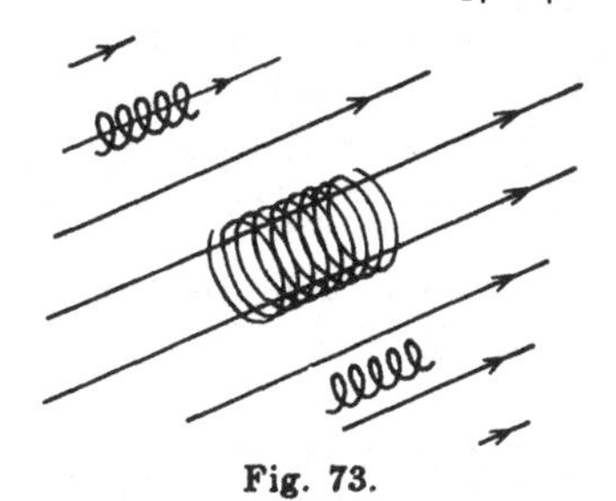

Fig. 73.

It seems useful to treat this pitch as a trivector quantity by giving it the orientation of the helix. Thus we get $\widehat{h} = -(2\pi m/q)\,\mathbf{v}\wedge\widehat{\mathbf{B}}/|\,\widehat{\mathbf{B}}\,|^2$. By virtue of (0.87), we write this as

$$\widehat{h} = \frac{2\pi m}{q}\mathbf{v}\wedge\widehat{\mathbf{B}}^{-1} .$$

In the particular case when the initial velocity is parallel to $\widehat{\mathbf{B}}$, the motion is flat and takes place in the plane of $\widehat{\mathbf{B}}$. The circular path of the particle in this motion is known as the *Larmor circle.*

Equation (6) can be used for determining the magnetic induction. One needs to inject charged particles with the velocities of various directions into the field. If, for one of the particles, the orbit becomes flat and closed, the plane of this orbit coincides with the plane of $\widehat{\mathbf{B}}$. Then we obtain from (6) $\widehat{\mathbf{B}} = -\frac{m}{q}\widehat{\boldsymbol{\omega}}$. Notice that, for positive charges, $\widehat{\mathbf{B}}$ and $\widehat{\boldsymbol{\omega}}$ have opposite directions. This way of determining $\widehat{\mathbf{B}}$ may be used also for nonuniform fields but then it is only approximate. For better approximation, one should use particle velocities which are as small as possible and charges as large as possible, then — as we see from (5a) — the radius of the (approximately) closed orbit becomes small and the field can be treated as uniform.

We now consider the motion in both electric and magnetic fields existing simultaneously. We do not assume any particular arrangement of the fields. Vector $\mathbf{E}$ may be oblique with respect to $\widehat{\mathbf{B}}$. We apply the nonrelativistic approximation $v \ll c$ to Eq. (1). Then $d\mathbf{p}/dt = m d\mathbf{v}/dt$ and the equation takes the form

$$\frac{d\mathbf{v}}{dt} = \frac{q}{m}(\mathbf{E} - \mathbf{v}\cdot\widehat{\mathbf{B}}) .$$

We decompose the velocity $\mathbf{v} = \mathbf{v}_{\|}+\mathbf{v}_{\perp}$ and the electric field $\mathbf{E} = \mathbf{E}_{\|}+\mathbf{E}_{\perp}$ into components parallel and perpendicular to $\widehat{\mathbf{B}}$. Since the vector $\mathbf{v}\cdot\widehat{\mathbf{B}} = \mathbf{v}_{\|}\cdot\widehat{\mathbf{B}}$ is parallel to $\widehat{\mathbf{B}}$, our equation separates into the following two:

$$\frac{d\mathbf{v}_{\|}}{dt} = \frac{q}{m}(\mathbf{E}_{\|} - \mathbf{v}_{\|}\cdot\widehat{\mathbf{B}}) , \tag{7}$$

$$\frac{d\mathbf{v}_{\perp}}{dt} = \frac{q}{m}\mathbf{E}_{\perp} , \tag{8}$$

in which the components $\mathbf{v}_{\|}$ and $\mathbf{v}_{\perp}$ do not influence each other.

Equation (8) has the known solution

$$\mathbf{v}_{\perp}(t) = \frac{q}{m}\mathbf{E}_{\perp}\,t + \mathbf{v}_{0\perp} \tag{9}$$

describing the uniform increase. We then pass to Eq. (7). We rewrite it with the Clifford product instead of the inner product:

$$\frac{d\mathbf{v}_{||}}{dt} = \frac{q}{m}(\mathbf{E}_{||} - \mathbf{v}_{||}\,\widehat{\mathbf{B}}) \,. \tag{10}$$

This is the inhomogeneous linear equation for the unknown function $\mathbf{v}_{||}$. Its solution is a sum of the general integral $e^{-\frac{1}{2}\hat{\omega}t}\,\mathbf{b}\,e^{\frac{1}{2}\hat{\omega}t} = e^{-\hat{\omega}t}\,\mathbf{b}$ of the homogeneous equation (here $\mathbf{b}$ is a constant vector parallel to $\widehat{\mathbf{B}}$) with a particular integral of the inhomogeneous equation. The latter may be found by choosing $\dot{\mathbf{v}} = 0$. Denoting $\mathbf{v}_{||} = \mathbf{v}_d$ in this case, then $\mathbf{v}_d\,\widehat{\mathbf{B}} = \mathbf{E}_{||}$. We find from this: $\mathbf{v}_d = \mathbf{E}_{||}\widehat{\mathbf{B}}^{-1}$ (the associativity of the Clifford product and the reversibility of nonzero multivectors are essential here). We may also write

$$\mathbf{v}_d = \mathbf{E}_{||}\cdot\widehat{\mathbf{B}}^{-1} = \mathbf{E}\cdot\widehat{\mathbf{B}}^{-1}$$

due to the properties of the inner product. In this manner, we obtain the solution to (10):

$$\mathbf{v}_{||}(t) = e^{-\hat{\omega}t}\,\mathbf{b} + \mathbf{E}\cdot\widehat{\mathbf{B}}^{-1} \,. \tag{11}$$

This is a rotation in the plane of $\widehat{\mathbf{B}}$, however, not around the origin in the velocity space but around the value $\mathbf{E}\cdot\widehat{\mathbf{B}}^{-1}$.

Further integration of Eqs. (11) and (9) yields the motion of the particle

$$\mathbf{r}_{||}(t) = e^{-\hat{\omega}t}\,(\mathbf{b}\cdot\hat{\omega}^{-1}) + (\mathbf{E}\cdot\widehat{\mathbf{B}}^{-1})t + \mathbf{r}_{0||} \,, \tag{12a}$$

$$\mathbf{r}_{\perp}(t) = \frac{q}{2m}\,\mathbf{E}_{\perp}\,t^2 + \mathbf{v}_{0\perp}\,t + \mathbf{r}_{0\perp} \,. \tag{12b}$$

We see that, in the direction orthogonal to $\widehat{\mathbf{B}}$, this is the uniformly accelerated motion. Whereas in the plane of $\widehat{\mathbf{B}}$, this is a superposition of the rotary motion with the rectilinear motion having velocity $\mathbf{v}_d = \mathbf{E}\cdot\widehat{\mathbf{B}}^{-1} = -\mathbf{E}\cdot\widehat{\mathbf{B}}/|\,\widehat{\mathbf{B}}\,|^2$, called the *electric drift velocity*. It is perpendicular to the electric field and independent of the initial conditions and of the charge q. In order to maintain the nonrelativistic approximation we have to assume $|\,\mathbf{E}\cdot\widehat{\mathbf{B}}^{-1}\,| = E_{||}/B \ll c$.

The trajectory for (12a) is a curve called *trochoid* or *generalized cycloid*. Depending on whether $|\,\mathbf{b}\,|$ is greater or less than $|\,\mathbf{v}_d\,|$, it takes the shape shown in Figs. 74 and 75 (see Problem 1). In particular, for $\mathbf{b} = 0$, that is, when the initial velocity satisfies $\mathbf{v}_{0||} = \mathbf{v}_d$, the motion parallel to $\widehat{\mathbf{B}}$ is uniform rectilinear. This corresponds to the situation when the magnetic part of the Lorentz force is in balance with the parallel component of the electric force.

This is exactly the aforenamed particular integral of Eq. (10). Finally, if $|\mathbf{b}| = |\mathbf{v}_d|$ the trajectory for (12a) is the *cycloid* (Fig. 76).

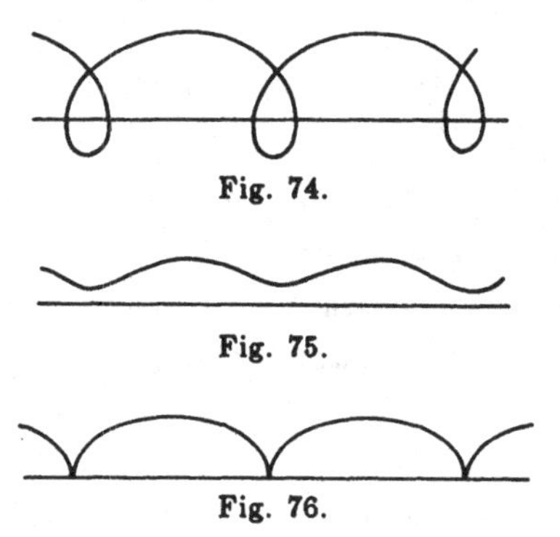

Fig. 74.

Fig. 75.

Fig. 76.

The spatial motion is, of course, the superposition of the described motion (12a) with the uniformly accelerated motion (12b). If $\mathbf{E} \parallel \widehat{\mathbf{B}}$, then for the particular initial velocity parallel to $\widehat{\mathbf{B}}$ the motion is flat and takes place over the named trochoid. Such a situation arises in the so-called *Hall effect* occurring with the current in a conductor immersed in a uniform magnetic field $\widehat{\mathbf{B}}$ parallel to the current. The curved trajectories of charge carriers imply that an electric charge accumulates on one side wall of the conductor.

Another special case is for $\mathbf{E} \perp \widehat{\mathbf{B}}$. Then $\mathbf{E}$ is absent in expression (12a) and the motion of the particle is a superposition of the uniformly accelerated motion (12b) along $\mathbf{E}$ with the circular motion (5a) parallel to $\widehat{\mathbf{B}}$. Thus the resulting path of the particle is a helix with variable pitch (see Fig. 77).

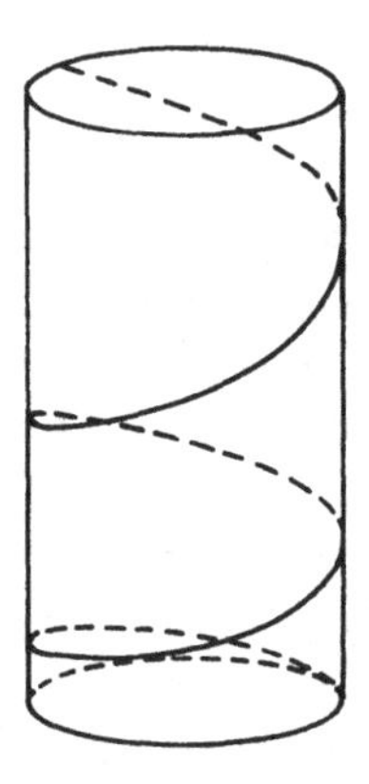

Fig. 77.

In the limit $\mathbf{E} \to 0$, the solution (12) reduces to the previously found solution (5). The solution (12) is not, however, suitable for the limiting transition $\hat{\mathbf{B}} \to 0$, since $\hat{\mathbf{B}}^{-1}$ occurs in it. Moreover, the condition $E_{\|}/B \ll c$ for the nonrelativistic approximation would be broken in such a limit. We shall return to this point — and generally to the motion of charges in uniform fields — again in Chap. 7 with considerations based on special relativity. In that chapter, we shall also cover the case $|\mathbf{E}_{\|}| > c|\hat{\mathbf{B}}|$ which we have omitted here.

§2. System of Charges in an External Field

Let us consider a system of charges placed in an external electrostatic field described by a scalar potential φ. The potential energy of a charge placed at the point $\mathbf{r}_s$ is $q_s\varphi(\mathbf{r}_s)$, so the total potential energy of the system is

$$U = \sum_s q_s \varphi(\mathbf{r}_s) .$$

We choose the origin of coordinates somewhere between the charges. We decompose the energy U in the power series with respect to $\mathbf{r}_s$:

$$U = \sum_s q_s \varphi(0) + \sum_s q_s \mathbf{r}_s \cdot \nabla\varphi(0) + \dots . \tag{13}$$

Assume that the electric field changes slowly in the region occupied by the charges, i.e., is approximately uniform. This assumption allows us to omit further terms in decomposition (13). Its first term is $Q\varphi(0)$ where $Q = \Sigma_s q_s$. In this zeroth approximation, the potential energy of the system is the same as if all the charges were concentrated at the origin. The second term of (13) yields

$$U^{(1)} = \mathbf{d} \cdot \nabla\varphi(0) ,$$

where $\mathbf{d} = \Sigma_s q_s \mathbf{r}_s$ is the dipole moment of the system of charges. After introducing $\mathbf{E} = -\nabla\varphi$, we write $U^{(1)} = -\mathbf{d} \cdot \mathbf{E}(0)$.

Now let $\mathbf{a}$ be a translation vector of the whole system of charges without destroying its configuration. In such a case, the potential energy in the chosen approximation becomes the following function of $\mathbf{a}$:

$$U(\mathbf{a}) = Q\varphi(\mathbf{a}) - \mathbf{d} \cdot \mathbf{E}(\mathbf{a}) .$$

The net force acting on the system is given by the minus gradient of this function with respect to $\mathbf{a}$, taken at the point $\mathbf{a} = 0$, that is

$$\mathbf{F} = -Q\nabla_a \varphi(\mathbf{a})\Big|_{\mathbf{a}=0} + \nabla_a[\mathbf{d} \cdot \mathbf{E}(\mathbf{a})]\Big|_{\mathbf{a}=0} = Q\mathbf{E}(0) + \nabla_r[\mathbf{d} \cdot \mathbf{E}(\mathbf{r})]\Big|_{\mathbf{r}=0} .$$

Due to Problem 0.28, we have $\nabla(\mathbf{d}\cdot\mathbf{E}) = (\mathbf{d}\cdot\nabla)\mathbf{E} - \mathbf{d}\cdot(\nabla\wedge\mathbf{E})$. By virtue of the Maxwell equation (1.17a), the second term vanishes for electrostatics, so in the approximation up to the first power in $\mathbf{r}_s$, we have

$$\mathbf{F} = q\mathbf{E} + (\mathbf{d}\cdot\nabla)\mathbf{E}\,, \tag{14}$$

i.e. the force is determined by the electric field and its first derivative in the direction of $\mathbf{d}$. For the electric dipole (that is, when $Q = 0$), the force depends only on the derivatives of $\mathbf{E}$. This means that only the nonuniform field acts, by a force, on an electric dipole.

We now calculate the net moment of force acting on the considered system of charges. For this purpose, we start with the forces acting on separate charges $\mathbf{F}_s = q_s\mathbf{E}(\mathbf{r}_s)$. Then the separate moments of force are $\widehat{\mathbf{K}}_s = \mathbf{r}_s\wedge\mathbf{F}_s$ and, in the approximation of first powers of $\mathbf{r}_s : \widehat{\mathbf{K}}_s = \mathbf{r}_s\wedge q\mathbf{E}(0)$. Thus the net moment of force is

$$\widehat{\mathbf{K}} = \sum_s q_s\mathbf{r}_s\wedge\mathbf{E}(0) = \mathbf{d}\wedge\mathbf{E}(0)\,. \tag{15}$$

We see that, in the chosen approximation, the moment of force is determined only by the electric field, not the derivatives.

We now consider a system of charges in a *bounded motion*, which means that the charges move in a bounded region of space and with bounded velocities. We also assume that the motion is *stationary*, that is periodically repeating. Let the system of charges be put in a constant and uniform external magnetic field. We shall look for the mean net force and the mean net moment of force. (By the mean value of a quantity F, we understand $\langle F\rangle = \frac{1}{T}\int_0^T F(t)dt$, where T is the period of the motion.)

The net force acting on our system of charges, due to (1), is

$$\mathbf{F} = -\sum_s q_s\mathbf{v}_s\cdot\widehat{\mathbf{B}} = -\sum_s q_s\frac{d\mathbf{r}_s}{dt}\cdot\widehat{\mathbf{B}} = -\frac{d}{dt}\sum_s q_s\mathbf{r}_s\cdot\widehat{\mathbf{B}}\,,$$

or

$$\mathbf{F}(t) = -\frac{d}{dt}[\mathbf{d}(t)\cdot\widehat{\mathbf{B}}]\,.$$

In this case, the mean net force is

$$\langle\mathbf{F}\rangle = -\frac{1}{T}[\mathbf{d}(T)\cdot\widehat{\mathbf{B}} - \mathbf{d}(0)\cdot\widehat{\mathbf{B}}] = 0\,,$$

since $t \to \mathbf{d}(t)$ is a periodic function with the period T. In this way, one shows that the mean value of the time derivative of any periodic function is zero.

The net moment of force exerted on the system of charges is

$$\hat{\mathbf{K}} = -\sum_s \mathbf{r}_s \wedge q_s (\mathbf{v}_s \cdot \hat{\mathbf{B}}) = -\sum_s q_s \mathbf{r}_s \wedge (\mathbf{v}_s \cdot \mathbf{B}) \,. \tag{16}$$

Notice the identity

$$\frac{d}{dt}[\mathbf{r}_s \wedge (\mathbf{r}_s \cdot \hat{\mathbf{B}})] = \dot{\mathbf{r}}_s \wedge (\mathbf{r}_s \cdot \hat{\mathbf{B}}) + \mathbf{r}_s \wedge (\dot{\mathbf{r}}_s \cdot \hat{\mathbf{B}}) = \mathbf{v}_s \wedge (\mathbf{r}_s \cdot \hat{\mathbf{B}}) + \mathbf{r}_s \wedge (\mathbf{v}_s \cdot \hat{\mathbf{B}}) \,,$$

which implies the equality

$$\frac{d}{dt}[\mathbf{r}_s \wedge (\mathbf{v}_s \cdot \hat{\mathbf{B}})] + [\mathbf{r}_s \wedge (\mathbf{v}_s \cdot \hat{\mathbf{B}}) - \mathbf{v}_s \wedge (\mathbf{r}_s \cdot \hat{\mathbf{B}})] = 2\,\mathbf{r}_s \wedge (\mathbf{v}_s \cdot \hat{\mathbf{B}}) \,.$$

By virtue of (0.32), the expression in the second square bracket is $(\mathbf{r}_s \wedge \mathbf{v}_s) \dot{\wedge} \hat{\mathbf{B}}$, so

$$\frac{d}{dt}[\mathbf{r}_s \wedge (\mathbf{r}_s \cdot \hat{\mathbf{B}})] + (\mathbf{r}_s \wedge \mathbf{v}_s) \dot{\wedge} \hat{\mathbf{B}} = 2\,\mathbf{r}_s \wedge (\mathbf{v}_s \cdot \hat{\mathbf{B}}) \,.$$

We employ this in (16):

$$\hat{\mathbf{K}} = -\frac{1}{2}\frac{d}{dt}\Big[\sum_s q_s \mathbf{r}_s \wedge (\mathbf{r}_s \cdot \hat{\mathbf{B}})\Big] - \frac{1}{2}\sum_s q_s (\mathbf{r}_s \wedge \mathbf{v}_s) \dot{\wedge} \hat{\mathbf{B}} \,.$$

The mean value of the first term is zero, so the mean net moment of force is

$$\langle \hat{\mathbf{K}} \rangle = -\frac{1}{2}\Big\langle \sum_s q_s (\mathbf{r}_s \wedge \mathbf{v}_s) \dot{\wedge} \hat{\mathbf{B}} \Big\rangle \,. \tag{17}$$

We remind the reader of the formula $\widehat{\mathbf{M}} = \frac{1}{2}\int dv\, \mathbf{r} \wedge \mathbf{j}(\mathbf{r})$ for the magnetic moment of currents. The moving charge q_s may be treated as a current with density $\mathbf{j}(\mathbf{r}) = q_s \mathbf{v}_s \delta^3(\mathbf{r} - \mathbf{r}_s)$ (compare with Eq. (1.37)), hence for such a charge

$$\widehat{\mathbf{M}}_s = \frac{1}{2} q_s \mathbf{r}_s \wedge \mathbf{v}_s \,. \tag{18}$$

We insert this into (17):

$$\langle \hat{\mathbf{K}} \rangle = -\Big\langle \sum_s \widehat{\mathbf{M}}_s \Big\rangle \dot{\wedge} \hat{\mathbf{B}} = -\langle \widehat{\mathbf{M}} \rangle \dot{\wedge} \hat{\mathbf{B}} \,. \tag{19}$$

We find that the mean net moment of forces acting on the stationary system of charges is given by the mingled product of the mean net magnetic moment with the magnetic induction. This formula is analogous to (15). The counterparts of Eqs. (15) and (19) written, by dint of (0.13) and (0.33), in terms of pseudovectors are even more similar:

$$\mathbf{K} = \mathbf{d} \times \mathbf{E}\,, \\ \langle \mathbf{K} \rangle = \langle \mathbf{M} \rangle \times \mathbf{B}\,. \tag{20}$$

The difference between expressions (19) and (20) is not only formal (volutors instead of pseudovectors), but also illustrates a different way of thinking about magnetic phenomena. Formula (20), written without the mean value sign: $\mathbf{K} = \mathbf{M} \times \mathbf{B}$, was found when the magnetic moment was treated in the same manner as the electric one, that is, as describing a dipole built of two magnetic charge poles. A physical model of such a dipole is a magnet in the shape of an elongated rod with the poles on the ends. Then the straight line of $\mathbf{M}$ coincides with the rod axis and the moment of force $\mathbf{K}$ juts out from the plane of rotation of the rod, as we show in Fig. 78. The same moment of force historically arose first as the moment of a couple of forces acting on two magnetic poles (see Fig. 79).

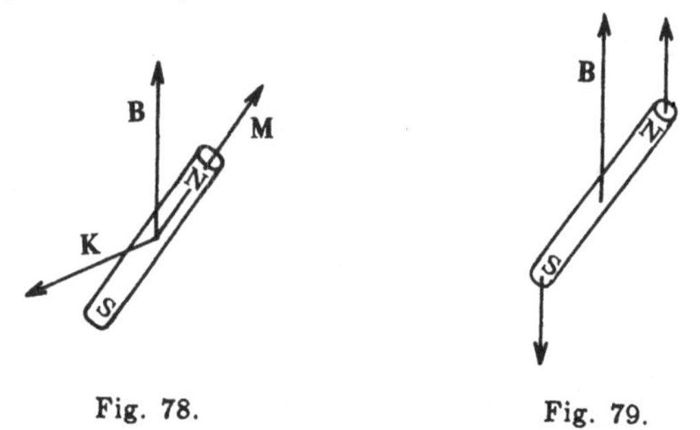

Fig. 78. Fig. 79.

On the other hand, formula (19) treats the magnetic moment two-dimensionally, as the magnetic shell. A physical model of such a dipole is the flat circuit. Then the plane of $\widehat{\mathbf{M}}$ coincides with the plane of the circuit (see Eq. (2.44)) and the moment of force $\widehat{\mathbf{K}}$ acting on the circuit coincides with its plane of rotation. We have shown, in Fig. 80, the circuit and the involved volutors as rectangles. We tried to draw them as dual quantities to the pseudovectors of Fig. 78. (The volutor $\widehat{\mathbf{K}}$ may be obtained also from $\widehat{\mathbf{B}}$ and $\widehat{\mathbf{M}}$ according to the prescription contained in Fig. 15.)

One more question — perhaps a naive one — is worth answering: Typically, a magnet is an elongated object, and the electric circuit is rather flat,

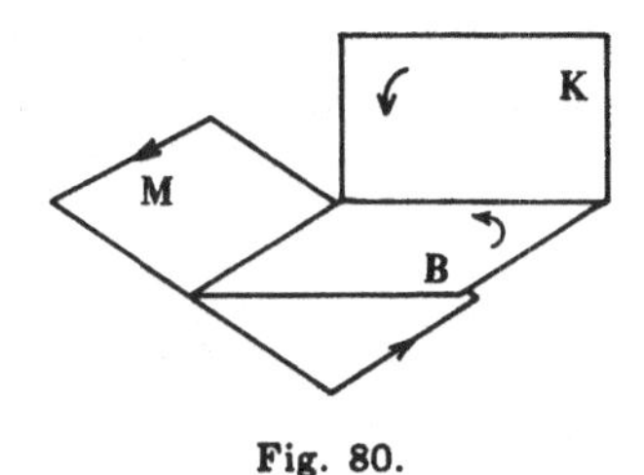

Fig. 80.

how then may one replace one by the other? Well, one should regard the magnet as a pile of magnetic shells, which we show in Fig. 81. This is a graphic illustration of a sentence from the beginning of Sec. 1.1: "Magnets are systems of vortical electric currents." The attempts of the early period of the science on magnetism to single out magnetic poles by cutting the magnets now seem rather funny.

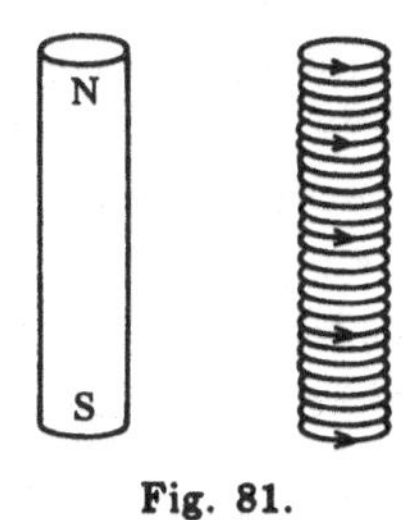

Fig. 81.

It may be useful at this point to make a digression about conceptual difficulties which may be surmounted by the volutorial description of magnetic moments. Let us, for this purpose, quote an excerpt from Weyl's book [17]:[1] "Ernst Mach tells of the intellectual shock he received when he learned as a boy that the magnetic needle is deflected in a certain sense, to the left or to the right, if suspended parallel to a wire through which an electric current is sent in a definite direction. Since the whole geometric and physical configuration, including the electric current and the south and north poles of the magnetic needle, to all appearances are symmetric with respect to the plane laid through the wire and the needle, the needle should react like Buridan's ass between equal bundles of hay and refuse to decide between left and right, just as scales of equal arms with equal weights neither go down on their left nor on their right side but stay horizontal." We show this "apparently

[1] Hermann Weyl, Symmetry. Copyright 1952, (c) 1980 by Princeton University Press. Excerpt reprinted with permission of Princeton University Press.

symmetric" initial situation in Fig. 82. Then, after switching on the current, forces emerge which rotate the needle up to the position shown in Fig. 83.

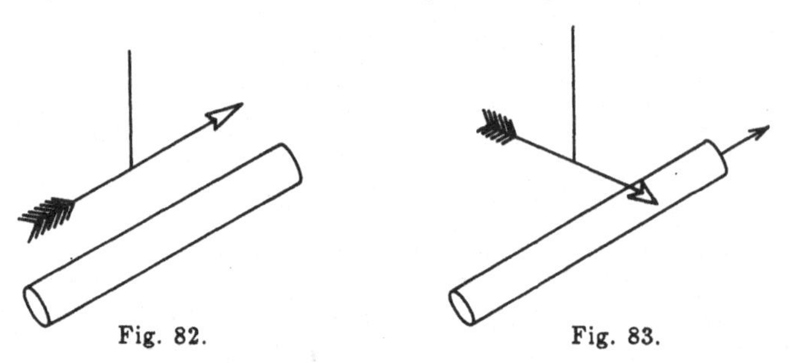

Fig. 82. Fig. 83.

Mach's difficulty lay in the apparent reflexion symmetry of the system in Fig. 82 with respect to the plane P passing the conductor and the magnetic needle. If we replace the arrow of the magnetic moment vector by a disc of the magnetic moment volutor, as in Fig. 84, we see, even in the absence of the current, that reflexion symmetry in P is lacking (the volutor $\widehat{\mathbf{M}}$ changes its sign). But any physical reason for tending to such symmetry is also lacking. Now, the flowing current releases forces which establish the reflexion symmetry by rotating the magnetic moment up to the position shown in Fig. 85. Such a mathematical description is more adequate to the physical nature of magnetism. One should realize that a needle is the magnet because it contains electrons performing circular motions in the atoms of the magnetic substance. After establishing the equilibrium of Fig. 85, the circular motions preserve their directions after a reflexion in P.

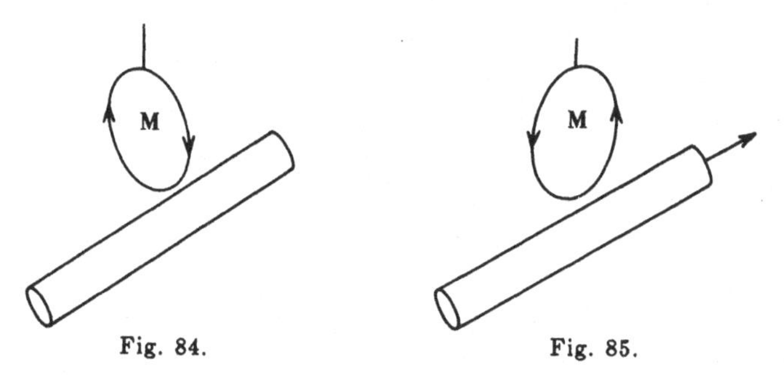

Fig. 84. Fig. 85.

Formula (19) (we write it here again:

$$\widehat{\mathbf{K}} = -\widehat{\mathbf{M}} \dot{\wedge} \widehat{\mathbf{B}} , \tag{21}$$

without the mean value sign) suggests a method for determining the direction of magnetic induction. There are two directions of $\widehat{\mathbf{M}}$, for which $\widehat{\mathbf{K}} = 0$; these are two opposite directions of parallel $\widehat{\mathbf{M}}$ and $\hat{\mathbf{B}}$. But only one of them — corresponding to identical orientations of $\widehat{\mathbf{M}}$ and $\hat{\mathbf{B}}$ — is the stable equilibrium position. Thus, when we have a small circuit with the directed area $\hat{\mathbf{S}}$, we place it in the magnetic field and let it rotate freely. After it attains the stable equilibruim position, we ascertain that the magnetic induction has the direction of $\hat{\mathbf{S}}$. This can be done practically by using small, round and flat magnets. If we put them in different points of the field, we may display in this manner the force surfaces (see Fig. 86). This photo should compete with the well known picture of iron filings thrown into the surroundings of a magnet. It may convince people that the force surfaces of the magnetic field have at least the same physical reality as the vectorial lines shown by the filings.

Fig. 86.

Let us now return to the main issue. If all charged particles of the system considered have equal ratio of charge q_s to the mass m_s, then due to (18), we may write

$$\widehat{\mathbf{M}} = \frac{1}{2}\sum_s q_s \mathbf{r}_s \wedge \mathbf{v}_s = \frac{q}{2m}\sum_s m_s \mathbf{r}_s \wedge \mathbf{v}_s \,.$$

Assuming that the velocities $\mathbf{v}_s$ are small in comparison with the velocity of light, then $m_s \mathbf{v}_s$ is the momentum and $m_s \mathbf{v}_s \wedge \mathbf{r}_s = \hat{\mathbf{L}}$ is the angular momentum of a particle. In that case,

$$\widehat{\mathbf{M}} = \frac{q}{2m}\sum_s \hat{\mathbf{L}}_s = \frac{q}{2m}\hat{\mathbf{L}} \,, \tag{22}$$

which means that the angular momentum of identical charged particles is proportional to their magnetic moment. The coefficient $q/2m$ is called the *giromagnetic ratio.*

Relation (22) allows us to find the function $t \to \widehat{\mathbf{M}}(t)$ if equality (21) holds for non-averaged quantities. The moment of force $\widehat{\mathbf{K}}$ is equal to the time derivative of the angular momentum, so

$$\frac{d}{dt}\hat{\mathbf{L}} = -\widehat{\mathbf{M}}\dot{\wedge}\widehat{\mathbf{B}}\,.$$

After using (22), we obtain

$$\frac{d}{dt}\widehat{\mathbf{M}} = \widehat{\mathbf{M}}\dot{\wedge}(-\frac{q}{2m}\widehat{\mathbf{B}})\,.$$

Comparison with Eq. (0.119) allows us to claim that

$$\widehat{\mathbf{M}}(t) = e^{-\frac{1}{2}\hat{\omega}_L t}\,\hat{\mathbf{M}}_0\,e^{\frac{1}{2}\hat{\omega}_L t}\,, \tag{23}$$

with $\hat{\omega}_L = -\frac{q}{2m}\widehat{\mathbf{B}}$, is the solution to the preceding equation. This means that the magnetic moment rotates in the plane of the external magnetic field with the angular frequency $\hat{\omega}_L$ which is called the *Larmor frequency*. Notice that it is twice as small as the cyclotron frequency. By dint of (22), it may also be written as $\hat{\omega}_L = \widehat{(M/L)}\,\widehat{\mathbf{B}}$. The rotation described by (23) which preserves the angle between $\widehat{\mathbf{M}}$ and $\widehat{\mathbf{B}}$ is called the *Larmor precession*. It resembles the behaviour of a top in the gravitational field and creates an additional average magnetic moment $\widehat{\mathbf{M}}_{ind}$ induced in this way by the external magnetic field. $\widehat{\mathbf{M}}_{ind}$ has the direction opposite to $\widehat{\mathbf{B}}$, that is, the induced magnetic moment weakens the external magnetic field.

We now comment on the neglected averaging in passing from (19) to (21). When the external magnetic field is weak enough, the Larmor frequency is small in comparison with the frequency of the assumed bounded motion. So the period T of the bounded motion is smaller than the period $2\pi/\omega_L$ of the Larmor precession. In this way, the mean value is calculated over T, and afterwards, the averaged quantities are subjected to much slower changes of the period $2\pi/\omega_L$.

§3. Lagrange and Hamilton Functions for a Charged Particle

The notions of *Lagrange function* or *Lagrangian* $L(x,\dot{x},t) = T(x,\dot{x},t) - V(x,t)$ and *Lagrange equations*

$$\frac{d}{dt}\frac{\partial L}{\partial \dot{x}} - \frac{\partial L}{\partial x} = 0 \quad \text{or} \quad \frac{d}{dt}\frac{\partial T}{\partial \dot{x}} - \frac{\partial T}{\partial x} = -\frac{\partial V}{\partial x}$$

as the equations of motion are known from classical mechanics. Here T denotes the kinetic energy, V the potential energy, x the collection of generalized coordinates (in the case of one particle one needs three coordinates) and $\dot{x}$ their time derivatives. The generalized forces $-\frac{\partial V}{\partial x}$ depend here only on the coordinates and time.

The Lagrange equation can be extended to the case of the so-called *generalized potential* $U(x, \dot{x}, t)$, when the forces also depend on the velocities. In this case, the Lagrange equations have the form

$$\frac{d}{dt}\frac{\partial T}{\partial \dot{x}} - \frac{\partial T}{\partial x} = \frac{d}{dt}\frac{\partial U}{\partial \dot{x}} - \frac{\partial U}{\partial x} .$$

The right-hand side is the force, also dependent on the velocities. This equation may also be written as

$$\frac{d}{dt}\frac{\partial L}{\partial \dot{x}} - \frac{\partial L}{\partial x} = 0 ,$$

where $L(x, \dot{x}, t) = T(x, \dot{x}, t) - U(x, \dot{x}, t)$.

An example of the nonpotential force,but having a generalized potential, is the Lorentz force $\mathbf{F} = q\,\mathbf{E} + q\,\widehat{\mathbf{B}} \cdot \mathbf{v}$. We are going to show that the expression

$$U = q\varphi - q\,\mathbf{A} \cdot \mathbf{v} ,$$

is the generalized potential for this force. Now,

$$-\frac{\partial U}{\partial x_i} = -q\,\frac{\partial \varphi}{\partial x_i} + q\,\frac{\partial}{\partial x_i}(\mathbf{A} \cdot \mathbf{v}) = -q\,\frac{\partial \varphi}{\partial x_i} + q\,\frac{\partial A_j}{\partial x_i}\,v_j ,$$

$$\frac{d}{dt}\frac{\partial U}{\partial \dot{x}_i} = -q\,\frac{d}{dt}\frac{\partial}{\partial v_i}(\mathbf{A} \cdot \mathbf{v}) = -q\,\frac{dA_i}{dt} .$$

Differentiation of the composed function $\mathbf{A}(\mathbf{r}(t), t)$ yields

$$\frac{d}{dt}\frac{\partial U}{\partial \dot{x}_i} = -q\,\frac{\partial A_i}{\partial x_j}\,v_j - q\,\frac{\partial A_i}{\partial t} ,$$

hence

$$\begin{aligned}\frac{d}{dt}\frac{\partial U}{\partial \dot{x}_i} - \frac{\partial U}{\partial x_i} &= -q\,\frac{\partial \varphi}{\partial x_i} - q\,\frac{\partial A_i}{\partial t} + q\left(\frac{\partial A_j}{\partial x_i} - \frac{\partial A_i}{\partial x_j}\right) v_j \\ &= q\,E_i + q\,B_{ij}\,v_j = q\,E_i + q\,(\widehat{\mathbf{B}} \cdot \mathbf{v})_i .\end{aligned}$$

We recognize in this the i-th coordinate of the Lorentz force which is what was to be shown. Therefore, we may write the Lagrangian of the charged particle in the electromagnetic field:

$$L = T - U = \frac{1}{2}mv^2 + q(\mathbf{A} \cdot \mathbf{v}) - q\varphi .$$

We now calculate the three components of the *generalized momentum* corresponding to the Lagrangian:

$$p_j = \frac{\partial L}{\partial \dot{x}_j} = \frac{\partial L}{\partial v_j} = m v_j + q A_j \quad \text{or} \quad \mathbf{p} = m\mathbf{v} + q\,\mathbf{A}\,.$$

The generalized momentum $\mathbf{p}$ is not, as we see, equal to the *mechanical momentum* $\boldsymbol{\pi} = m\mathbf{v}$, for the additional term $q\,\mathbf{A}$ occurs connected with the electromagnetic field. One may meet another terminology as well: $\mathbf{p}$ — *canonical momentum*, $\boldsymbol{\pi}$ — *kinetic momentum.* The relation between them is

$$\mathbf{p} = \boldsymbol{\pi} + q\,\mathbf{A}\,. \tag{24}$$

The canonical momentum is not uniquely determined, since the vector potential $\mathbf{A}$ is not unique — both quantities depend on the choice of gauge.

The *Hamilton function* or *Hamiltonian* is defined as follows by the Lagrangian and generalized momentum:

$$H = \sum_j p_j x_j - L\,.$$

The Hamiltonian for the charged particle is

$$H = \mathbf{p}\cdot\mathbf{v} - L = m\mathbf{v}^2 + q\,\mathbf{A}\cdot\mathbf{v} - \frac{1}{2}m\mathbf{v}^2 - q\,\mathbf{A}\cdot\mathbf{v} + q\,\varphi = \frac{1}{2}m\mathbf{v}^2 + q\varphi = \frac{\boldsymbol{\pi}^2}{2m} + q\varphi\,.$$

We see that the Hamiltonian is equal to the total energy which does not depend on the magnetic field ($\mathbf{A}$, containing all the information about it, is absent in the above expression). One cannot say, however, that the Hamiltonian does not describe the magnetic field, because the Hamilton function should be expressed in terms of the generalized momentum. We do this using (24):

$$H = \frac{(\mathbf{p} - q\,\mathbf{A})^2}{2m} + q\varphi\,. \tag{25}$$

The *Poisson bracket*

$$\{F, G\} = \sum_{j=1}^{3} \left(\frac{\partial F}{\partial x_j}\frac{\partial G}{\partial p_j} - \frac{\partial F}{\partial p_j}\frac{\partial G}{\partial x_j} \right),$$

for two functions F, G of the generalized coordinates and momenta, is an important tool in the Hamilton formalism. For particular functions, one obtains

$$\left.\begin{aligned} &\{p_l, p_j\} = 0\,, \qquad \{p_j, G(\mathbf{r}, t)\} = -\frac{\partial G}{\partial x_j}\,, \\ &\{F(\mathbf{r}, t),\, G(\mathbf{r}, t)\} = 0\,, \{F(\mathbf{r}, t), p_j\} = \frac{\partial F}{\partial x_j}\,, \\ &\{F, G_1 G_2\} = \{F, G_1\}G_2 + \{F, G_2\}G_1\,. \end{aligned}\right\} \tag{26}$$

The Poisson brackets are used for writing out the equation of motion

$$\frac{dG}{dt} = \{G, H\} + \frac{\partial G}{\partial t}\,, \tag{27}$$

for any physical quantity G (here H is the Hamiltonian).

We use Eq. (27) to find the constants of motion in the uniform magnetic field. Defining the vectors $\boldsymbol{\pi} = \mathbf{p} - q\,\mathbf{A}$, $\boldsymbol{\kappa} = \mathbf{p} + q\,\mathbf{A}$, we compute the following Poisson bracket using the rules (26):

$$\begin{aligned}\{\kappa_i, \pi_j\} &= \{p_i + q\,A_i, p_j - q\,A_j\} = q\{A_i, p_j\} - q\{p_i, A_j\} \\ &= q\Big(\frac{\partial A_i}{\partial x_j} + \frac{\partial A_j}{\partial x_i}\Big)\,.\end{aligned}$$

Choose one of the possible vector potentials for the uniform magnetic field, namely (2.33):

$$A_l = \frac{1}{2}\,x_k\,B_{kl}\,, \tag{28}$$

found for the field of the solenoid. Then

$$\frac{\partial A_i}{\partial x_j} + \frac{\partial A_j}{\partial x_i} = \frac{1}{2}\,(B_{ji} + B_{ij}) = 0\,.$$

Only potential (28) has such a property. It is not possessed by other potentials corresponding to the uniform magnetic field, considered in Chap. 2. Therefore, for gauge (28), the identity

$$\{\kappa_i, \pi_j\} = 0\,, \tag{29}$$

holds.

We now calculate the time derivative of $\boldsymbol{\kappa}$ according to (27):

$$\begin{aligned}\dot{\kappa}_i &= \{\kappa_i, H\} + \frac{\partial \kappa_i}{\partial t} = \{\kappa_i, H\} + q\,\frac{\partial A_i}{\partial t} = \{\kappa_i, \frac{\boldsymbol{\pi}^2}{2m}\} \\ &= \frac{1}{2m}\,\{\kappa_i, \pi_k\,\pi_k\} = \frac{1}{m}\{\kappa_i, \pi_k\}\pi_k = 0\,,\end{aligned}$$

by virtue of (29). In this way, an interesting result was obtained. For the constant uniform magnetic field, the vector

$$\boldsymbol{\kappa} = \mathbf{p} + q\,\mathbf{A} = m\mathbf{v} + 2q\,\mathbf{A}$$

is the constant of motion for the vector potential (28) with the axial symmetry.

This situation is similar to the motion in the constant electric field, where the total energy $\mathcal{E} = \frac{1}{2}m\mathbf{v}^2 + q\varphi$ is a constant of motion. In analogy with the decomposition of this energy into kinetic and potential parts, some physicists claim that $\boldsymbol{\kappa}$ is the *total momentum* which consists of the kinetic part $m\mathbf{v}$ and the *potential part* $2p\,\mathbf{A}$.

We now ponder whether a similar constant of the motion may be found for the arbitrary static magnetic field. Assume that the field is described by the vector potential $\mathbf{A}$ which does not satisfy the condition

$$\frac{\partial A_i}{\partial x_j} + \frac{\partial A_j}{\partial x_i} = 0 \quad \text{for any} \quad i, j \in \{1, 2, 3\} . \tag{30}$$

Does a gauge transformation $\mathbf{A}' = \mathbf{A} + \nabla\chi$ exist, after which a new potential $\mathbf{A}'$ satisfies (30)? Equation (30), applied to $\mathbf{A}'$, yields

$$\frac{\partial A'_i}{\partial x_j} + \frac{\partial A'_j}{\partial x_i} = \frac{\partial A_i}{\partial x_j} + \frac{\partial A_j}{\partial x_i} 2 \frac{\partial^2 \chi}{\partial x_i\, \partial x_j} = 0 .$$

This implies that the gauge function χ must satisfy the differential equation

$$\frac{\partial^2 \chi}{\partial x_i\, \partial x_j} = -\frac{1}{2}\left(\frac{\partial A_i}{\partial x_j} + \frac{\partial A_j}{\partial x_i}\right) \quad \text{for any} \quad i, j \in \{1, 2, 3\} , \tag{31}$$

in which the right-hand side is assumed to be known. The partial differential equations are soluble only when they satisfy the consistency conditions. One of them has the form

$$\frac{\partial}{\partial x_k}\left(\frac{\partial^2 \chi}{\partial x_i\, \partial x_j}\right) = \frac{\partial}{\partial x_i}\left(\frac{\partial^2 \chi}{\partial x_j \partial x_k}\right) .$$

Having used (31), we may write this as

$$\frac{\partial^2 A_i}{\partial x_k\, \partial x_j} + \frac{\partial^2 A_j}{\partial x_k\, \partial x_i} = \frac{\partial^2 A_j}{\partial x_i\, \partial x_k} + \frac{\partial^2 A_k}{\partial x_i\, \partial x_j} ,$$

or

$$\frac{\partial^2 A_i}{\partial x_k\, \partial x_i} = \frac{\partial^2 A_k}{\partial x_i\, \partial x_j} .$$

Due to the commutativity of partial derivatives, this may also be written as

$$\frac{\partial}{\partial x_j}\left(\frac{\partial A_i}{\partial x_k} - \frac{\partial A_k}{\partial x_i}\right) = 0 ,$$

or, by virtue of (2.1),

$$\frac{\partial}{\partial x_j} B_{ki} = 0 \quad \text{for any} \quad i, j, k \in \{1, 2, 3\} .$$

This means that the magnetic field has to be uniform if a vector potential is to exist such that Eq. (30) is satisfied. This observation implies that the expression $\boldsymbol{\kappa} = \mathbf{p} + q\,\mathbf{A}$ may be the constant of motion only in the uniform magnetic field and only for the particular "solenoidal" vector potential.

An interesting relation of Hamiltonian (25) with the harmonic oscillator is worth noticing in the case of the uniform magnetic field. We insert the "solenoidal" potential (28) into (25) and put $\varphi = 0$

$$H = \frac{1}{2m}\left(\mathbf{p} - \frac{1}{2}\,q\,\mathbf{r}\cdot\widehat{\mathbf{B}}\right)^2 = \frac{\mathbf{p}^2}{2m} - \frac{q}{2m}\,(\mathbf{p}\wedge\mathbf{r})\cdot\widehat{\mathbf{B}} + \frac{q^2}{8m}\,|\,\mathbf{r}\cdot\widehat{\mathbf{B}}\,|^2 .$$

Equation (0.25) was used in this derivation. We now introduce the Larmor frequency $\widehat{\boldsymbol{\omega}}_L = -\frac{q}{2m}\,\widehat{\mathbf{B}}$:

$$H = \frac{\mathbf{p}^2}{2m} + (\mathbf{p}\wedge\mathbf{r})\cdot\widehat{\boldsymbol{\omega}}_L + \frac{m}{2}\,|\,\mathbf{r}\cdot\widehat{\boldsymbol{\omega}}_L\,|^2 .$$

After employing the property $|\,\mathbf{r}\cdot\widehat{\boldsymbol{\omega}}_L\,| = r_{\|}\,\omega_L$, where $\mathbf{r}_{\|}$ is the component of $\mathbf{r}$ parallel to $\widehat{\mathbf{B}}$, we obtain

$$H = \frac{\mathbf{p}^2}{2m} - (\mathbf{r}\wedge\mathbf{p})\cdot\widehat{\boldsymbol{\omega}}_L + \frac{m}{2}\,r_{\|}^2\,\omega_L^2 .$$

We introduce the angular momentum $\widehat{\mathbf{L}} = \mathbf{r}\wedge\mathbf{p}$:

$$H = \frac{\mathbf{p}^2}{2m} + \frac{m}{2}\,\omega_L^2\,r_{\|}^2 - \widehat{\mathbf{L}}\cdot\widehat{\boldsymbol{\omega}}_L .$$

The second term on the right-hand side does not contain the component $\mathbf{r}_{\perp}$, and the third term chooses only the component of $\widehat{\mathbf{L}}$ parallel to $\widehat{\mathbf{B}}$. In this manner, the motion in the direction orthogonal to $\widehat{\mathbf{B}}$ is free. Therefore, the problem may be restricted to the plane of $\widehat{\mathbf{B}}$; then the Hamiltonian takes the form

$$H_{\|} = \frac{\mathbf{p}_{\|}^2}{2m} + \frac{1}{2}\,\omega_L^2\,\mathbf{r}_{\|}^2 - \widehat{\mathbf{L}}\cdot\widehat{\boldsymbol{\omega}}_L ,$$

that is, it becomes the sum of the plane isotropic oscillator Hamiltonian with the additional term $-\widehat{\mathbf{L}}\cdot\widehat{\boldsymbol{\omega}}_L$. It is known from mechanics that the angular momentum (in fact, only its component parallel to $\hat{\mathbf{B}}$, after our restriction) is a constant of the motion for the plane oscillator. In that case, by virtue of (27), the projection of $\hat{\mathbf{L}}$ on the plane of $\hat{\mathbf{B}}$ is the constant of the motion.

In this manner, besides the three constants of motion conatined in $\boldsymbol{\kappa}$, we have found the fourth one. We should, however, remember that $\hat{\mathbf{L}}$ is the "canonical" angular momentum, which means that $\hat{\mathbf{L}}$ depends, through $\mathbf{p}$, on the chosen gauge:

$$\hat{\mathbf{L}} = \mathbf{r} \wedge \mathbf{p} = \mathbf{r} \wedge \boldsymbol{\pi} + \frac{q}{2} \mathbf{r} \wedge (\mathbf{r} \cdot \hat{\mathbf{B}}) ,$$

which after restriction to $\mathbf{r}_{\|}$, is

$$\hat{\mathbf{L}}_{\|} = \mathbf{r}_{\|} \wedge \boldsymbol{\pi}_{\|} + \frac{q r_{\|}^2}{2} \hat{\mathbf{B}} .$$

Problems

1. Justify the importance of inequalities $|\mathbf{b}| \lesseqgtr |\mathbf{E} \cdot \hat{\mathbf{B}}^{-1}|$ for the trochoid discussed in Sec. 1.
2. Demonstrate that in the flat motion described by (12a) for different initial velocities of the same magnitude and for the same initial positions, all possible trochoids pass through another common point as shown in Fig. 87.

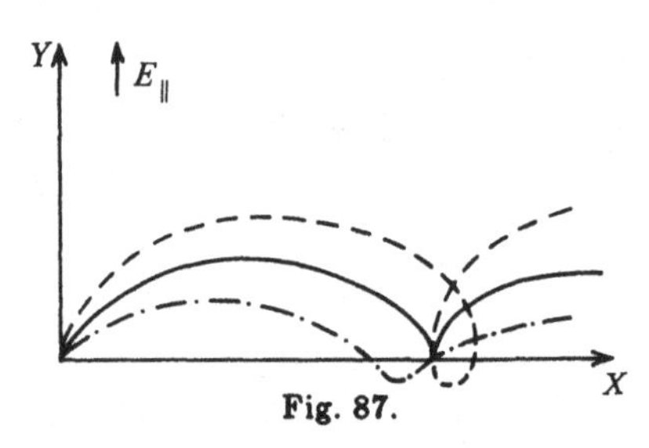

Fig. 87.

3. A particle with mass m and charge q moves in the electric field of a charge Q at rest in the origin of coordinates and in the uniform magnetic field parallel to plane $\mathbf{e}_1 \mathbf{e}_2$. The charge q has the initial velocity $\mathbf{v}_0 = v_0 \, \mathbf{e}_2$ and the initial position $\mathbf{r}_0 = x_0 \, \mathbf{e}_1$. For which value of the magnetic induction does the charge q move over a circle?
4. An electric charge moves with the initial velocity $\mathbf{v}_0 = v_{0y} \, \mathbf{e}_2 + v_{0z} \, \mathbf{e}_3$ in the magnetic field given by the formula $\hat{\mathbf{B}}(\mathbf{r}) = \mathbf{e}_2 \wedge \mathbf{e}_3 \, B_0 \sin kz$ with a

constant k. Does the charge move permanently in a plane parallel to $\mathbf{e}_2\,\mathbf{e}_3$? Find the motion in the approximation $v_y \ll v_z$.

5. Two parallel resistanceless slide-bars separated by a distance d are set up under the angle α to the horizon (see Fig. 88). The bars are connected at the top by resistance R. A resistanceless cross-piece of mass m slips down, thus closing the electric circuit. The whole is placed in a uniform magnetic field of horizontal induction $\hat{\mathbf{B}}$ (as a volutor). What limiting velocity does the cross-piece achieve?

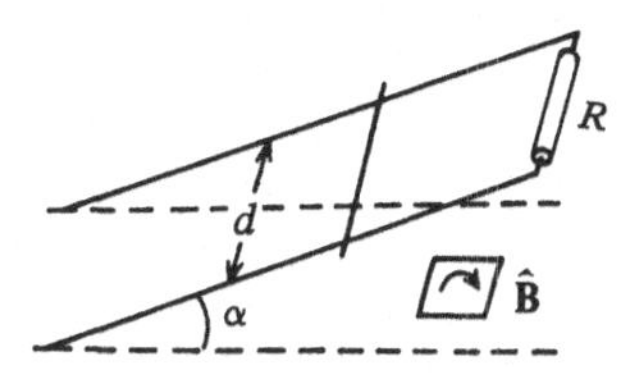

Fig. 88.

6. Starting with Eq. (21), find the energy of the magnetic dipole with the moment $\widehat{\mathbf{M}}$ placed in a constant uniform magnetic field of induction $\hat{\mathbf{B}}$. Express it as a function of the angle between $\widehat{\mathbf{M}}$ and $\hat{\mathbf{B}}$. Compare the result with the solution to Problem 2.14.
7. Apply the equation of motion (27) to the kinetic momentum $\boldsymbol{\pi}$ when the pure electric field is described by the potentials $A' = 0, \varphi' = f(\mathbf{r})$, and afterwards by the potentials given in Problem 2.7. Compare the equations.
8. Find the value of $\boldsymbol{\kappa}$ from the explicit function $t \to \mathbf{r}(t)$ known for the charge in the unform magnetic field.
9. Show that $\{\pi_i, \pi_j\} = q\,B_{ij}, \{\kappa_i, \kappa_j\} = -q\,B_{ij}$.

Chapter 4

PLANE ELECTROMAGNETIC FIELDS

§1. Introducing Plane Electromagnetic Fields

The reader may have noticed in Sec. 1.8 that the two linear solutions to the Maxwell equation

$$f(\mathbf{r},t) = [\omega t - k(\mathbf{r}\cdot\mathbf{n})\mathbf{n}]\mathbf{m} \quad \text{and} \quad f(\mathbf{r},t) = [\omega t\mathbf{n} - k(\mathbf{r}\cdot\mathbf{n})]\mathbf{m}$$

consist of two factors playing different roles. The first factor, placed in square brackets, guarantees fulfilment of the differential equation, whereas the other factor causes f to be the electromagnetic cliffor (i.e. the sum of a vector and a volutor).

We now build more general solutions to the Maxwell equation out of the two cliffors:

$$X(\mathbf{r},t) = \omega t - k\mathbf{n}(\mathbf{n}\cdot\mathbf{r}) \quad \text{and} \quad Y(\mathbf{r},t) = k(\mathbf{n}\cdot\mathbf{r}) - \omega t\mathbf{n}\,. \tag{1}$$

Both X and Y are combinations of a scalar and a vector parallel to $\mathbf{n}$, thus both commute with $\mathbf{n}$:

$$X\mathbf{n} = \mathbf{n}X\,, \qquad Y\mathbf{n} = \mathbf{n}Y\,, \tag{2}$$

and the identities

$$X\mathbf{n} = -Y\,, \qquad Y\mathbf{n} = -X\,, \tag{3}$$

are satisfied.

Since the linear combinations of scalars and vectors parallel to $\mathbf{n}$ form a subalgebra, it is obvious that the powers X^m and Y^m for any natural m are of the form $\varphi + \mathbf{n}\,\psi$, with scalars φ and ψ.

Lemma 1. The cliffor-valued functions X^m and Y^m satisfy the homogeneous Maxwell equation (1.24) if, and only if, $\sqrt{\varepsilon\mu}\omega = k$.

Proof. Notice that

$$\frac{\partial X}{\partial t} = \omega\,, \quad \frac{\partial X}{\partial x_i} = -kn_i\,\mathbf{n}\,, \quad i \in \{1,2,3\}\,,$$

where $n_i = \mathbf{n}\cdot\mathbf{e}_i$. We now calculate the derivatives of X^m:

$$\frac{\partial}{\partial t}X^m = \sum_{j=1}^{m} X^{j-1}\frac{\partial X}{\partial t}X^{m-j} = m\omega\,X^{m-1}\,,$$

$$\frac{\partial}{\partial x_i}X^m = \sum_{j=1}^{m} X^{j-1}\frac{\partial X}{\partial x_i}X^{m-j} = -k\,n_i\sum_{j=1}^{m} X^{j-1}\,\mathbf{n}\,X^{m-j}\,.$$

Equation (2) implies

$$\frac{\partial}{\partial x_i}X^m = -m\,k\,n_i\,\mathbf{n}\,X^{m-1}\,,$$

and

$$\nabla X^m = \mathbf{e}_i\frac{\partial}{\partial x_i}X^m = -m\,k\,\mathbf{n}^2\,X^{m-1} = -mk\,X^{m-1}\,.$$

Thus we obtain

$$\not{D}\,X^m = \left(\sqrt{\varepsilon\mu}\frac{\partial}{\partial t} + \nabla\right)X^m = m(\sqrt{\varepsilon\mu}\omega - k)X^{m-1}\,.$$

Therefore, $\not{D}X^m = 0$ if, and only if, $\sqrt{\varepsilon\mu}\omega = k$.

Similar calculations may be performed for Y^m. ■

If we substitute $\mathbf{k} = k\,\mathbf{n}$ and $x_0 = ut = t/\sqrt{\varepsilon\mu}$, we obtain for the functions X and Y the expressions

$$X = kx_0 - \mathbf{n}\,(\mathbf{k}\cdot\mathbf{r})\,, \qquad Y = \mathbf{k}\cdot\mathbf{r} - \mathbf{k}\,x_0\,. \tag{4}$$

Let $\mathcal{A}(\mathbf{n})$ denote the *subalgebra generated* by $\mathbf{n}$ and I, that is, the set of all linear combinations of $\mathbf{n}$ and I and their products. This, of course, is commutative algebra. The functions X and Y have their values in $\mathcal{A}(\mathbf{n})$. It follows from Lemma 1 that the power series

$$g(X) = \sum_{m=0}^{\infty} a_m X^m = \sum_{m=0}^{\infty} X^m a_m \quad \text{or} \quad g^*(Y) = \sum_{m=0}^{\infty} a_m Y^m = \sum_{m=0}^{\infty} Y^m a_m \,, \tag{5}$$

with $a_m \in \mathcal{A}(\mathbf{n})$ satisfy the homogeneous Maxwell equation.[1)] The functions (5) take their values in $\mathcal{A}(\mathbf{n})$, that is

$$g = g_1 + g_2 \mathbf{n} + g_3 I \mathbf{n} + g_4 I \,, \tag{6}$$

where $g_1, \ldots, g_4$ are scalars.

We now wish to find solutions to the Maxwell equation which are electromagnetic cliffors. We seek them in the form $f = gN$ where N is a constant cliffor and g is given by (6). We represent N generally as $N = \lambda + \mathbf{c} + \widehat{\mathbf{C}} + \tau I$, whence

$$\begin{aligned}
gN =& (g_1 + g_2 \mathbf{n} + g_3 I \mathbf{n} + g_4 I)\,(\lambda + \mathbf{c} + \widehat{\mathbf{C}} + \tau I) \\
=& [g_1 \lambda + g_2 (\mathbf{n} \cdot \mathbf{c}) + g_3 I (\mathbf{n} \wedge \widehat{\mathbf{C}}) - g_4 \tau] \\
&+ [g_1 \mathbf{c} + g_2 \lambda \mathbf{n} + g_2 (\mathbf{n} \cdot \widehat{\mathbf{C}}) + g_3 I (\mathbf{n} \wedge \mathbf{c}) - g_3 \tau \mathbf{n} + g_4 I \widehat{\mathbf{C}}] \\
&+ [g_1 \widehat{\mathbf{C}} + g_2 \tau \mathbf{n} I + g_2 \mathbf{n} \wedge \mathbf{c} + g_3 \lambda \mathbf{n} I + g_3 I (\mathbf{n} \cdot \widehat{\mathbf{C}}) + g_4 I \mathbf{c}] \\
&+ [g_1 \tau I + g_2 \mathbf{n} \wedge \widehat{\mathbf{C}} + g_3 I (\mathbf{n} \cdot \mathbf{c}) + g_4 \lambda I] \,.
\end{aligned}$$

We have grouped multivectors of definite ranks in separate square brackets. The scalar and trivector parts have to vanish, so

$$\begin{aligned}
g_1 \lambda + g_2 (\mathbf{n} \cdot \mathbf{c}) + g_3 I (\mathbf{n} \wedge \widehat{\mathbf{C}}) - g_4 \tau = 0 \,, \\
g_1 \tau I + g_2 \mathbf{n} \wedge \widehat{\mathbf{C}} + g_3 I (\mathbf{n} \cdot \mathbf{c}) + g_4 \lambda I = 0 \,.
\end{aligned}$$

The functions $g_1, \ldots, g_4$ may be linearly independent, hence their coefficients must be zero separately:

$$\lambda = 0, \quad \tau = 0, \quad \mathbf{n} \cdot \mathbf{c} = 0, \quad \mathbf{n} \wedge \widehat{\mathbf{C}} = 0 \,.$$

[1)] For simplicity of reasoning, let us assume that the power series of the derivatives $\sum_{m=1}^{\infty} a_m \, m X^{m-1}$ is convergent.

The first two equalities indicate that $N = \mathbf{c} + \widehat{\mathbf{C}}$, and the two last equalities mean that

$$\mathbf{c} \perp \mathbf{n}, \quad \widehat{\mathbf{C}} \parallel \mathbf{n}, \tag{7}$$

which occurs if, and only if, $\mathbf{n} \wedge \mathbf{c} = \mathbf{n}\,\mathbf{c}$, $\mathbf{n} \cdot \widehat{\mathbf{C}} = \mathbf{n}\,\widehat{\mathbf{C}}$ and equivalently,

$$\mathbf{n}\,N = -N\,\mathbf{n}\,. \tag{8}$$

In this manner, we may state that

$$\begin{aligned} gN = &\,(g_1\,\mathbf{c} + g_2\,\mathbf{n}\,\widehat{\mathbf{C}} + g_3\,I\,\mathbf{n}\,\mathbf{c} + g_4\,I\,\widehat{\mathbf{C}}) \\ &+ (g_1\,\widehat{\mathbf{C}} + g_2\,\mathbf{n}\,\mathbf{c} + g_3\,I\,\mathbf{n}\,\widehat{\mathbf{C}} + g_4\,I\,\mathbf{c})\,, \end{aligned}$$

is the electromagnetic cliffor if, and only if, the condition (7) or, equivalently, (8) is satisfied.

Thus we obtain the conclusion that the functions

$$f = gN = \Big(\sum_{m=0}^{\infty} a_m X^m\Big)N \quad \text{and} \quad f^* = gN = \Big(\sum_{m=0} a_m Y^m\Big)N \tag{9}$$

satisfying the homogeneous Maxwell equation are the electromagnetic cliffors if, and only if, $N = \mathbf{c} + \widehat{\mathbf{C}}$ anticommutes with $\mathbf{n}$ or alternatively, if, and only if, (7) is satisfied.

Since g commutes with $\mathbf{n}$, the condition (8) implies

$$\mathbf{n}\,f = -f\,\mathbf{n}\,, \tag{10}$$

which, in turn, is equivalent to

$$\mathbf{e} \perp \mathbf{n}, \quad \widehat{\mathbf{b}} \parallel \mathbf{n}\,, \tag{11}$$

if f, of the form (9), is decomposed as $f = \mathbf{e} + \widehat{\mathbf{b}}$. The same holds for f^* in place of f.

In this manner, we have obtained quite a rich family of the electromagnetic field functions. Their common feature is that they depend on $\mathbf{r}$ through the scalar product $\mathbf{n} \cdot \mathbf{r}$, as it is visible from (1). This means that, for fixed time, they are constant on the planes described by the equations $\mathbf{n} \cdot \mathbf{r} = \text{const}$. That is why we call them *plane electromagnetic fields*.

Physicists usually refer to time-dependent solutions of the homogeneous Maxwell equations as *electromagnetic waves*. Let us reserve the word "wave" for solutions with some elements of periodicity in time dependence. Since we do not assume anything of this kind for our functions g, we have chosen a more general name: plane electromagnetic fields.

Conditions (11) written in the traditional language have the form $\mathbf{E} \perp \mathbf{n}$, $\mathbf{B} \perp \mathbf{n}$ which is known as the *transversality condition* for plane electromagnetic fields. This condition is usually mentioned simultaneously and on equal footing with the *perpendicularity condition* $\mathbf{E} \perp \mathbf{B}$. We have not obtained here any condition relating $\mathbf{c}$ with $\widehat{\mathbf{C}}$, which means that the electric part $\mathbf{e}$ and the magnetic part $\widehat{\mathbf{b}}$ of the plane field f may be chosen independently of each other (of course, with the fulfilment of (11)). Hence the transversality condition is more fundamental than that of perpendicularity. We shall see later which special solutions do indeed satisfy the latter condition.

One may ask whether, for a given solution f or f^* of the form (9), a function g' can be found such that $f = g'N'$ and g' is scalar-valued. In general, the answer to this question is negative, but for a specific value of the argument X or Y, that is for specific time t_0 and position $\mathbf{r}_0$ where g is invertible, this can be achieved by redefining the quantities as follows:

$$g'(\mathbf{r},t) = g(\mathbf{r},t)g^{-1}(\mathbf{r}_0,t_0), \quad N' = g(\mathbf{r}_0,t_0)N .$$

Since $g(\mathbf{r}_0,t_0)$ belongs to $\mathcal{A}(\mathbf{n})$, it commutes with $\mathbf{n}$, hence N' satisfies condition (8) as well as N. Thus the primed quantities satisfy all the relevant conditions found previously for the unprimed quantities and, additionally, $g'(\mathbf{r}_0,t_0) = 1$ which obviously is a scalar.

We shall find the energy-momentum cliffor (2.63) for plane fields (9). Making use of the second property (7), we write $\widehat{\mathbf{C}} = \mathbf{n}\mathbf{c}'$ where $\mathbf{c} \perp \mathbf{n}$, and compute the expression

$$NN^\dagger = (\mathbf{c} + \mathbf{n}\,\mathbf{c}')\,(\mathbf{c} + \mathbf{c}'\,\mathbf{n}) = \mathbf{c}^2 + \mathbf{n}\,\mathbf{c}'\,\mathbf{c} + \mathbf{c}\,\mathbf{c}'\,\mathbf{n} + \mathbf{c}'^2 .$$

Anticommutativity of $\mathbf{c}, \mathbf{c}'$ with $\mathbf{n}$ implies

$$\begin{aligned} NN^\dagger &= \mathbf{c}^2 + \mathbf{c}'^2 + \mathbf{n}\,(\mathbf{c}'\,\mathbf{c} + \mathbf{c}\,\mathbf{c}') , \\ NN^\dagger &= |\,N\,|^2 + 2\mathbf{n}\,(\mathbf{c}\cdot\mathbf{c}') . \end{aligned} \tag{12}$$

We calculate $ff^\dagger$ for $f = gN$:

$$ff^\dagger = gNN^\dagger\, g^\dagger = g[|\,N\,|^2 + 2\mathbf{n}\,(\mathbf{c}\cdot\mathbf{c}')]\, g^\dagger .$$

Since g commutes with $\mathbf{n}$,

$$ff^\dagger = [|\,N\,|^2 + 2\mathbf{n}\,(\mathbf{c}\cdot\mathbf{c}')]\, gg^\dagger . \tag{13}$$

By dint of (6), functions (5) can be represented as

$$g = \varphi + \mathbf{n}\psi \,, \tag{14}$$

with "complex" scalars $\varphi = g_1 + g_4 I$, $\psi = g_2 + g_3 I$, hence

$$ff^\dagger = [|N|^2 + 2\mathbf{n}(\mathbf{c}\cdot\mathbf{c}')][\varphi\varphi^\dagger + \mathbf{n}(\varphi^\dagger\psi + \varphi\psi^\dagger) + \psi\psi^\dagger] \,.$$

We introduce the notation $\langle\varphi,\psi\rangle = \frac{1}{2}(\varphi\psi^\dagger + \varphi^\dagger\psi)$ and observe that $\varphi\varphi^\dagger + \psi\psi^\dagger = |g|^2$. In this manner, we obtain

$$\frac{1}{2}ff^\dagger = \frac{1}{2}|g|^2|N|^2 + 2\langle\varphi,\psi\rangle\,\mathbf{c}\cdot\mathbf{c}' + \mathbf{n}\left(\langle\varphi,\psi\rangle|N|^2 + |g|^2\mathbf{c}\cdot\mathbf{c}'\right) \,. \tag{15}$$

This implies that the energy density and the Poynting vector are

$$w = \frac{1}{2}|g|^2|N|^2 + 2\langle\varphi,\psi\rangle\,\mathbf{c}\cdot\mathbf{c}' \,, \tag{16a}$$

$$\mathbf{S} = \mathbf{n}\,u(\langle\varphi,\psi\rangle|N|^2 + |g|^2\mathbf{c}\cdot\mathbf{c}') \,. \tag{16b}$$

Equation (16b) indicates that the energy propagates parallel to $\mathbf{n}$ in the positive or negative direction, depending on the expression in the bracket. The planes of constant value of the field are perpendicular to the direction of energy transport.

The calculations can be repeated for the field f^* in place of f, leading to the same conclusions.

§2. Travelling Plane Fields

It would be interesting to establish when the velocity of the energy transport (2.65) is maximal for plane electromagnetic fields (9). We insert (16) into (2.65) such that

$$v = \frac{|\mathbf{S}|}{w} = \frac{2u|\langle\varphi,\psi\rangle|N|^2 + |g|^2\mathbf{c}\cdot\mathbf{c}'|}{|g|^2|N|^2 + 4\langle\varphi,\psi\rangle\,\mathbf{c}\cdot\mathbf{c}'} \,.$$

As we know from §2.6, the maximal possible value is $v = u$. in this case,

$$\pm 2(\langle\varphi,\psi\rangle|N|^2 + |g|^2\mathbf{c}\cdot\mathbf{c}') = |g|^2|N|^2 + 4\langle\varphi,\psi\rangle\,\mathbf{c}\cdot\mathbf{c}' \,. \tag{17}$$

After simple manipulations, we get

$$|g|^2|N|^2 \pm 2\langle\varphi,\psi\rangle|N|^2 \pm 2|g|^2\mathbf{c}\cdot\mathbf{c}' + 4\langle\varphi,\psi\rangle\,\mathbf{c}\cdot\mathbf{c}' = 0 \,,$$

which is equivalent to

$$(|g|^2 \pm 2\langle\varphi,\psi\rangle)\,(|N|^2 \pm 2\mathbf{c}\cdot\mathbf{c}') = 0\,.$$

The first bracket cannot vanish identically for all $\mathbf{r}$ and t, so $N^2 \pm 2\mathbf{c}\cdot\mathbf{c}' = 0$, which can be written as $(\mathbf{c}\pm\mathbf{c}')^2 = 0$. In this manner, we obtain

$$\mathbf{c}' = \pm\mathbf{c}\,, \tag{18}$$

or, equivalently, $\widehat{C} = \mathbf{n}\,\mathbf{c}$, whence

$$N = \mathbf{c}\pm\mathbf{n}\,\mathbf{c} = (1\pm\mathbf{n})\,\mathbf{c} \tag{19}$$

is the necessary and sufficient condition for the maximal velocity of the energy transport in the whole space filled by the fields (9). Since $\mathbf{n}$ commutes with g, we obtain, in this case,

$$f = (1\pm\mathbf{n})g\mathbf{c}\,, \tag{20}$$

where $\mathbf{c}\perp\mathbf{n}$ and g is a function of X or Y. We call such solutions the *travelling plane fields.* They coincide with what is traditionally called the *plane radiation field.*

Vector $\mathbf{n}$ may be treated as a linear operator in the space of cliffors, acting as the multiplication from the left. Due to the associativity of the Clifford product, this operator satisfies $\mathbf{n}^2 = 1$, since we have chosen $\mathbf{n}$ as the unit vector. Let us introduce the operator $P_\mathbf{n} = \frac{1}{2}(1+\mathbf{n})$ and compute its square:

$$P_\mathbf{n}^2 = \frac{1}{2}(1+\mathbf{n})\frac{1}{2}(1+\mathbf{n}) = \frac{1}{4}(1+2\mathbf{n}+\mathbf{n}^2) = \frac{1}{4}(2+2\mathbf{n}) = \frac{1}{2}(1+\mathbf{n}) = P_\mathbf{n}\,.$$

The operator satisfying $P_\mathbf{n}^2 = P_\mathbf{n}$ is referred to as the *idempotent.* $P_\mathbf{n}$ can also be considered as the *projection operator*, by which we understand that for any cliffor Z, $P_\mathbf{n}Z$ is not changed under $P_\mathbf{n}$:

$$P_\mathbf{n}(P_\mathbf{n}Z) = P_\mathbf{n}Z\,.$$

Due to the identity $\mathbf{n}\,P_\mathbf{n} = P_\mathbf{n}$ (easy to check), we ascertain that this operator projects onto a subspace of *eigencliffors* of the operator $\mathbf{n}$, corresponding to the eigenvalue $+1$:

$$\mathbf{n}\,(P_\mathbf{n}Z) = P_\mathbf{n}Z\,.$$

Similarly, $P_{-\mathbf{n}} = \frac{1}{2}(1-\mathbf{n})$ is the projection operator onto eigencliffors of $\mathbf{n}$, corresponding to the eigenvalue -1. It follows from $P_\mathbf{n} + P_{-\mathbf{n}} = 1$ and

$P_{\mathbf{n}} P_{-\mathbf{n}} = 0$ that any cliffor can be uniquely decomposed into the sum of two eigencliffors of n, corresponding to the eigenvalues ± 1, respectively

$$Z = P_{\mathbf{n}} Z + P_{-\mathbf{n}} Z = Z_{+} + Z_{-} \ .$$

When an electromagnetic cliffor $f = \mathbf{e} + \widehat{\mathbf{b}}$ satisfies (11), then $\mathbf{n} f$ is also an electromagnetic cliffor, that is, $\mathbf{n} f$ contains only the vector and volutor parts

$$\mathbf{n} f = \mathbf{n} \wedge \mathbf{e} + \mathbf{n} \cdot \widehat{\mathbf{b}} \ .$$

Thus we state that the operators $P_{\mathbf{n}}$ and $P_{-\mathbf{n}}$, when acting on the electromagnetic cliffor f satisfying (11), give new electromagnetic cliffors

$$f_{+} = P_{\mathbf{n}} f \quad \text{and} \quad f_{-} = P_{-\mathbf{n}} f \ , \tag{21}$$

which satisfy

$$\mathbf{n} f_{+} = f_{+} \ , \qquad \mathbf{n} f_{-} = -f_{-} \ .$$

Since f's of the form (20) are the eigencliffors of $\mathbf{n}$, we arrive at the following observation: the travelling plane fields are the eigencliffors of $P_{\mathbf{n}}$ or $P_{-\mathbf{n}}$ to the eigenvalue $+1$ and the eigencliffors of $\mathbf{n}$ to the eigenvalues ± 1, respectively.

The reader may check (see Problem 1) that the reverse is also true: if the electromagnetic cliffor (9) is the eigencliffor of $\mathbf{n}$, it describes the travelling field and, in this case, the energy propagates with the maximal velocity.

When the eigenequation $\mathbf{n} f = \pm f$ or, equivalently, $\mathbf{n} N = \pm N$ is satisfied, then the product gN may be simplified, since the vector $\mathbf{n}$ standing in (6) can be replaced by the numbers ± 1, respectively. In this case, one may write $g = (g_1 \pm g_2) + (g_4 \pm g_3) I$ instead of (6) and, therefore, put $\psi = 0$ in expression (14). This observation, along with (18), simplifies the energy-momentum cliffor (15):

$$\frac{1}{2} f f^{\dagger} = \frac{1}{2} |\, g \,|^2 |\, N \,|^2 \pm \mathbf{n} |\, g \,|^2 |\, \mathbf{c} \,|^2 \ .$$

We have also $|\, \mathbf{c} \,|^2 = \frac{1}{2} |\, N \,|^2$ in this case, hence

$$\frac{1}{2} f f^{\dagger} = \frac{1}{2} (1 \pm \mathbf{n}) |\, g \,|^2 |\, N \,|^2 = P_{\pm \mathbf{n}} |\, g \,|^2 |\, N \,|^2 \ .$$

Equations (16) should now be replaced by

$$w = \frac{1}{2} |\, g \,|^2 |\, N \,|^2 \ , \tag{22a}$$

$$\mathbf{S} = \pm\frac{1}{2}\mathbf{n}\,u|\,g\,|^2|\,N\,|^2 = \pm\mathbf{n}\,uw\,. \tag{22b}$$

Consequently, we propose the following definition: the plane electromagnetic field f changing along the straight line of vector $\mathbf{n}$ (this means $\mathbf{n}\,f = -f\,\mathbf{n}$) is *travelling in the direction* $+\mathbf{n}$ if $\mathbf{n}\,f = f$, and is *travelling in the direction* $-\mathbf{n}$ if $\mathbf{n}\,f = -f$. Thus our general solutions (9) are superpositions of two fields (21) travelling in opposite directions. Due to (22), the energy-momentum cliffor for the travelling field is

$$\frac{1}{2}f_\pm f_\pm^\dagger = (1 \pm \mathbf{n})w\,. \tag{23}$$

Lemma 1. A plane electromagnetic field f is the travelling field if, and only if, $f^2 = 0$.

Proof. The necessary condition. Let f be the plane field travelling along $\mathbf{n}$, that is, $\mathbf{n}\,f = \lambda f$ where $\lambda = \pm 1$. Take first $\lambda = 1$:

$$f^2 = ff = f\,\mathbf{n}\,f = -\mathbf{n}\,ff = -ff = -f^2 \quad \text{hence } f^2 = 0\,.$$

Take now $\lambda = -1$:

$$f^2 = f(-\mathbf{n}\,f) = -f\,\mathbf{n}\,f = \mathbf{n}\,ff = -f^2\,,$$

hence again $f^2 = 0$.

The sufficient condition. Let $f^2 = 0$. This means

$$(\mathbf{e} + \widehat{\mathbf{b}})\,(\mathbf{e} + \widehat{\mathbf{b}}) = \mathbf{e}^2 + \mathbf{e}\,\widehat{\mathbf{b}} + \widehat{\mathbf{b}}\,\mathbf{e} + \widehat{\mathbf{b}}^2 = 0\,.$$

Using (0.71) and (0.79),

$$|\,\mathbf{e}\,|^2 - |\,\widehat{\mathbf{b}}\,|^2 + 2\,\mathbf{e} \wedge \widehat{\mathbf{b}} = 0\,.$$

Equating separately the scalar and trivector parts to zero gives $|\,\mathbf{e}\,| = |\,\widehat{\mathbf{b}}\,|$ and $\widehat{\mathbf{b}} \parallel \mathbf{e}$. The latter condition, along with (11), yields $\widehat{\mathbf{b}} = \lambda\,\mathbf{n} \wedge \mathbf{e}$ and the former yields $\lambda = \pm 1$, so $\widehat{\mathbf{b}} = \pm\mathbf{n} \wedge \mathbf{e} = \pm\mathbf{n}\,\mathbf{e}$. This implies $\mathbf{n}\,f = \mathbf{n}\,\mathbf{e} + \mathbf{n}\,\widehat{\mathbf{b}} = \mathbf{n}\,\mathbf{e} \pm \mathbf{n}^2\mathbf{e} = \pm\widehat{\mathbf{b}} \pm \mathbf{e} = \pm f$, so f is the travelling field in the direction $\pm\mathbf{n}$. ■

The condition $|\,\mathbf{e}\,| = |\,\widehat{\mathbf{b}}\,|$ obtained in the course of the proof can be interpreted as the *synchronicity* of the electric and magnetic field changes. On the other hand, the condition $\widehat{\mathbf{b}} = \pm\mathbf{n} \wedge \mathbf{e}$ means that $\widehat{\mathbf{b}}$ and $\mathbf{e}$ are coplanar quantities. In the traditional language, one writes this as $\mathbf{b} = \pm\mathbf{n} \times \mathbf{e}$.

We recognize this as the *perpendicularity* condition for electromagnetic waves. Now we see that the subclass of plane electromagnetic fields satisfying the perpendicularity is the subclass of travelling fields.

It is interesting to know how the energy-momentum cliffor $\frac{1}{2}ff^{\dagger}$ of the field $f = f_+ + f_- = P_{\mathbf{n}}\,f + P_{-\mathbf{n}}\,f$ can be expressed by the cliffors (23). The following calculations lead to this end. Due to (10), we have $P_{\pm\mathbf{n}}\,f = fP_{\mp\mathbf{n}}$, whence

$$f_+f_-^{\dagger} = (P_{\mathbf{n}}\,f)\,(P_{-\mathbf{n}}\,f)^{\dagger} = fP_{-\mathbf{n}}(fP_{\mathbf{n}})^{\dagger} = fP_{-\mathbf{n}}\,P_{\mathbf{n}}\,f^{\dagger} = 0\,,$$

because $P_{\mathbf{n}}\,P_{-\mathbf{n}} = 0$. Similarly $f_-f_+^{\dagger} = 0$, therefore

$$ff^{\dagger} = (f_+ + f_-)(f_+ + f_-)^{\dagger} = f_+f_+^{\dagger} + f_-f_-^{\dagger}\;.$$

In this manner, we arrive at the relation

$$\frac{1}{2}ff^{\dagger} = \frac{1}{2}f_+f_+^{\dagger} + \frac{1}{2}f_-f_-^{\dagger}\;, \tag{24}$$

stating that for the plane field, which is a superposition of two plane fields travelling in opposite directions, the energy density and the energy flux are sums of the respective quantities connected separately to the travelling fields. Since the constituent energy fluxes are opposite and, at the same time, the energy density is nonnegative, it is not strange that for the superposed field, the energy flux has its magnitude smaller than the product $uw = u(w_+ + w_-)$ and the velocity of the energy transport is not maximal.

If the Poynting vector $\mathbf{S}$ is zero for the field (9), we are allowed to call it the *standing field.* In this case, by dint of (24), the Poynting vectors $\mathbf{S}_{\pm}$ of the constituent travelling fields have equal magnitudes.

For the travelling fields of the form (20) also the constant N expressed by (19) is the eigencliffor of $\mathbf{n}$. This implies that

$$XN_{\pm} = [\omega\,t - k(\mathbf{n}\cdot\mathbf{r})\,\mathbf{n}\,]N_{\pm} = [\omega\,t \mp k(\mathbf{n}\cdot\mathbf{r})]N_{\pm}\;,$$

and

$$YN_{\pm} = [k(\mathbf{n}\cdot\mathbf{r}) - \omega\,t\,\mathbf{n}\,]N_{\pm} = [k(\mathbf{n}\cdot\mathbf{r}) \mp \omega\,t]N_{\pm}\;.$$

We now apply the identity $P_{\mathbf{n}} + P_{-\mathbf{n}} = 1$ to the product gN:

$$gN = (P_{\mathbf{n}} + P_{-\mathbf{n}})gN = g(P_{\mathbf{n}} + P_{-\mathbf{n}})N = gN_+ + gN_-\;.$$

Since $\mathbf{n}^m N_\pm = (\pm 1)^m N_\pm$ for any natural m, we can replace the vector $\mathbf{n}$ standing in X or Y in respective places in the expressions (9) by ± 1. We may, therefore, write

$$f(\mathbf{r},t) = g(X)N = g(\omega t - \mathbf{k}\cdot\mathbf{r})N_+ + g(\omega t + \mathbf{k}\cdot\mathbf{r})N_- , \tag{25}$$

and

$$f^*(\mathbf{r},t) = g(Y)N = g(\mathbf{k}\cdot\mathbf{r} - \omega t)N_+ + g(\mathbf{k}\cdot\mathbf{r} + \omega t)N_- . \tag{26}$$

Thus we see that the plane fields defined as the functions of the cliffor argument X or Y are reduced to the sums of two functions of the scalar arguments. This result allows us to extend our definition of plane electromagnetic fields to cases where g is not expressible by a power series. In such cases, formulae (25) and (26) may be regarded merely as definitions of $g(X)N$ and $g(Y)N$.

In the case of travelling fields, only one term is present in sums (25) and (26). We now see that the whole spacetime dependence of the electromagnetic field is blended into the single scalar argument $\pm(\omega t - \mathbf{k}\cdot\mathbf{r})$ or $\omega t + \mathbf{k}\cdot\mathbf{r}$ of the function g. Let us call it the *phase* of the travelling field. The space changes of the field can be cancelled by its time changes, so that the phase can be kept constant. This means physically that the planes of constant field move uniformly with the time flow. We state that the electromagnetic field *propagates* according to the equation $\omega t \mp \mathbf{k}\cdot\mathbf{r} = \text{const}$. One gets, after differentiation,

$$\mathbf{n}\cdot d\mathbf{r} = \pm\frac{\omega}{k}dt ,$$

hence the velocity normal to the plane of constant phase is

$$\mathbf{u} = \pm\frac{\omega}{k}\mathbf{n} = \pm\omega\,\mathbf{k}^{-1} . \tag{27}$$

This is called the *phase velocity*. We obtain $u = 1/\sqrt{\varepsilon\mu}$ from Lemma 1 of the previous section for the magnitude of this velocity. In this manner, we ascertain that the energy propagation velocity coincides with the phase velocity of the travelling fields.

It can be seen from the right-hand formulae (1) and (9) that the value of solution f^* at time $t = 0$ — we call this the *initial value* of f^* — is of the form

$$f^*(\mathbf{r},0) = g(\mathbf{k}\cdot\mathbf{r})N .$$

This observation allows us to state that, for any initial electromagnetic field function of the form $g(\mathbf{k}\cdot\mathbf{r})N$ with $\mathbf{n}N = -N\mathbf{n}$, one automatically gets the time-dependent solution by the substitution

$$\mathbf{k}\cdot\mathbf{r} \to Y = \mathbf{k}\cdot\mathbf{r} - \mathbf{k}\,u\,t . \tag{28}$$

We illustrate the procedure by the following example. Let the initial electromagnetic field be purely the magnetic field discussed in Sec. 1.6.5 and illustrated in Fig. 40. The vector $\mathbf{n}$ is now the unit vector perpendicular to the plates carrying the surface currents. Let us choose $k = 1$. If the origin of coordinates is chosen exactly half-way between the plates, the function g describing the initial field can be chosen as the *step function:*

$$g_a(x) = \begin{cases} 0 & \text{for} \quad x < -\frac{a}{2} \,, \\ 1 & \text{for} \quad -\frac{a}{2} \leq x \leq \frac{a}{2} \,, \\ 0 & \text{for} \quad x > \frac{a}{2} \,, \end{cases}$$

where a is the distance between the plates, and the field itself is described by the function $f^*(\mathbf{r},0) = g_a(\mathbf{n}\cdot\mathbf{r})\,\widehat{\mathbf{b}}_0$, where $\widehat{\mathbf{b}}_0 = \sqrt{\mu}\,\widehat{\mathbf{H}}_0$ for $\widehat{\mathbf{H}}_0$ — the initial magnetic field between the plates.

If the surface currents "confining" the magnetic field between the plates are suddenly switched off at time $t = 0$ (this is possible for superconducting plates at least as a thought experiment) the electromagnetic field should fulfil the homogeneous Maxwell equation — thus we may apply our previous observations which lead to solution (26):

$$f^*(\mathbf{r},t) = g_a(\mathbf{n}\cdot\mathbf{r} - u\,t)N_+ + g_a(\mathbf{n}\cdot\mathbf{r} + u\,t)N_- \;,$$

where $N_\pm = \frac{1}{2}\,(1 \pm \mathbf{n})\,\widehat{\mathbf{b}}_0$. In this manner, we obtain

$$f^*(\mathbf{r},t) = \frac{1}{2}\,g_a(\mathbf{n}\cdot\mathbf{r} - u\,t)\,(1+\mathbf{n})\,\widehat{\mathbf{b}}_0 + \frac{1}{2}\,g_a(\mathbf{n}\cdot\mathbf{r} + u\,t)\,(1-\mathbf{n})\,\widehat{\mathbf{b}}_0 \;.$$

The first term in this sum, $f_+^* = \frac{1}{2}\,g_a(\mathbf{n}\cdot\mathbf{r} - u\,t)\,(1+\mathbf{n})\,\widehat{\mathbf{b}}_0$, has a constant nonzero value for $-\frac{a}{2} + u\,t \leq \mathbf{n}\cdot\mathbf{r} \leq \frac{a}{2} + u\,t$ with the magnetic part $\widehat{\mathbf{b}}_+ = \frac{1}{2}\,\widehat{\mathbf{b}}_0$ and the electric part $\mathbf{e}_+ = \frac{1}{2}\,\mathbf{n}\,\widehat{\mathbf{b}}_0$. Therefore it is a step-like "bunch" of the electromagnetic field travelling in the direction $+\mathbf{n}$. The second term f_-^* is nonzero for $-\frac{a}{2} - u\,t \leq \mathbf{n}\cdot\mathbf{r} \leq \frac{a}{2} - u\,t$ with $\widehat{\mathbf{b}}_- = \frac{1}{2}\,\widehat{\mathbf{b}}_0$, $\mathbf{e}_- = -\frac{1}{2}\,\mathbf{n}\cdot\widehat{\mathbf{b}}_0$ and it describes a similar "bunch" moving in the direction $-\mathbf{n}$.

We see that the electromagnetic field splits into two step-like "bunches" travelling in opposite directions from the region of initial nonzero magnetic field. The two "bunches" contain both electric and magnetic parts of equal magnitudes $|\,\mathbf{e}_\pm\,| = |\,\widehat{\mathbf{b}}_\pm\,|$, but in the regions where the "bunches" overlap, the electric part of the field cancels out. Therefore, for the times $t > \frac{a}{u}$, the electric field is nonzero in the same regions where the magnetic field is nonzero.

It is visible from the left-hand formulae (1) and (9) that the value of the solution f on the distinguished plane $\mathbf{n}\cdot\mathbf{r}=0$ — we call it the *boundary value* of f — is of the form

$$f(0,t)=g(\omega t)N\,.$$

This shows that a kind of boundary value problem can be solved. When the field is given on the plane $\mathbf{n}\cdot\mathbf{r}=0$ as the function of time only, of the form $g(\omega t)N$ for $\mathbf{n}N=-N\mathbf{n}$, then the space-time dependent solution is obtained automatically by the substitution

$$\omega t\to X=\omega t-\mathbf{n}\,(\mathbf{k}\cdot\mathbf{r})\,. \tag{29}$$

Let us illustrate the procedure by another example. Let the boundary electromagnetic field be exponentially decaying: $f(0,t)=e^{-\omega t}N$, then

$$f(\mathbf{r},t)=e^{-X}N=e^{-\omega t+n(\mathbf{k}\cdot\mathbf{r})}N\,. \tag{30}$$

For $N=\mathbf{c}$, by virtue of (0.97), the division into the electric and magnetic parts yields

$$\mathbf{e}\,(\mathbf{r},t)=e^{-\omega t}\,\mathbf{c}\,\cosh\mathbf{k}\cdot\mathbf{r}\,,$$
$$\widehat{\mathbf{b}}\,(\mathbf{r},t)=e^{-\omega t}\,\mathbf{n}\,\mathbf{c}\,\sinh\mathbf{k}\cdot\mathbf{r}\,.$$

A similar solution can be obtained if $N=\widehat{\mathbf{C}}$, that is, if only the magnetic field exists on the boundary plane. The energy-momentum cliffor for (30) is

$$\frac{1}{2}ff^{\dagger}=\frac{1}{2}e^{-2\omega t}\,e^{2\mathbf{n}\,(\mathbf{k}\cdot\mathbf{r})}\,|\,N\,|^{2}\,,$$

from which the Poynting vector

$$\mathbf{S}=\frac{u}{2}e^{-2\omega t}\,\sinh(2\mathbf{k}\cdot\mathbf{r})|\,N\,|^{2}\,\mathbf{n}$$

is obtained. At all times, the energy flows outwards from the distinguished plane $\mathbf{n}\cdot\mathbf{r}=0$. This particular solution is physically non-interesting because it grows too fast for $r\to\infty$.

Let us, nevertheless, apply decomposition (25) to (30):

$$f(\mathbf{r},t)=e^{-\omega t+\mathbf{k}\cdot\mathbf{r}}\frac{1}{2}(1+\mathbf{n})N+e^{-\omega t-\mathbf{k}\cdot\mathbf{r}}\frac{1}{2}(1-\mathbf{n})N\,.$$

This is the superposition of two exponentially sloped fields travelling with velocities $\pm u\,\mathbf{n}$ in the directions of their spatial growth. This explains why the field decreases at the distinguished plane.

§3. Reflection and Refraction

When considering the behaviour of plane electromagnetic fields at an interface between two media, it is convenient to consider the travelling fields in order to deal with the energy incoming only from one side of the interface.

Let the plane interface separate two media with different constants ε and μ. We choose the Cartesian coordinates such that the $\mathbf{e}_1\mathbf{e}_2$ - plane coincides with the interface. We denote by subscript 1 the medium for $x_3 > 0$ and the other one by 2. Let two plane electromagnetic fields exist in the half-space $x_3 > 0$:

the *incident field*

$$f(\mathbf{r}, t) = g(\omega\, t - \mathbf{k}\cdot\mathbf{r})\,(1+\mathbf{n})\,\mathbf{c}\;, \tag{31a}$$

the *reflected field*

$$f^{(r)}(\mathbf{r}, t) = g(\omega\, t - \mathbf{k}^{(r)}\cdot\mathbf{r})\,(1+\mathbf{n}^{(r)})\,\mathbf{c}^{(r)}\;, \tag{31b}$$

whereas the *penetrating field*,

$$f^{(p)}(\mathbf{r}, t) = g(\omega\, t - \mathbf{k}^{(p)}\cdot\mathbf{r})\,(1+\mathbf{n}^{(p)})\,\mathbf{c}^{(p)}\;, \tag{31c}$$

exists for $x_3 < 0$, where g is a function containing only the scalar and trivector parts — it is the same for all three fields. We assume here various vectors $\mathbf{k}, \mathbf{k}^{(r)}, \mathbf{k}^{(p)}$ differing by their directions $\mathbf{n}, \mathbf{n}^{(r)}, \mathbf{n}^{(p)}$ and magnitudes. Also other features described by the initial vectors $\mathbf{c}, \mathbf{c}^{(r)}, \mathbf{c}^{(p)}$ may be different. An important assumption is that the time coefficient ω is the same for all plane fields, because the rhythm of time changes, enforced by the incident field, should be the same for the secondary fields $f^{(r)}$ and $f^{(p)}$.

The same premise implies that the spatial variation of all fields must be compatible at $x_3 = 0$. Consequently, we must have the arguments of g-function all equal at $x_3 = 0$:

$$\mathbf{k}\cdot\mathbf{r} = \mathbf{k}^{(r)}\cdot\mathbf{r} = \mathbf{k}^{(p)}\cdot\mathbf{r} \;\text{ for }\; x_3 = 0\;. \tag{32}$$

It follows from this that the equalities for the coordinates

$$k_1 = k_1^{(r)} = k_1^{(p)}\,, \qquad k_2 = k_2^{(r)} = k_2^{(p)}\;, \tag{33}$$

have to be satisfied, which means that all vectors $\mathbf{k}, \mathbf{k}^{(r)}, \mathbf{k}^{(p)}$ have equal projections on the interface — they may differ only in the third component. The plane determined by $\mathbf{k}$ and the normal vector $\mathbf{e}_3$ to the interface is called the *plane of incidence*. Equations (33) show that the vectors $\mathbf{k}^{(r)}$ and $\mathbf{k}^{(p)}$ also lie in this plane. We choose the basis vector $\mathbf{e}_1$ such that $k_2 = k_2^{(r)} = k_2^{(p)} = 0$. Then the plane of incidence coincides with that of $\mathbf{e}_1\mathbf{e}_3$.

We know from Lemma 1 in Sec. 1 that the relations

$$\frac{\omega}{k} = \frac{1}{\sqrt{\varepsilon_1\mu_1}} = u_1\,, \quad \frac{\omega}{k^{(r)}} = \frac{1}{\sqrt{\varepsilon_1\mu_1}} = u_1\,, \quad \frac{\omega}{k^{(p)}} = \frac{1}{\sqrt{\varepsilon_2\mu_2}} = u_2\,, \tag{34}$$

are fulfilled. They imply that the angle of incidence θ_1 is equal to the angle of reflexion θ_1' (see Fig. 89). Indeed, the magnitudes of $\mathbf{k}$ and $\mathbf{k}^{(r)}$ are equal due to (34); their first components are equal due to (33), and the third components must be opposite in order that the reflected field propagates back to medium 1. This equality of two angles is known as the *reflection law* for the light.

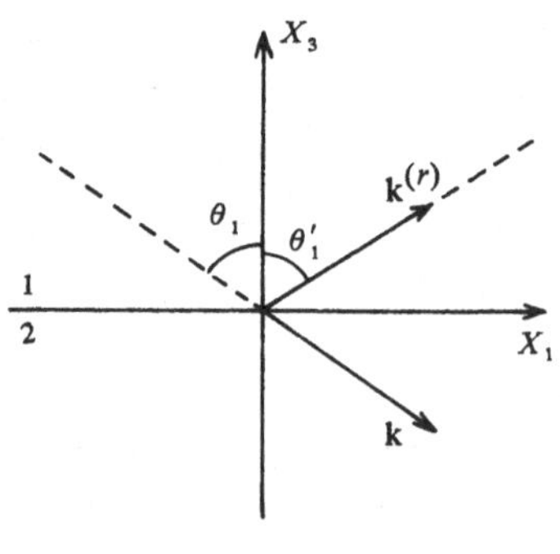

Fig. 89.

Vectors $\mathbf{k}$ and $\mathbf{k}^{(p)}$ have equal first components and the same sign of the third components (in order that the field $f^{(p)}$ moves into medium 2) and, by virtue of (34), have different magnitudes, hence the angle of penetration θ_2 must be different from θ_1 (see Fig. 90). This phenomenon is known as the *refraction* for the light and θ_2 is also called the *angle of refraction.* Of course $k_1/k = \sin\theta_1, k_1^{(p)}/k^{(p)} = \sin\theta_2$ which, due to (33), implies

$$\frac{k^{(p)}}{k} = \frac{\sin\theta_1}{\sin\theta_2}\,.$$

Relations (34) now yield

$$\frac{u_1}{u_2} = \frac{\sin\theta_1}{\sin\theta_2}\,. \tag{35}$$

We ascertain that the ratio of sines of the angles of incidence and of refraction is independent of the angle of incidence. It depends only on the properties of the two media. This fact is called *Snell's law* in the case of light. Ratio (35), denoted $n_{12} = u_1/u_2$, is called the *refraction index* of medium 2 with respect to medium 1. In this manner, we obtain

$$n_{12} = \sqrt{\frac{\varepsilon_2\mu_2}{\varepsilon_1\mu_1}} . \tag{36}$$

The penetrating field is also called the *refracted field.*

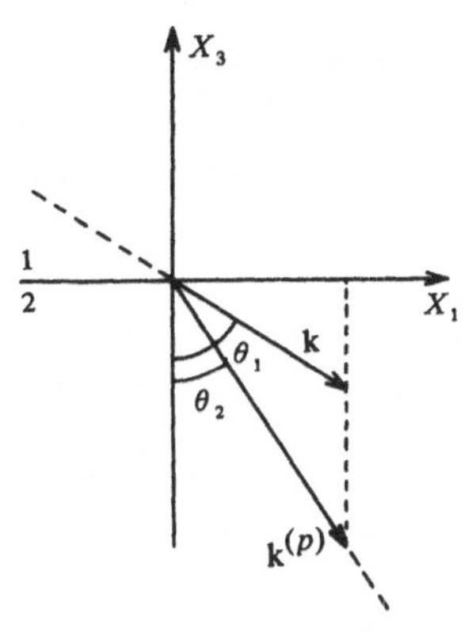

Fig. 90.

Up to now, only the condition of identical rhythms of the three fields has been used for finding the relations between the three quantities $\mathbf{k}, \mathbf{k}^{(r)}$ and $\mathbf{k}^{(p)}$. Their consequences (i.e. the reflection law and Snell's law) are called the *kinematic properties* of the phenomenon.

We shall now invoke the boundary conditions of §1.4. They are easy to apply if the function g is purely a scalar function, so we assume that g is a scalar.[1)] In this case, the separation of cliffor f into the vector and bivector parts depends only on the constant $(1+\mathbf{n})\mathbf{c}$ in (31a) and on respective constants in (31b,c).

In terms of fields (31), the continuity of the tangent components of $\mathbf{E}$ and $\hat{\mathbf{B}}$ leads to the conditions

$$\frac{1}{\sqrt{\varepsilon_1}}(\mathbf{c} + \mathbf{c}^{(r)}) \wedge \mathbf{e}_3 = \frac{1}{\sqrt{\varepsilon_2}}\mathbf{c}^{(p)} \wedge \mathbf{e}_3 , \tag{37a}$$

[1)] If this is not the case, we redefine g according to the prescription mentioned in §1 and conclude that g is a scalar at some point on the interface. In fact, there are more such points — they form a line perpendicular to the plane of incidence. Afterwards, we consider the boundary conditions at this set of points.

$$\sqrt{\mu_1}\,(\mathbf{n}\,\mathbf{c}+\mathbf{n}^{(r)}\,\mathbf{c}^{(r)})\wedge\mathbf{e}_3=\sqrt{\mu_2}\,(\mathbf{n}^{(p)}\,\mathbf{c}^{(p)})\wedge\mathbf{e}_3\ , \tag{37b}$$

and the continuity of the normal components of $\mathbf{D}$ and $\hat{\mathbf{H}}$ leads to

$$\sqrt{\varepsilon_1}\,(\mathbf{c}+\mathbf{c}^{(r)})\cdot\mathbf{e}_3=\sqrt{\varepsilon_2}\,\mathbf{c}^{(p)}\cdot\mathbf{e}_3\ , \tag{37c}$$

$$\frac{1}{\sqrt{\mu_1}}\,(\mathbf{n}\,\mathbf{c}+\mathbf{n}^{(r)}\,\mathbf{c}^{(r)})\cdot\mathbf{e}_3=\frac{1}{\sqrt{\mu_2}}\,(\mathbf{n}^{(p)}\,\mathbf{c}^{(p)})\cdot\mathbf{e}_3\ . \tag{37d}$$

It is convenient to consider two separate situations: one in which the incident field, with the electric part as a vector and the magnetic part as a volutor, is perpendicular to the plane of incidence (see Fig. 91) and the other in which the incident field is parallel to the plane of incidence (see Fig. 92).

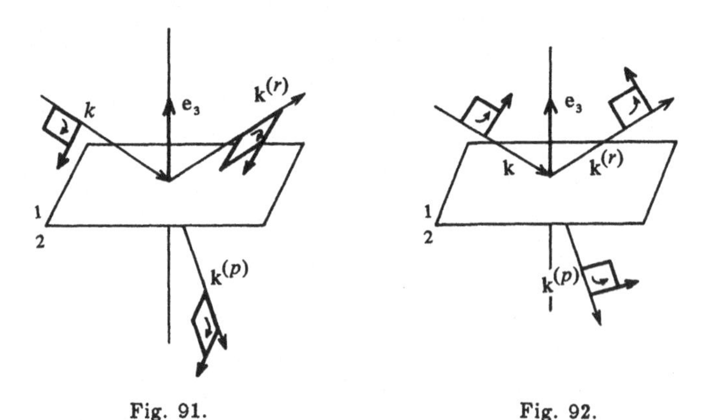

Fig. 91. Fig. 92.

In the first situation, the electric fields represented in (37) by vectors $\mathbf{c},\mathbf{c}^{(r)},\mathbf{c}^{(p)}$ are all perpendicular to $\mathbf{e}_3$, hence the outer product in (37a) can be replaced by the Clifford product and the common factor $\mathbf{e}_3$ can be divided out:

$$\frac{1}{\sqrt{\varepsilon_1}}\,(\mathbf{c}+\mathbf{c}^{(r)})=\frac{1}{\sqrt{\varepsilon_2}}\,\mathbf{c}^{(p)}\ . \tag{38}$$

For the same reason, it is implied that both sides of Eq. (37c) are zero. Equation (37d) leads to

$$\frac{1}{\sqrt{\mu_1}}\left[\mathbf{c}\,(\mathbf{n}\cdot\mathbf{e}_3)+\mathbf{c}^{(r)}(\mathbf{n}^{(r)}\cdot\mathbf{e}_3)\right]=\frac{1}{\sqrt{\mu_2}}\,\mathbf{c}^{(p)}(\mathbf{n}^{(p)}\cdot\mathbf{e}_3)\ ,$$

i.e.,

$$\frac{1}{\sqrt{\mu_1}}\,(-\mathbf{c}\,\cos\theta_1+\mathbf{c}^{(r)}\cos\theta_1)=-\frac{1}{\sqrt{\mu_2}}\,\mathbf{c}^{(p)}\cos\theta_2\ . \tag{39}$$

Equation (37b), after using Snell's law, gives again condition (38). Thus we are left with the system of the two Eqs. (38,39). One may solve it for two unknowns $\mathbf{c}^{(r)}$ and $\mathbf{c}^{(p)}$

$$\left.\begin{aligned} \mathbf{c}^{(r)} &= \frac{\sqrt{\mu_2/\mu_1}\cos\theta_1 - \sqrt{\varepsilon_2/\varepsilon_1}\cos\theta_2}{\sqrt{\mu_2/\mu_1}\cos\theta_1 + \sqrt{\varepsilon_2/\varepsilon_1}\cos\theta_2}\,\mathbf{c}\,, \\ \mathbf{c}^{(p)} &= \frac{2n_{12}\cos\theta_1}{\sqrt{\mu_2/\mu_1}\cos\theta_1 + \sqrt{\varepsilon_2/\varepsilon_1}\cos\theta_2}\,\mathbf{c}\,, \end{aligned}\right\} \tag{40}$$

where notation (36) is employed.

In the second situation, i.e. when the incident field is parallel to the plane of incidence, the magnetic fields represented by the volutors $\widehat{\mathbf{C}} = \mathbf{n}\,\mathbf{c}$, $\widehat{\mathbf{C}}^{(r)} = \mathbf{n}^{(r)}\,\mathbf{c}^{(r)}$ and $\widehat{\mathbf{C}}^{(p)} = \mathbf{n}^{(p)}\,\mathbf{c}^{(p)}$ are parallel to $\mathbf{e}_3$, hence both sides of (37b) are zero, whereas the inner products in (37d) can be replaced by the Clifford product and the common factor $\mathbf{e}_3$ can be divided out:

$$\frac{1}{\sqrt{\mu_1}}(\widehat{\mathbf{C}} + \widehat{\mathbf{C}}^{(r)}) = \frac{1}{\sqrt{\mu_2}}\widehat{\mathbf{C}}^{(p)}\,. \tag{41}$$

The two other Eqs. (37a,c) lead to the same condition

$$\frac{1}{\sqrt{\varepsilon_1}}(\widehat{\mathbf{C}} - \widehat{\mathbf{C}}^{(r)})\cos\theta_1 = \frac{1}{\sqrt{\varepsilon_2}}\widehat{\mathbf{C}}^{(p)}\cos\theta_2\,. \tag{42}$$

The system of Eqs. (41,42) can be solved for $\widehat{\mathbf{C}}^{(r)}$ and $\widehat{\mathbf{C}}^{(p)}$ as follows:

$$\left.\begin{aligned} \widehat{\mathbf{C}}^{(r)} &= \frac{\sqrt{\varepsilon_2/\varepsilon_1}\cos\theta_1 - \sqrt{\mu_2/\mu_1}\cos\theta_2}{\sqrt{\varepsilon_2/\varepsilon_1}\cos\theta_1 + \sqrt{\mu_2/\mu_1}\cos\theta_2}\,\widehat{\mathbf{C}}\,, \\ \widehat{\mathbf{C}}^{(p)} &= \frac{2n_{12}\cos\theta_1}{\sqrt{\varepsilon_2/\varepsilon_1}\cos\theta_1 + \sqrt{\mu_2/\mu_1}\cos\theta_2}\,\widehat{\mathbf{C}}\,. \end{aligned}\right\} \tag{43}$$

Relations (40) and (43) are known in optics as *Fresnel formulae*.

For a *normal incidence*, that is for $\theta_1 = \theta_2 = 0$, both situations coincide and can be expressed in terms of the electric parts

$$\mathbf{c}^{(r)} = \frac{\sqrt{\mu_2/\mu_1} - \sqrt{\varepsilon_2/\varepsilon_1}}{\sqrt{\mu_2/\mu_1} + \sqrt{\varepsilon_2/\varepsilon_1}}\,\mathbf{c}\,. \tag{44a}$$

$$\mathbf{c}^{(p)} = \frac{2n_{12}}{\sqrt{\mu_2/\mu_1} + \sqrt{\varepsilon_2/\varepsilon_1}}\,\mathbf{c}\,. \tag{44b}$$

In this case, the field $f + f^{(r)}$ existing in the first medium is a superposition of two plane fields travelling in opposite directions, so we can write it as the sum $f_+ + f_-$:

$$f + f^{(r)} = g(\omega t - \mathbf{k}\cdot\mathbf{r})\,(1+\mathbf{n})\,\mathbf{c} + g(\omega t + \mathbf{k}\cdot\mathbf{r})\,(1-\mathbf{n})\,\mathbf{c}^{(r)}\ ,$$

where $\mathbf{n} = -\mathbf{e}_3$ and $\mathbf{k}^{(r)} = -\mathbf{k}$. We insert vector $\mathbf{n}$ inside the argument of functions g:

$$f + f^{(r)} = g[\omega t - \mathbf{n}\,(\mathbf{k}\cdot\mathbf{r})]\,(1+\mathbf{n})\,\mathbf{c} + g[\omega t - \mathbf{n}\,(\mathbf{k}\cdot\mathbf{r})]\,(1-\mathbf{n})\,\mathbf{c}^{(r)}\ ,$$

and substitute (44a),

$$f + f^{(r)} = g[\omega t - \mathbf{n}\,(\mathbf{k}\cdot\mathbf{r})]\left[(1+\mathbf{n})\,\mathbf{c} + \frac{\sqrt{\mu_2/\mu_1} - \sqrt{\varepsilon_2/\varepsilon_1}}{\sqrt{\mu_2/\mu_1} + \sqrt{\varepsilon_2/\varepsilon_1}}\,(1-\mathbf{n})\,\mathbf{c}\right]\ ,$$

$$f + f^{(r)} = g[\omega t - \mathbf{n}\,(\mathbf{k}\cdot\mathbf{r})]\,2\,\frac{\sqrt{\mu_2/\mu_1} + \sqrt{\varepsilon_2/\varepsilon_1}\,\mathbf{n}}{\sqrt{\mu_2/\mu_1} + \sqrt{\varepsilon_2/\varepsilon_1}}\,\mathbf{c}\ . \qquad (45)$$

In this manner, we have obtained the specific example of the plane electromagnetic field of the form (9): $f + f^{(r)} = g(X)N$, where

$$N = 2\frac{\sqrt{\mu_2/\mu_1} + \sqrt{\varepsilon_2/\varepsilon_1}\,\mathbf{n}}{\sqrt{\mu_2/\mu_1} + \sqrt{\varepsilon_2/\varepsilon_1}}\,\mathbf{c}$$

is not the eigencliffor of $\mathbf{n}$. We see that, by the perpendicular reflection, it is possible to obtain only the left-hand form of plane solutions (9).

By virtue of (22b), the densities of the energy flux for the incident, reflected and penetrating fields are, respectively,

$$\mathbf{S} = \mathbf{n}\,u_1|\,g\,|^2|\,\mathbf{c}\,|^2\ ,$$
$$\mathbf{S}^{(r)} = -\mathbf{n}\,u_1|\,g\,|^2|\,\mathbf{c}^{(r)}\,|^2\ ,$$
$$\mathbf{S}^{(p)} = \mathbf{n}\,u_2|\,g\,|^2|\,\mathbf{c}^{(p)}\,|^2\ .$$

The ratio $T = S^{(p)}/S$ is called the *transmission coefficient* of the interface. We obtain, with the aid of (44b),

$$T = \frac{u_2|\,\mathbf{c}^{(p)}\,|^2}{u_1|\,\mathbf{c}\,|^2} = \frac{4n_{12}}{(\sqrt{\mu_2/\mu_1} + \sqrt{\varepsilon_2/\varepsilon_1})^2}\ . \qquad (46)$$

We see that it is independent of the incident field — it is the characteristics of the two media. It is easy to check that $T \le 1$ and $T = 1$ if, and only if,

$\varepsilon_2/\varepsilon_1 = \mu_2/\mu_1$. The ratio $R = S^{(r)}/S$ is called the *reflection coefficient* of the interface. By dint of (44a), we obtain

$$R = \frac{|\mathbf{c}^{(r)}|^2}{|\mathbf{c}|^2} = \frac{(\sqrt{\mu_2/\mu_1} - \sqrt{\varepsilon_2/\varepsilon_1})^2}{(\sqrt{\mu_2/\mu_1} + \sqrt{\varepsilon_2/\varepsilon_1})^2} \,. \tag{47}$$

One may easily check that $R + T = 1$, which means that the energy flux of the incident field divides exactly between the reflected and penetrating fields. Of course, for $\varepsilon_2/\varepsilon_1 = \mu_2/\mu_1$, the reflected field is absent.

§4. Plane Electromagnetic Waves

We turn now to the plane electromagnetic fields which are periodic functions of the time variable. As we know from the substitutions (28) and (29), this implies that the fields are also periodic in the space dependence. We shall call them *plane electromagnetic waves*. The magnitude $|\mathbf{S}|$ of the energy flux is called the *intensity* of the wave.

One can apply the two procedures of Sec. 2, namely choosing the initial-value field or the boundary-value field for finding the plane waves. If the initial-value field is of the form

$$f^*(\mathbf{r}, 0) = g(\mathbf{k} \cdot \mathbf{r})N$$

with a periodic function g, one obtains the solution in the form (26) which means that the electromagnetic field is the superposition of two waves of the same spatial "shape" g[1] travelling in opposite directions $\pm\mathbf{n}$. Of course, only one travelling wave is present if N is an eigencliffor of $\mathbf{n}$.

If the boundary-value field on the plane $\mathbf{n} \cdot \mathbf{r} = 0$ has the form

$$f(0, t) = g(\omega\, t)N$$

with the periodic function g, the solution (25) is obtained which is the superposition of two waves with the same "shape" of time changes[1] travelling in opposite directions $\pm\mathbf{n}$. Of course, only one travelling wave is present if $\mathbf{n}N = \pm N$.

In the case of scalar functions g, for $N = \mathbf{c} + \widehat{\mathbf{C}}$, the electric part of the solutions has the form $\mathbf{e}(\mathbf{r}, t) = g(\mathbf{r}, t)\,\mathbf{c}$. This means that the tip point of the electric vector in a fixed point $\mathbf{r}$ traces a linear segment parallel to $\mathbf{c}$ — in such a case, we say that the electromagnetic wave is *linearly polarized*.

[1] We should remark that the preservation of shape takes place only for nondispersive media, i.e. when ε and μ do not depend on the frequency.

We now restrict ourselves to the *harmonic waves*, that is, to functions which can be expressed by trigonometric ones (sine and cosine) of the same period. Let the initial field be given by the sine function

$$f^*(\mathbf{r},0) = (\sin \mathbf{k}\cdot\mathbf{r})N ,$$

where $\mathbf{n}\,N = -N\,\mathbf{n}$. Then

$$f^*(\mathbf{r},t) = (\sin Y)N = \sin(\mathbf{k}\cdot\mathbf{r} - \mathbf{n}\,\omega\, t)N , \tag{48a}$$

and, due to Problem 0.24,

$$f^*(\mathbf{r},t) = (\sin \mathbf{k}\cdot\mathbf{r}\cos \mathbf{n}\,\omega\, t - \sin \mathbf{n}\,\omega\, t \cos \mathbf{k}\cdot\mathbf{r})N ,$$

or

$$f^*(\mathbf{r},t) = (\sin \mathbf{k}\cdot\mathbf{r}\cos \omega\, t - \mathbf{n}\sin \omega\, t \cos \mathbf{k}\cdot\mathbf{r})N . \tag{48b}$$

Equation (26) gives the following representation in terms of the travelling waves:

$$f^*(\mathbf{r},t) = \sin(\mathbf{k}\cdot\mathbf{r} - \omega\, t)N_+ + \sin(\mathbf{k}\cdot\mathbf{r} + \omega\, t)N_- .$$

Of course, only one term is present if N is the eigencliffor of $\mathbf{n}$. The parameters $\mathbf{k}$ and ω obtain their names for the harmonic waves: $\mathbf{k}$ is the *wave vector*, its magnitude is the *wave number*, ω is the *angular* or *circular frequency*. They are connected with the *period* $T = 2\pi/\omega$ of the wave and with the *wave length* $\lambda = 2\pi/|\,\mathbf{k}\,|$.

For $N = \mathbf{c} + \widehat{\mathbf{C}}$, we extract from (48b) the vector and bivector parts

$$\mathbf{e}\,(\mathbf{r},t) = \mathbf{c}\cos\omega\, t \sin \mathbf{k}\cdot\mathbf{r} - \mathbf{n}\,\widehat{\mathbf{C}}\sin\omega\, t\cos\mathbf{k}\cdot\mathbf{r} ,$$
$$\widehat{\mathbf{b}}\,(\mathbf{r},t) = \widehat{\mathbf{C}}\cos\omega\, t \sin \mathbf{k}\cdot\mathbf{r} - \mathbf{n}\,\mathbf{c}\sin\omega\, t\cos\mathbf{k}\cdot\mathbf{r} .$$

The time dependence of the electric field is the superposition of two harmonic oscillations $\mathbf{a}\cos\omega\, t + \mathbf{a}'\sin\omega\, t$ (where $\mathbf{a} = \mathbf{c}\sin\mathbf{k}\cdot\mathbf{r}, \mathbf{a}' = \mathbf{n}\,\widehat{\mathbf{C}}\cos\mathbf{k}\cdot\mathbf{r}$), hence the tip of the electric vector traces an ellipse when the vectors $\mathbf{c}$ and $\mathbf{n}\,\widehat{\mathbf{C}}$ are not parallel. Only in the points $\mathbf{k}\cdot\mathbf{r} = m\pi/2$, for integer m, one term only is present and the ellipse reduces to a linear segment. Of course the oscillations of $\mathbf{e}$ are linear everywhere when N is the eigencliffor of $\mathbf{n}$, since $\mathbf{n}\,\widehat{\mathbf{C}}$ and $\mathbf{c}$ are parallel in such a case.

Now let the boundary field be given on the plane $\mathbf{n}\cdot\mathbf{r} = 0$ by the sine function: $f(0,t) = (\sin\omega\, t)N$ with $\mathbf{n}\,N = -N\,\mathbf{n}$. Then

$$f(\mathbf{r},t) = (\sin X)N \tag{49a}$$

and, due to Problem 0.24,

$$f(\mathbf{r},t) = (\sin\omega t \cos\mathbf{k}\cdot\mathbf{r} - \mathbf{n}\cos\omega t \sin\mathbf{k}\cdot\mathbf{r})N \ . \tag{49b}$$

The decomposition (25) yields

$$f(\mathbf{r},t) = \sin(\omega t - \mathbf{k}\cdot\mathbf{r})N_+ + \sin(\omega t + \mathbf{k}\cdot\mathbf{r})N_- \ .$$

One may similarly discuss the time dependence of the electric vector — the observations are the same as above.

If $N = N_+ = \mathbf{c} + \mathbf{n}\,\mathbf{c}$, the solutions (48) and (49) can be written in the same form:

$$f(\mathbf{r},t) = \sin(\mathbf{k}\cdot\mathbf{r} - \omega t)\,(\mathbf{c} + \mathbf{n}\,\mathbf{c}) \ . \tag{50}$$

This is the *travelling plane harmonic wave* with the *linear polarization.* With time flow, the electric vector $\mathbf{e}$ at a fixed point changes only its magnitude and orientation, the same is true for the magnetic volutor $\hat{\mathbf{b}}$. As concerns the space changes, the tip points of $\mathbf{e}$ at a fixed time draw the sinusoid along a straight line parallel to $\mathbf{n}$. The volutors $\hat{\mathbf{b}}$ lie in the plane determined by $\mathbf{n}$ and $\mathbf{c}$, which we show in Fig. 93. It would be natural to call it the polarization plane. This name, however, had already been used in the last century for the plane in which the magnetic pseudovector oscillates, that is, for the plane perpendicular to the one distinguished in Fig. 93. Afterwards, another name was coined for the plane of $\mathbf{e}$ and $\hat{\mathbf{b}}$, namely the *plane of light oscillations.*

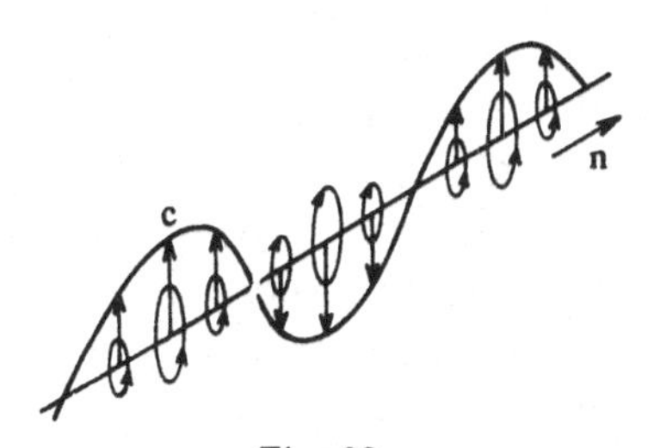

Fig. 93.

In the traditional presentations, using only vectors and pseudovectors, the magnetic field $\mathbf{H}$ for the plane travelling harmonic wave is perpendicular to $\mathbf{E}$ and the choice of the plane of oscillations for the electromagnetic field seems to be a matter of pure convention. Now, in the multivector language, both quantities $\mathbf{E}$ and $\hat{\mathbf{H}}$ oscillate in the same plane, so the discrimination of this plane to describe the polarized wave is less conventional. This particular plane is more important also because of physical reasons: All the forces exerted on test charges placed in the field of the wave act in this plane.

The superposition of the incident and reflected waves according to (45),

$$\sin[\omega t - n(\mathbf{k}\cdot\mathbf{r})]\,\frac{\sqrt{\mu_2/\mu_1}+\sqrt{\varepsilon_2/\varepsilon_1}\,\mathbf{n}}{\sqrt{\mu_2/\mu_1}+\sqrt{\varepsilon_2/\varepsilon_1}}\,2\mathbf{c}\,, \tag{51}$$

is an easily realizable example of the plane harmonic wave (49). Since $\mathbf{n}\widehat{\mathbf{C}} \parallel \mathbf{c}$ in this case, we may call it the *plane harmonic wave* with *linear polarization* (this time it is not the running wave).

The energy-momentum cliffor for the wave (48) is

$$\begin{aligned}\frac{1}{2}ff^\dagger &= \frac{1}{2}NN^\dagger(\sin\mathbf{k}\cdot\mathbf{r}\cos\omega t - \mathbf{n}\sin\omega t\cos\mathbf{k}\cdot\mathbf{r})^2\\ &= \frac{1}{2}NN^\dagger(\sin^2\mathbf{k}\cdot\mathbf{r}\cos^2\omega t + \cos^2\mathbf{k}\cdot\mathbf{r}\sin^2\omega t\\ &\qquad - \frac{\mathbf{n}}{2}\sin 2\mathbf{k}\cdot\mathbf{r}\sin 2\omega t)\,.\end{aligned}$$

The same expression is obtained for the wave (49). Its time average (i.e. the integral over the period divided by the period length T) is

$$\left\langle\frac{1}{2}ff^\dagger\right\rangle = \frac{1}{2}NN^\dagger\left(\frac{1}{2}\sin^2\mathbf{k}\cdot\mathbf{r} + \frac{1}{2}\cos^2\mathbf{k}\cdot\mathbf{r}\right) = \frac{1}{4}NN^\dagger\,.$$

Hence, by virtue of (12),

$$\left\langle\frac{1}{2}ff^\dagger\right\rangle = \frac{1}{4}|N|^2 + \frac{1}{2}\mathbf{n}(\mathbf{c}\cdot\mathbf{c}')\,.$$

We see that the averaged energy flux can be parallel or antiparallel to $\mathbf{n}$ depending on the sign of the scalar product $\mathbf{c}\cdot\mathbf{c}'$. Of course, for the travelling waves, the intensity of the wave is maximal among the waves with the same energy density. The zero average intensity wave is treated as the *standing wave*. For $\mathbf{c}' = 0$, that is, for $N = \mathbf{c}$, we obtain such a wave. The electric and magnetic parts of expression (48) are, in this case,

$$\mathbf{e}(\mathbf{r},t) = \mathbf{c}\sin\mathbf{k}\cdot\mathbf{r}\cos\omega t, \quad \widehat{\mathbf{b}}(\mathbf{r},t) = \mathbf{n}\mathbf{c}\cos\mathbf{k}\cdot\mathbf{r}\sin\omega t\,.$$

Both are products of the form $\psi(\mathbf{r})\varphi(t)$ which is characteristic for the standing waves.

The examples with the cosine function instead of sine in (48a) and (49a) are not interesting because of the identity $\cos X = \sin(X + \frac{\pi}{2})$ which follows from Problem 0.24.

Another option for the periodic changes of the initial or boundary field exists, the rotation around $\mathbf{n}$. As we know from (0.104), for N anticommuting with $\mathbf{n}$, such a rotation by angle β is described by $N \rightarrow e^{-I\mathbf{n}\beta}N$. Let, therefore, the initial electromagnetic field be

$$f^*(\mathbf{r},0) = e^{-I\,\mathbf{n}\,(\mathbf{k}\cdot\mathbf{r})}\,N\,. \tag{52}$$

The Taylor expansion of the function $g(\mathbf{k}\cdot\mathbf{r}) = e^{-I\,\mathbf{n}\,(\mathbf{k}\cdot\mathbf{r})}$ has its coefficients in $\mathcal{A}(\mathbf{n})$, so we may apply the formalism developed in Secs. 1 and 2. Thus substitution (28) with the aid of (3) gives the space-time dependence

$$f^*(\mathbf{r},t) = e^{-I\,\mathbf{n}\,Y}N = e^{IX}N\,, \tag{53a}$$

which may also be written as

$$f^*(\mathbf{r},t) = e^{I[\omega\,t-\mathbf{n}\,(\mathbf{k}\cdot\mathbf{r})]}N\,. \tag{53b}$$

By virtue of (0.94), the exponential function is a combination of the trigonometric functions, therefore we are allowed to call (53) the plane harmonic wave.

It is visible from the form $f^* = e^{-I\,\mathbf{n}\,(\mathbf{k}\cdot\mathbf{r})}\,(e^{I\omega\,t}N)$ that for each fixed time the space changes of the field are rotations. Hence the tip point of the electric vector traces a right-hand helix along any line parallel to $\mathbf{n}$. Therefore we call (53) the plane harmonic wave with the *right spiral polarization*. The form $f^* = e^{I\omega\,t}[e^{-I\,\mathbf{n}\,(\mathbf{k}\cdot\mathbf{r})}N]$ demonstrates that for a fixed point $\mathbf{r}$ the time dependence of the field is the Larmor-Reinich transformation (1.27). Having denoted $f_\mathbf{r} = e^{-I\,\mathbf{n}\,(\mathbf{k}\cdot\mathbf{r})}N = \mathbf{e}_\mathbf{r} + \widehat{\mathbf{b}}_\mathbf{r}$, we obtain

$$\begin{aligned} f^* = e^{I\omega\,t}f_\mathbf{r} = (\mathbf{e}_\mathbf{r}\,\cos\omega\,t - \mathbf{b}_\mathbf{r}\sin\omega\,t) \\ + I(\mathbf{b}_\mathbf{r}\cos\omega\,t + \mathbf{e}_\mathbf{r}\sin\omega\,t)\,. \end{aligned} \tag{54}$$

The electric vector in the first bracket traces an ellipse if $\mathbf{b}_\mathbf{r}$ is not parallel to $\mathbf{e}_\mathbf{r}$. In the particular case when $\mathbf{b}_\mathbf{r} \parallel \mathbf{e}_\mathbf{r}$, it reduces to a linear segment. We show in Fig. 94 the space-time variations of the electric vector under discussion. The ellipse in it illustrates the time changes of $\mathbf{e}$ in a fixed point. The helix representing the space changes rotates and alternatively swells and shrinks simultaneously on its whole length.

We decompose (53a) according to (26):

$$f^* = e^{-I\,\mathbf{n}\,(\mathbf{k}\cdot\mathbf{r}-\omega\,t)}N_+ + e^{-I\,\mathbf{n}\,(\mathbf{k}\cdot\mathbf{r}+\omega\,t)}N_-\,. \tag{55}$$

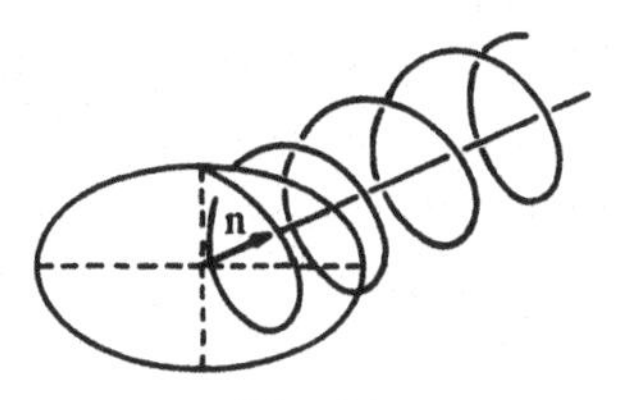

Fig. 94.

Both terms describe rotation by the angle $\mathbf{k}\cdot\mathbf{r}-\omega t$ and $\mathbf{k}\cdot\mathbf{r}+\omega t$, respectively, around $\mathbf{n}$. Therefore the two constituent travelling waves forming (55) are *circularly polarized*. In order to decide which is the sense of the circular polarization, let us quote Robson [18, Pg. 7]: "The tip of the E-vector will appear to trace out an ellipse when viewed along the direction in which the light propagates. If the E-vector rotates around the ellipse in a clockwise (anticlockwise) direction when viewed by an observer who receives the beam of light, the ray is said to be right-handed (left-handed) elliptically polarized light... If the ellipse becomes a circle, the light is called right-handed (left-handed) circularly polarized light."

We remind the reader of the relative directions of $\mathbf{n}$ and $\hat{\mathbf{n}} = I\,\mathbf{n}$ in Fig. 95. The first term in (55) has the propagation direction $\mathbf{n}$ and the direction of time rotations $-\hat{\mathbf{n}}$, therefore we state that its circular polarization is right-handed. The second term in (55) propagates in the direction $-\mathbf{n}$ but the time rotations take place in the direction $\hat{\mathbf{n}}$, hence its circular polarization is also right-handed. In this way, we find that the two travelling waves forming (53) have the same *right circular polarization.*

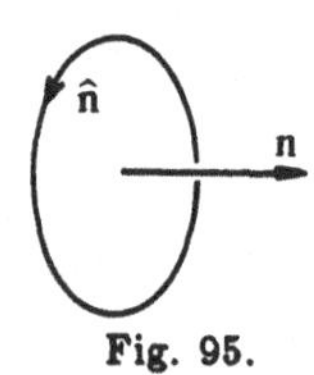

Fig. 95.

The eigenequations $\mathbf{n}\,N_\pm = \pm N_\pm$ imply $e^{I\,\mathbf{n}\beta}N_\pm = e^{\pm I\beta}N_\pm$ which means that the Larmor-Reinich transformation is equivalent to the ordinary rotation for the travelling waves. Hence we can write (55) alternatively as

$$f^*(\mathbf{r},t) = e^{I(\omega t-\mathbf{k}\cdot\mathbf{r})}N_+ + e^{I(\omega t+\mathbf{k}\cdot\mathbf{r})}N_- \ .$$

If, instead of (52), we start with the opposite space rotation $f^*(\mathbf{r},0) = e^{I\,\mathbf{n}(\mathbf{k}\cdot\mathbf{r})}N$, we obtain the plane wave

$$f^*(\mathbf{r},t) = e^{I\,\mathbf{n}Y}N = e^{-IX}N \tag{56}$$

which, as the space picture of electric oscillations, has the left-handed helix. The time changes of $\mathbf{e}$ again form an ellipse. Figure 96 illustrates the space-time variations of the electric vector for this wave. We call it the plane harmonic wave with the *left spiral polarization*. Its decomposition into the travelling waves

$$f^* = e^{-I\,\mathbf{n}\,(\omega t - \mathbf{k}\cdot\mathbf{r})}N_+ + e^{I\,\mathbf{n}\,(\omega t + \mathbf{k}\cdot\mathbf{r})}N_- \,, \tag{57}$$

shows that it consists of two travelling waves with the same *left circular polarization*.

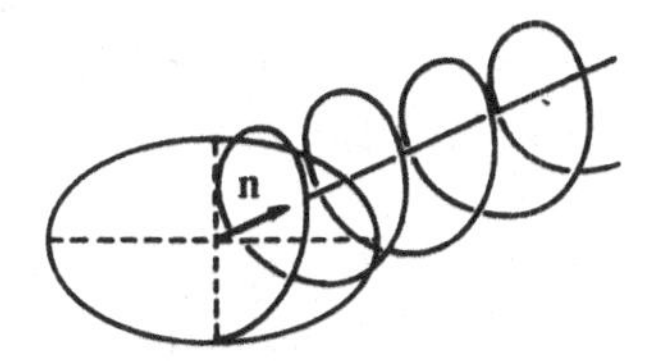

Fig. 96.

The energy-momentum cliffor for the waves (53) and (56) is

$$\frac{1}{2}f^* f^{*\dagger} = \frac{1}{2}\,e^{\pm IX}NN^\dagger e^{\mp IX} \;.$$

We use (12)

$$\frac{1}{2}f^* f^{*\dagger} = \frac{1}{2}\,e^{\pm IX}[|\,N\,|^2 + 2\,\mathbf{n}\,(\mathbf{c}\cdot\mathbf{c}')]e^{\mp IX} \;.$$

Since $\mathbf{n}$ commutes with the exponent and the whole exponential, we obtain

$$\frac{1}{2}f^* f^{*\dagger} = \frac{1}{2}|\,N\,|^2 + \mathbf{n}\,(\mathbf{c}\cdot\mathbf{c}') \;. \tag{58}$$

The energy density and flux do not depend on time nor position, which is typical for the circularly polarized waves. Thus the spiral polarization is, in this sense, a generalization of the circular one.

For $\mathbf{c} \perp \mathbf{c}'$, we obtain the standing wave. In such a case, $\widehat{\mathbf{C}} = \mathbf{n}\,\mathbf{c}' \perp \mathbf{c}$. This implies that $\mathbf{e}_\mathbf{r} \perp \widehat{\mathbf{b}}_\mathbf{r}$ or, equivalently, $\mathbf{e}\,(\mathbf{r},t) \parallel \mathbf{b}\,(\mathbf{r},t)$. We see that, for this particular standing wave with the spiral polarization, the perpendicularity condition is highly broken — the magnetic pseudovector $\mathbf{B}$ is parallel to $\mathbf{E}$ for all times and positions.

Let the boundary field now be gives on the plane $\mathbf{n}\cdot\mathbf{r}=0$ by

$$f(0,t)=e^{-I\mathbf{n}\omega t}N\,. \tag{59}$$

This, due to (29), gives us the space-time dependence

$$f(\mathbf{r},t)=e^{-I\mathbf{n}X}N=e^{IY}N\,, \tag{60a}$$

which may also be written as

$$f(\mathbf{r},t)=e^{I(\mathbf{k}\cdot\mathbf{r}-\mathbf{n}\omega t)}N\,. \tag{60b}$$

The form $f=e^{-I\mathbf{n}\omega t}(e^{I\mathbf{k}\cdot\mathbf{r}}N)$ shows that, for a fixed point $\mathbf{r}$, the time dependence of the field is the rotation in the direction $\hat{\mathbf{n}}=I\mathbf{n}$, hence we call (60) the plane harmonic wave with a *round polarization in the direction* $\hat{\mathbf{n}}$.[1)] For the alternative dependence with the plus sign in (59), the solution has a *round polarization in the direction* $-\hat{\mathbf{n}}$.

From the form $f=e^{I\mathbf{k}\cdot\mathbf{r}}(e^{-I\mathbf{n}\omega t}N)$, we see that for each fixed time the space change of the field is the Larmor-Reinich transformation, hence, in general, the curve joining the tips of $\mathbf{e}(\mathbf{r},t)$ for $t=$ constant along a line parallel to $\mathbf{n}$ is a helix with an elliptic base. In particular cases, it reduces to a sinusoid or to a helix with a circular base.

One may decompose (60a) according to (25) into the travelling waves:

$$f(\mathbf{r},t)=e^{-I\mathbf{n}(\omega t-\mathbf{k}\cdot\mathbf{r})}N_+ + e^{-I\mathbf{n}(\omega t+\mathbf{k}\cdot\mathbf{r})}N_-\,. \tag{61}$$

This demonstrates that the plane harmonic wave (60) is the superposition of two waves travelling in opposite directions and with the opposite circular polarizations (right-handed for the direction $-\mathbf{n}$ and left-handed for the direction $+\mathbf{n}$). By virtue of the eigenequations, Eq. (61) can also be written as

$$f(\mathbf{r},t)=e^{-I(\omega t-\mathbf{k}\cdot\mathbf{r})}N_+ + e^{I(\omega t+\mathbf{k}\cdot\mathbf{r})}N_-\,.$$

For the alternative dependence with the plus sign in (59), the solution can be decomposed as

$$\begin{aligned} e^{I\mathbf{n}X}N &= e^{I\mathbf{n}(\omega t-\mathbf{k}\cdot\mathbf{r})}N_+ + e^{I\mathbf{n}(\omega t+\mathbf{k}\cdot\mathbf{r})}N_- \\ &= e^{I(\omega t-\mathbf{k}\cdot\mathbf{r})}N_+ + e^{-I(\omega t+\mathbf{k}\cdot\mathbf{r})}N_-\,. \end{aligned} \tag{62}$$

[1)]The words "circular polarization" would be better in this context, but they have been used for a long time to describe only the travelling waves. Traditionally, the circular polarizations are divided into the left-handed and right-handed. However, these adjectives make no sense here for the round polarization. They are appropriate rather for the spiral polarization, as we used them earlier in this section.

This is the superposition of two travelling waves with the opposite circular polarizations (left-handed for the direction $-\mathbf{n}$ and right-handed for the direction $+\mathbf{n}$).

The energy-momentum cliffor for the waves (60) and (62) is

$$\frac{1}{2}ff^{\dagger} = \frac{1}{2}e^{\pm IY}NN^{\dagger}e^{\mp IY} = \frac{1}{2}|N|^2 + \mathbf{n}(\mathbf{c}\cdot\mathbf{c}') .$$

This expression is independent of $\mathbf{r}$ and t, therefore we are allowed to claim that the round polarization is another generalization of the circular one.

For $\mathbf{c} \perp \mathbf{c}'$, the standing waves are obtained and the condition $\mathbf{e}(\mathbf{r},t) \perp \widehat{\mathbf{b}}(\mathbf{r},t)$ is satisfied as well. This demonstrates that standing waves with round polarization exist for which $\mathbf{E}$ is parallel to $\mathbf{B}$.

When looking at the exponents in solutions (53b) and (60b), one tends to call the expressions $X = \omega t - \mathbf{n}(\mathbf{k}\cdot\mathbf{r})$ and $Y = \mathbf{k}\cdot\mathbf{r} - \mathbf{n}\omega t$, the phases. Can one also speak about phase velocity in this case? The answer to this question has to be negative because one speaks about phase velocity when the time changes of the term ωt can cancel out the space changes of $\mathbf{k}\cdot\mathbf{r}$ to keep their combination constant. This cannot happen in the general case because the changes named cannot cancel each other within X or Y since they take place in different "directions" in the linear space of cliffors. Therefore we must conclude that for plane harmonic waves, the concept of phase velocity in general makes no sense. Only for the travelling waves, i.e. for separate terms in sums in (55), (57), (61) and (62), does phase velocity become meaningful — it is, of course, equal to $\pm\mathbf{n}\,u$, as derived in §2.

§5. Description of the Polarization

Expression (51) is not the only possible plane harmonic wave for which the electric oscillations take place on a linear segment. One may write a more general expression

$$\sin[\omega t - \mathbf{n}(\mathbf{k}\cdot\mathbf{r})](\alpha + \beta\mathbf{n})\mathbf{c} = (\sin X)(\alpha + \beta\mathbf{n})\mathbf{c} ,$$

(here $\mathbf{c} \perp \mathbf{n}$ and α, β are scalars) which can also be called a plane harmonic wave with *linear polarization*. Expressions like $(\sin X)(\mathbf{c} + \widehat{\mathbf{C}})$, where $\mathbf{c}$ and $\widehat{\mathbf{C}}$ are arbitrary anticommuting with $\mathbf{n}$, do not give the linear oscillations of $\mathbf{e}$, therefore they cannot be called linearly polarized. They may, however, be obtained as superpositions of two waves linearly polarized in perpendicular directions:

$$(\sin X)N = (\sin X)(\alpha + \beta\mathbf{n})\mathbf{c}_1 + (\sin X)(\gamma + \delta\mathbf{n})\mathbf{c}_2 ,$$

where $\mathbf{c}_1 \perp \mathbf{c}_2$, $\mathbf{c}_1, \mathbf{c}_2 \perp \mathbf{n}$ and $\alpha, \beta, \gamma, \delta$ are scalars, since the decomposition $N = (\alpha + \beta\,\mathbf{n})\,\mathbf{c}_1 + (\gamma + \delta\,\mathbf{n})\,\mathbf{c}_2$ is possible for any N satisfying (8).

One may easily check that our generalizations of circularly polarized waves can be expressed as the following combinations of the sine-type waves (Problem 6):
The right spiral polarization:

$$e^{IX}N = (\cos X)N + (\sin X)IN = \sin(X + \frac{\pi}{2})N + (\sin X)IN \,. \qquad (63a)$$

The left spiral polarization:

$$e^{-IX}N = \sin(X + \frac{\pi}{2})N - (\sin X)IN \,. \qquad (63b)$$

The round polarization in direction $\widehat{\mathbf{n}} = I\mathbf{n}$:

$$e^{IY}N = \sin(Y + \frac{\pi}{2})N + (\sin Y)IN \,. \qquad (63c)$$

The round polarization in direction $-\widehat{\mathbf{n}}$:

$$e^{-IY}N = \sin(Y + \frac{\pi}{2})N - (\sin Y)N \,. \qquad (63d)$$

The reverse relations:

$$(\sin X)N = \frac{1}{2I}(e^{IX}N + e^{-IX}N) \,, \qquad (64a)$$

$$(\sin Y)N = \frac{\mathbf{n}}{2I}(e^{IY}N + e^{-IY}N) \qquad (64b)$$

can also be established. We see that, among plane harmonic waves, some are "linearly" dependent, where the word "linearly" stands for the combinations with the coefficients in the subalgebra $\mathcal{A}(\mathbf{n})$.

We shall look for some physical quantities appropriate for characterizing the round and spiral polarizations. We start with the former.

It is useful to introduce volutor quantity $\mathbf{e} \wedge \frac{\partial \mathbf{e}}{\partial t}$ describing the plane motion of the electric vector at a fixed point. It is possible to check that $\partial\,\mathbf{e}/\partial t = \mathbf{e}\,I\,\mathbf{n}\,\omega = \mathbf{e}\widehat{\boldsymbol{\omega}}$ for the wave (60) where $\widehat{\boldsymbol{\omega}} = I\mathbf{n}\,\omega$ is the volutor of angular velocity. Since $\mathbf{e}$ is parallel to $\widehat{\boldsymbol{\omega}}$, we have $\mathbf{e}\widehat{\boldsymbol{\omega}} = \mathbf{e}\cdot\widehat{\boldsymbol{\omega}}$ and, because $\mathbf{e}\cdot\widehat{\boldsymbol{\omega}}$ is perpendicular to $\mathbf{e}$,

$$\mathbf{e} \wedge \frac{\partial\,\mathbf{e}}{\partial t} = \mathbf{e}\,(\mathbf{e}\widehat{\boldsymbol{\omega}}) = \mathbf{e}^2\widehat{\boldsymbol{\omega}} = |\,\mathbf{e}\,|^2\widehat{\boldsymbol{\omega}} \,.$$

Thus the direction of the proposed volutor is the same as that of $\widehat{\omega}$.

Similarly, the identity $\frac{\partial\,\mathbf{b}}{\partial t} = \mathbf{b}\widehat{\omega}$ holds for the wave (60), hence $\mathbf{b}\wedge\partial\,\mathbf{b}/\partial t = |\,\mathbf{b}\,|^2\widehat{\omega}$. By dint of the identity (0.33), we may write this as

$$-\widehat{\mathbf{b}}\,\dot{\wedge}\,\frac{\partial\,\widehat{\mathbf{b}}}{\partial t} = |\,\widehat{\mathbf{b}}\,|^2\widehat{\omega}\,.$$

It is useful to consider one half of the sum of the two last expressions:

$$\frac{1}{2}\left(\mathbf{e}\wedge\frac{\partial\,\mathbf{e}}{\partial t} - \widehat{\mathbf{b}}\,\dot{\wedge}\,\frac{\partial\,\widehat{\mathbf{b}}}{\partial t}\right) = \frac{1}{2}(|\,\mathbf{e}\,|^2 + |\,\widehat{\mathbf{b}}\,|^2)\widehat{\omega} = w\widehat{\omega}\,. \tag{65}$$

We introduce its quotient with the energy density w and call it *roundness* $\widehat{\Omega}$:

$$\widehat{\Omega} = \frac{1}{2w}\left(\mathbf{e}\wedge\frac{\partial\,\mathbf{e}}{\partial t} - \widehat{\mathbf{b}}\,\dot{\wedge}\,\frac{\partial\,\widehat{\mathbf{b}}}{\partial t}\right)\,. \tag{66}$$

Result (65) gives, for the wave (60), the identity $\widehat{\Omega} = \widehat{\omega}$. Similarly, we obtain $\widehat{\Omega} = -\widehat{\omega}$ for the alternative to (60) with the opposite exponent. We summarize the observations: The roundness for the wave with the round polarization is constant and equal to the angular velocity of the electromagnetic quantities $\mathbf{e}$ and $\widehat{\mathbf{b}}$.

We now pass to the wave with spiral polarization. A trivector quantity would be more adequate in this case. We propose the expression:

$$\frac{1}{2}\left(\mathbf{e}\wedge\frac{\partial\,\mathbf{e}}{\partial\xi}\wedge\mathbf{n} + \mathbf{b}\wedge\frac{\partial\,\mathbf{b}}{\partial\xi}\wedge\mathbf{n}\right)\,,$$

where $\xi = \mathbf{n}\cdot\mathbf{r}$. One may verify that $\partial\,\mathbf{e}/\partial\xi = \mathbf{e}\,I\,\mathbf{n}\,k, \partial\,\mathbf{b}/\partial\xi = \mathbf{b}\,I\,\mathbf{n}\,k$ for the wave (53), hence

$$\frac{1}{2}\left(\mathbf{e}\wedge\frac{\partial\,\mathbf{e}}{\partial\xi} - \widehat{\mathbf{b}}\,\dot{\wedge}\,\frac{\partial\,\widehat{\mathbf{b}}}{\partial\xi}\right)\wedge\mathbf{n} = \frac{1}{2}(|\,\mathbf{e}\,|^2 + |\,\widehat{\mathbf{b}}\,|^2)\,(I\,\mathbf{n}\,k)\wedge\mathbf{n} = wkI\,.$$

We introduce its quotient with the energy density and call it *spirality* $\widehat{s}$:

$$\widehat{s} = \frac{1}{2w}\left(\mathbf{e}\wedge\frac{\partial\,\mathbf{e}}{\partial\xi} - \widehat{\mathbf{b}}\,\dot{\wedge}\,\frac{\partial\,\widehat{\mathbf{b}}}{\partial\xi}\right)\wedge\mathbf{n}\,. \tag{67}$$

The above calculations demonstrate that for the wave (53) with the right spiral polarization, the identity $\widehat{s} = kI$ holds. For its alternative (56) with the left spiral polarization, one obtains $\widehat{s} = -kI$.

The spirality, like all trivectors, has two relevant features: the amplitude and the orientation. The physical meaning of the orientation of spirality is just the screw sense of the helix being the space picture of electric oscillations. The magnitude of spirality has also a physical meaning — it is merely $k = 2\pi/\lambda$, where λ is the pitch of the helix. When the pitch tends to infinity, the helix becomes a straight line (one may say: the helix disappears) and naturally the spirality tends to zero. Thus the spirality for the wave with spiral polarization is a kind of "directed wave number".

Let us transform expression (67). First of all, due to the perpendicularity of the multiplied factors, we can replace the products $\wedge$ and $\dot{\wedge}$ by the Clifford product and use the anticommutativity of the factors:

$$\hat{s} = \frac{1}{2w}\left(\mathbf{e}\frac{\partial\,\mathbf{e}}{\partial\xi} - \widehat{\mathbf{b}}\frac{\partial\,\widehat{\mathbf{b}}}{\partial\xi}\right)\mathbf{n} = \frac{1}{2w}\left(\widehat{\mathbf{b}}\mathbf{n}\frac{\partial\,\widehat{\mathbf{b}}}{\partial\xi} - \mathbf{e}\,\mathbf{n}\frac{\partial\,\mathbf{e}}{\partial\xi}\right),$$

and afterwards, we return to the inner and outer products:

$$\hat{s} = \frac{1}{2w}\left[\widehat{\mathbf{b}}\wedge\left(\mathbf{n}\cdot\frac{\partial\,\widehat{\mathbf{b}}}{\partial\xi}\right) - \mathbf{e}\wedge\left(\mathbf{n}\wedge\frac{\partial\,\mathbf{e}}{\partial\xi}\right)\right].$$

For the plane waves, the identities $\mathbf{n}\wedge\frac{\partial\,\mathbf{e}}{\partial\xi} = \nabla\wedge\mathbf{e}, \mathbf{n}\cdot\frac{\partial\,\widehat{\mathbf{b}}}{\partial\xi} = \nabla\cdot\widehat{\mathbf{b}}$ are satisfied, therefore we obtain

$$\hat{s} = \frac{1}{2w}[\widehat{\mathbf{b}}\wedge(\nabla\cdot\widehat{\mathbf{b}}) - \mathbf{e}\wedge(\nabla\wedge\mathbf{e})]\,. \tag{68}$$

This formula — as distinct from (67) — does not contain the explicit dependence on the direction $\mathbf{n}$. It follows from the homogeneous Maxwell equation in uniform medium that $\nabla\cdot\widehat{\mathbf{b}} = \frac{1}{u}\frac{\partial\,\mathbf{e}}{\partial t}, \nabla\wedge\mathbf{e} = -\frac{1}{u}\frac{\partial\,\widehat{\mathbf{b}}}{\partial t}$, thus we may write (68) as

$$\hat{s} = \frac{1}{2uw}\left(\mathbf{e}\wedge\frac{\partial\,\widehat{\mathbf{b}}}{\partial t} - \mathbf{b}\wedge\frac{\partial\,\mathbf{e}}{\partial t}\right).$$

This is the transformed version of (67).

It is useful to have a compact expression for the travelling harmonic plane wave with *elliptic polarization*. We shall demonstrate that

$$f(\mathbf{r},t) = (1+\mathbf{n})\{\mathbf{n}\wedge[\mathbf{h}\,e^{\hat{\mathbf{i}}(\mathbf{k}\cdot\mathbf{r}-\omega t)}]\}\,, \tag{69}$$

where $\hat{\mathbf{i}}$ is a unit volutor of arbitrary direction, $\mathbf{h}\,||\,\hat{\mathbf{i}}$ and $\mathbf{h}\perp\mathbf{n}$ describes the wave named. The condition $\mathbf{h}\,||\,\hat{\mathbf{i}}$ implies that the mapping $\mathbf{h}\to\mathbf{h}\,e^{\hat{\mathbf{i}}\varphi}$ is the rotation by the angle φ in the direction $\widehat{\mathbf{i}}$, hence the changes of the volutor $\widehat{\mathbf{b}}$ contained in (69)

$$\widehat{\mathbf{b}}(\mathbf{r},t)=\mathbf{n}\wedge[\mathbf{h}\,e^{\widehat{\mathbf{i}}(\mathbf{k}\cdot\mathbf{r}-\omega t)}]$$

are rotations around $\mathbf{n}$ and, possibly, dilations if the vectors $\mathbf{h}\,e^{\widehat{\mathbf{i}}\varphi}$ for $\varphi\neq 0$ are not perpendicular to $\mathbf{n}$ (see Fig. 97). The same observations concern the vector $\mathbf{e}(\mathbf{r},t)=\mathbf{n}\widehat{\mathbf{b}}(\mathbf{r},t)$. Additionally, $\mathbf{e}$ is always perpendicular to $\mathbf{n}$. Therefore the tip of $\mathbf{e}$ in a fixed point orbits an *ellipse* obtained by projecting the circle $\varphi\rightarrow\mathbf{h}\,e^{\widehat{\mathbf{i}}\varphi}$ onto the plane orthogonal to $\mathbf{n}$ (this ellipse is denoted by C in Fig. 97). The major semiaxis of the ellipse coincides with $\mathbf{h}$.

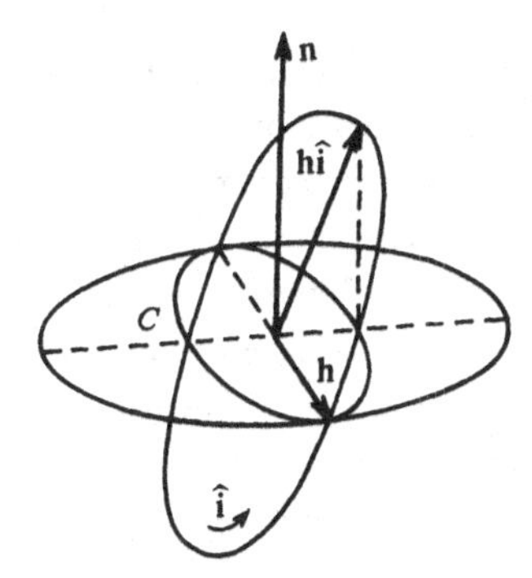

Fig. 97.

Having checked (see Problem 7) that function (69) satisfies the Maxwell equation, we state that this function describes the travelling harmonic plane wave with elliptic polarization. Different directions of $\widehat{\mathbf{i}}$ correspond to different kinds of the elliptic polarization. So for $\widehat{\mathbf{i}}\perp\mathbf{n}$, the outer product (69) can be replaced by the Clifford product and the polarization becomes circular. If the angle between $\mathbf{n}$ and $\widehat{\mathbf{i}}$ is not right, the ellipse C is not a circle and its major semiaxis is parallel to $\mathbf{n}\cdot\widehat{\mathbf{i}}$; whilst for $\widehat{\mathbf{i}}\parallel\mathbf{n}$, the ellipse C reduces to the linear segment parallel to $\mathbf{n}\cdot\widehat{\mathbf{i}}$.

Various relative directions of the volutor $\widehat{\mathbf{i}}$ with respect to $\mathbf{n}$ can be described by various relative directions of the vector $\mathbf{i}=-I\widehat{\mathbf{i}}$ with respect to $\mathbf{n}$. If $\mathbf{n}$ is fixed, the possible directions of $\mathbf{i}$ form a sphere with $\mathbf{n}$ as the north pole. The directions $\mathbf{i}=\pm\mathbf{n}$, i.e. the points on both poles, correspond to the circular polarizations (the north pole — to the right-handed one) whereas the points on the equator correspond to the linear polarization (the line of electric oscillations is orthogonal to $\mathbf{i}$). Figure 98 illustrates the sphere and the possible elliptic polarizations for several points on it.

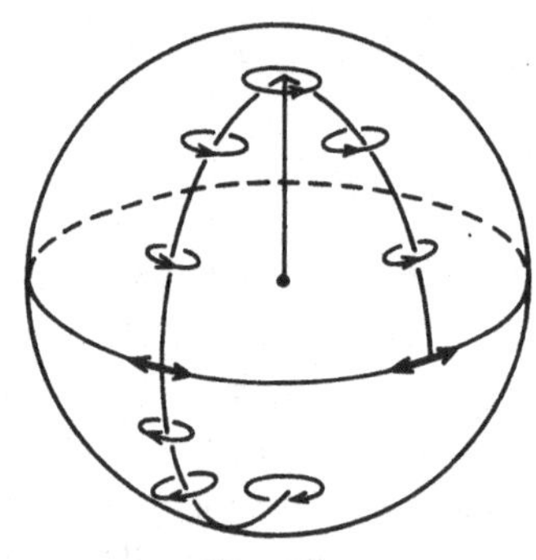

Fig. 98.

Readers familiar with the *Poincaré sphere* (a notion serving to describe elliptic polarizations) will notice that the unit sphere characterized is something different, because two points lying on our sphere symmetrically with respect to north-south axis correspond to the same elliptic polarization. Loosely speaking, our sphere has "twice as many" points as the Poincaré sphere. However, the points on our sphere have a more direct relation to the shape of the ellipse: it is sufficient to draw the great circle perpendicular to the radius vector $\mathbf{i}$ of a given point and project it onto the equator plane.

One may check (see Problem 7) that the energy-momentum cliffor for the wave (69) is

$$\frac{1}{2} f f^{\dagger} = \frac{1}{2}(1+\mathbf{n})|\,\mathbf{h}\,|^2[1+\cos^2\theta + \sin^2\theta\cos 2(\mathbf{k}\cdot\mathbf{r} - \omega\, t)]\ ,$$

where θ is the angle between $\mathbf{i}$ and $\mathbf{n}$. Thus the time-average value of this cliffor is

$$\frac{1}{2}\langle f f^{\dagger}\rangle = \frac{1}{2}(1+\mathbf{n})|\,\mathbf{h}\,|^2(1+\cos^2\theta)\ .$$

The quantities determining the spirality and roundness are

$$\frac{1}{2u}\left(\mathbf{e}\wedge\frac{\partial\,\widehat{\mathbf{b}}}{\partial t} - \widehat{\mathbf{b}}\wedge\frac{\partial\,\mathbf{e}}{\partial t}\right) = k|\,\mathbf{h}\,|^2\mathbf{n}\wedge\widehat{\mathbf{i}} = k|\,\mathbf{h}\,|^2 I\cos\theta\ ,$$

and

$$\frac{1}{2}\left(\mathbf{e}\wedge\frac{\partial\,\mathbf{e}}{\partial t} - \widehat{\mathbf{b}}\,\dot{\wedge}\,\frac{\partial\,\widehat{\mathbf{b}}}{\partial t}\right) = -\omega|\,\mathbf{h}\,|^2\mathbf{n}\,(\mathbf{n}\wedge\widehat{\mathbf{i}}) = -\omega|\,\mathbf{h}\,|^2\widehat{\mathbf{n}}\cos\theta\ ,$$

respectively. We see that they are zero for $\theta = \frac{\pi}{2}$, that is, for the linear polarization, and are maximal for $\theta = 0$, that is, for the circular polarization.

Problems

1. Let the electromagnetic cliffor of the form (9) be the eigencliffor of $P_{\mathbf{n}} = \frac{1}{2}(1+\mathbf{n})$ to the eigenvalue one or zero. Show that $|\mathbf{S}|/w = 1/\sqrt{\varepsilon\mu}$. Here $\mathbf{S}$ is the Poynting vector and w is the energy density of the electromagnetic field (9).
2. Show that, for the wave (53), the explicit functions $\mathbf{E}(\mathbf{r},t)$ and $\mathbf{B}(\mathbf{r},t)$ are

$$\begin{aligned}\mathbf{E}(\mathbf{r},t) &= (\mathbf{E}_0 \cos\mathbf{k}\cdot\mathbf{r} + \mathbf{n}\times\mathbf{E}_0 \sin\mathbf{k}\cdot\mathbf{r})\cos\omega t \\ &\quad - \frac{1}{\sqrt{\varepsilon\mu}}(\mathbf{B}_0 \cos\mathbf{k}\cdot\mathbf{r} + \mathbf{n}\times\mathbf{B}_0 \sin\mathbf{k}\cdot\mathbf{r})\sin\omega t\,, \\ \mathbf{B}(\mathbf{r},t) &= \sqrt{\varepsilon\mu}(\mathbf{E}_0 \cos\mathbf{k}\cdot\mathbf{r} + \mathbf{n}\times\mathbf{E}_0 \sin\mathbf{k}\cdot\mathbf{r})\sin\omega t \\ &\quad + (\mathbf{B}_0 \cos\mathbf{k}\cdot\mathbf{r} + \mathbf{n}\times\mathbf{B}_0 \sin\mathbf{k}\cdot\mathbf{r})\cos\omega t\,.\end{aligned}$$

3. Find the explicit functions $\mathbf{E}(\mathbf{r},t)$ and $\mathbf{B}(\mathbf{r},t)$ for the waves (56) and (60).
4. Check that $\mathbf{A}(\mathbf{r},t) = (-1/\omega\sqrt{\varepsilon})e^{I\,\mathbf{n}\,(\mathbf{k}\cdot\mathbf{r}-\omega t)}I\,\mathbf{n}\,\mathbf{c}, \varphi(\mathbf{r},t) = 0$ are the electromagnetic potentials for the travelling wave $e^{I\,\mathbf{n}\,(\mathbf{k}\cdot\mathbf{r}-\omega t)}(1+\mathbf{n})\,\mathbf{c}$ with $\mathbf{c}\perp\mathbf{n}$. Is the Lorentz condition satisfied?
5. Show that $\mathbf{A}(\mathbf{r},t) = (1/\omega\sqrt{\varepsilon})\,\mathbf{c}\sin(\mathbf{k}\cdot\mathbf{r}-\omega t), \varphi = 0$ are the electromagnetic potentials for the travelling wave $(1+\mathbf{n})\,\mathbf{c}\cos(\mathbf{k}\cdot\mathbf{r}-\omega t)$. Check the Lorentz condition.
6. Prove the identities (63) and (64).
7. Check that expression (69) satisfies the homogeneous Maxwell equation. Find the energy-momentum cliffor, spirality and roundness for the wave (69).
8. Two travelling harmonic plane waves of equal frequencies polarized linearly in perpendicular directions propagate in the same direction. The phase difference is δ. Determine the polarization of the superposed wave.
9. Two harmonic waves with linear polarizations are travelling in different directions. Let $\mathbf{S}$ be the Poynting vector for the superposed field and $\mathbf{S}_i$ — for the constituent waves. Is the equality $\mathbf{S} = \mathbf{S}_1 + \mathbf{S}_2$ satisfied?
10. Deduce that the identities $\frac{\partial\mathbf{E}}{\partial t} = \pm\omega\mathbf{n}\times\mathbf{E}, \frac{\partial\mathbf{B}}{\partial t} = \pm\omega\mathbf{n}\times\mathbf{B}$ are valid for two kinds of the waves with round polarization.
11. Show that the identities $\nabla\times\mathbf{E} = \pm k\,\mathbf{E}, \nabla\times\mathbf{B} = \pm k\,\mathbf{B}$ are satisfied for the waves with spiral polarization, the plus sign corresponding to the left polarization and the minus to the right one.

Chapter 5

VARIOUS KINDS OF ELECTROMAGNETIC WAVES

§1. The Plane Harmonic Wave in a Conducting Medium

A conductor is characterized by its *resistance* R occurring in *Ohm's law* which for stationary currents has the form $\varphi_2 - \varphi_1 = RJ$ where J is the current and $\varphi_2 - \varphi_1$ is the electric potential difference along the conductor. The resistance R for a cylindrical conductor of length l and cross section area s is $R = l/s\sigma$, where σ, called *conductivity*, depends only on the conducting medium. Ohm's law can be transformed into the so-called *local* or *differential form*

$$\mathbf{j} = \sigma \mathbf{E} \, . \tag{1}$$

We assume that the reader is familiar with the derivation.

Substitution of (1) in the Maxwell equation (1.23) yields

$$\mathcal{D} f = -\sqrt{\mu}\mathbf{j} = -\sqrt{\mu}\sigma \mathbf{E} = -\sqrt{\frac{\mu}{\varepsilon}}\sigma \mathbf{e} \, .$$

This can also be written as

$$\nabla f + \frac{1}{u}\frac{\partial f}{\partial t} + \sqrt{\frac{\mu}{\varepsilon}}\sigma \mathbf{e} = 0 \, . \tag{2}$$

We are interested in finding a plane wave solution to this equation. We restrict ourselves to a harmonic wave of the following form:

$$f(\mathbf{r},t) = e^{-\widehat{\mathbf{n}}\,\omega\, t} g(\mathbf{n}\cdot\mathbf{r})\ , \tag{3}$$

where $\widehat{\mathbf{n}} = I\mathbf{n}$ and g is a cliffor-valued function. In order for f to be an electromagnetic cliffor, g has to contain only the vector and bivector parts: $g = \mathbf{g}_v + \widehat{\mathbf{g}}_b$ and, moreover, the conditions $\mathbf{n} \perp \mathbf{g}_v$ and $\mathbf{n}\|\widehat{\mathbf{g}}_b$ must be satisfied. This signifies that g anticommutes with $\mathbf{n}$ and with $\widehat{\mathbf{n}}$, so we can write

$$f(\mathbf{r},t) = g(\mathbf{n}\cdot\mathbf{r}) e^{\widehat{\mathbf{n}}\,\omega\, t}\ . \tag{4}$$

In this manner, the time changes of the field are rotations and we see that the wave (3) has *round polarization* in direction $\widehat{\mathbf{n}}$.

The electric part of (4) is $\mathbf{e} = \mathbf{g}_v\, e^{\widehat{\mathbf{n}}\,\omega\, t}$, so Eq. (2) takes the form

$$(\boldsymbol{\nabla} g) e^{\widehat{\mathbf{n}}\,\omega\, t} - \frac{1}{u}\widehat{\mathbf{n}}\,\omega\, g\, e^{\widehat{\mathbf{n}}\,\omega\, t} + \sqrt{\frac{\mu}{\varepsilon}}\,\sigma\,\mathbf{g}_v\, e^{\widehat{\mathbf{n}}\,\omega\, t} = 0\ .$$

We multiply both sides by $e^{-\widehat{\mathbf{n}}\,\omega\, t}$ from the right:

$$\boldsymbol{\nabla} g - \frac{1}{u}\widehat{\mathbf{n}}\,\omega\, g + \sqrt{\frac{\mu}{\varepsilon}}\,\sigma\,\mathbf{g}_v = 0\ .$$

Function g depends on $\mathbf{r}$ only through $\xi = \mathbf{n}\cdot\mathbf{r}$, hence $\boldsymbol{\nabla} g = \mathbf{n}\, g'$ where $g' = dg/d\xi$. In this way, we get the equation

$$\mathbf{n}\, g' = \frac{1}{u} I\,\mathbf{n}\,\omega\, g - \sqrt{\frac{\mu}{\varepsilon}}\,\sigma\,\mathbf{g}_v$$

which, after introducing the notation

$$k_0 = \frac{\omega}{u}\ , \qquad \sqrt{\frac{\mu}{\varepsilon}}\,\sigma = k_0\,\kappa\ , \tag{5}$$

can be altered to the form

$$g' = k_0\, I g - k_0\,\kappa\,\mathbf{n}\,\mathbf{g}_v\ .$$

We equate separately the vector and the bivector parts:

$$\mathbf{g}_v' = k_0\, I\,\widehat{\mathbf{g}}_b\ , \quad \widehat{\mathbf{g}}_b' = k_0\, I\,\mathbf{g}_v - k_0\,\kappa\,\mathbf{n}\,\mathbf{g}_v\ . \tag{6}$$

This is a system of two first-order ordinary differential equations. The standard method of solving it is by finding the *characteristic values* of its matrix

$$A = \begin{pmatrix} 0 & k_0 I \\ k_0 I - k_0 \kappa \mathbf{n} & 0 \end{pmatrix} ,$$

that is, solutions λ_i to the *characteristic equation* $\det(A - \lambda \mathbb{1}) = 0$, and then finding the eigenvectors of A, that is, columns X_i such that $AX_i = \lambda_i X_i$. Thereafter, the solution to (6) is written (in column form):

$$\begin{pmatrix} \mathbf{g}_v \\ \widehat{\mathbf{g}}_b \end{pmatrix} = \sum_i e^{\lambda_i \xi} X_i .$$

In our case, the matrix A has its elements in the subalgebra $\mathcal{A}(\mathbf{n})$, therefore we seek its eigenvalues also in $\mathcal{A}(\mathbf{n})$. The characteristic equation has the form

$$\lambda^2 = -k_0^2 (1 + \kappa \widehat{\mathbf{n}}) .$$

Since $\widehat{\mathbf{n}}^2 = -1$, we can treat this equation formally as in the field of complex numbers with $\widehat{\mathbf{n}}$ playing the role of an imaginary unit. The usual way of finding the square roots of a complex number yields the characteristic values

$$\lambda_{1,2} = \pm \frac{k_0}{\sqrt{2}} \left(\sqrt{\sqrt{1+\kappa^2} - 1} - \sqrt{\sqrt{1+\kappa^2} + 1}\, \widehat{\mathbf{n}} \right) .$$

Having introduced the notation

$$\gamma = \frac{k_0}{\sqrt{2}} \sqrt{\sqrt{1+\kappa^2} - 1} , \qquad k = \frac{k_0}{\sqrt{2}} \sqrt{\sqrt{1+\kappa^2} + 1} , \tag{7}$$

we get the characteristic values in the form

$$\lambda_{1,2} = \pm(\gamma - k \widehat{\mathbf{n}}) .$$

The eigenvectors

$$X_1 = \begin{pmatrix} k_0 \mathbf{c}_1 \\ -(\gamma I + k \mathbf{n}) \mathbf{c}_1 \end{pmatrix} , \qquad X_2 = \begin{pmatrix} k_0 \mathbf{c}_2 \\ (\gamma I + k \mathbf{n}) \mathbf{c}_2 \end{pmatrix}$$

correspond to these values, where $\mathbf{c}_1, \mathbf{c}_2$ are arbitrary vectors orthogonal to $\mathbf{n}$.

In this manner, we have found the solution to (6) in the following column form

$$\begin{pmatrix} \mathbf{g}_v \\ \widehat{\mathbf{g}}_b \end{pmatrix} = e^{(\gamma - k\widehat{\mathbf{n}})\,\mathbf{n}\cdot\mathbf{r}} \begin{pmatrix} k_0\,\mathbf{c}_1 \\ (-\gamma\,I - k\,\mathbf{n})\,\mathbf{c}_1 \end{pmatrix} + e^{-(\gamma - k\widehat{\mathbf{n}})\,\mathbf{n}\cdot\mathbf{r}} \begin{pmatrix} k_0\,\mathbf{c}_2 \\ (\gamma I + k\,\mathbf{n})\,\mathbf{c}_2 \end{pmatrix},$$

which may be rewritten in the cliffor form

$$g = \mathbf{g}_v + \widehat{\mathbf{g}}_b = e^{(\gamma - k\widehat{\mathbf{n}})\,\mathbf{n}\cdot\mathbf{r}}[k_0\,\mathbf{c}_1 - (\gamma\,I + k\,\mathbf{n})\,\mathbf{c}_1] \\ + e^{(-\gamma + k\widehat{\mathbf{n}})\,\mathbf{n}\cdot\mathbf{r}}[k_0\,\mathbf{c}_2 + (\gamma\,I + k\,\mathbf{n})\,\mathbf{c}_2]\,.$$

We insert this in (3) with the substitution $\mathbf{k} = k\,\mathbf{n}$:

$$f = e^{\gamma\,\mathbf{n}\cdot\mathbf{r}}\,e^{-\widehat{\mathbf{n}}\,(\mathbf{k}\cdot\mathbf{r}+\omega\,t)}(k_0 - \gamma\,I - \mathbf{k})\,\mathbf{c}_1 + e^{-\gamma\,\mathbf{n}\cdot\mathbf{r}}\,e^{\widehat{\mathbf{n}}\,(\mathbf{k}\cdot\mathbf{r}-\omega\,t)}(k_0 + \gamma\,I + \mathbf{k})\,\mathbf{c}_2\,. \tag{8}$$

The two terms of this function can be interpreted as two travelling waves, both containing the periodic and nonperiodic dependence on the space variable $\xi = \mathbf{n}\cdot\mathbf{r}$. The periodic dependence determined by factors $e^{-\widehat{\mathbf{n}}\,(\mathbf{k}\cdot\mathbf{r}+\omega\,t)}$ and $e^{\widehat{\mathbf{n}}\,(\mathbf{k}\cdot\mathbf{r}-\omega\,t)}$ justifies calling the waves "travelling". The planes of constant phase run in the directions $-\mathbf{n}$ and $+\mathbf{n}$, respectively, both with the wave number k given, according to (5) and (7), by

$$k = \frac{\omega}{\sqrt{2}\,u}\sqrt{\sqrt{1 + \left(\frac{\sigma}{\varepsilon\,\omega}\right)^2} + 1}\,.$$

Thus the *phase velocity*

$$\frac{\omega}{k} = \frac{\sqrt{2}\,u}{\sqrt{\sqrt{1 + (\sigma/\varepsilon\,\omega)^2} + 1}}$$

is smaller than in a nonconducting medium (that is, for $\sigma = 0$). Moreover, it depends on the angular frequency — this phenomenon is referred to as *dispersion.*

The nonperiodic dependence is determined by the scalar factors $e^{\gamma\,\mathbf{n}\cdot\mathbf{r}}$ and $e^{-\gamma\,\mathbf{n}\cdot\mathbf{r}}$. They influence only the magnitude of the field and show that in both terms, it decreases in the direction of propagation. Thus the electromagnetic wave entering a conductor is damped — this phenomenon is called *skin effect.* It is described quantitatively by the so-called *skin* or *penetration depth*

$$\delta = \frac{1}{\gamma} = \frac{\sqrt{2}\,u}{\omega\sqrt{\sqrt{1 + (\sigma/\varepsilon\,\omega)^2} - 1}}$$

defined as the distance at which the wave is damped to $1/e = 0.369$ of its initial amplitude.

Let us discuss the particular solution for $\mathbf{c}_1 = 0$:

$$f = e^{-\gamma\, \mathbf{n}\cdot\mathbf{r}}\, e^{\widehat{\mathbf{n}}\,(\mathbf{k}\cdot\mathbf{r}-\omega\, t)}(k_0 + \gamma\, I + \mathbf{k})\,\mathbf{c}_2 \; . \tag{9}$$

This can be interpreted as a left-handed circularly polarized damped wave propagating only in one direction $+\mathbf{n}$. The two exponential factors do not mix the vector and bivector parts of the remainder (the first factor influences the magnitudes, the second one performs the rotation), which can be considered as the initial electromagnetic cliffor. So the initial magnetic part

$$\widehat{\mathbf{b}}_0 = (\gamma\, I + \mathbf{k})\mathbf{c}_2 \tag{10}$$

is not parallel to the initial electric part $\mathbf{e}_0 = k_0\, \mathbf{c}_2$. We may transform (10) as follows

$$\widehat{\mathbf{b}}_0 = (k + \gamma\, I\mathbf{n})\,\mathbf{n}\,\mathbf{c}_2 = (k + \gamma\,\widehat{\mathbf{n}})\,\mathbf{n}\,\mathbf{c}_2 \; .$$

If one writes the even cliffor $k + \gamma\,\widehat{\mathbf{n}}$ in the form $k_0\, a\, e^{\widehat{\mathbf{n}}\,\alpha}$ with real α, a and $a > 0$, one obtains

$$a = \frac{1}{k_0}\sqrt{k^2 + \gamma^2} = \sqrt[4]{1 + \kappa^2}\; , \tag{11}$$

$$\alpha = \text{arc tan}\frac{\gamma}{k} = \text{arc tan}\sqrt{\frac{\sqrt{1+\kappa^2} - 1}{\sqrt{1+\kappa^2} + 1}}\; . \tag{12}$$

We observe that the magnetic part $\widehat{\mathbf{b}} = a\, e^{\widehat{\mathbf{n}}\,\alpha}\mathbf{n}\,\mathbf{e}$ has a magnitude a times larger than $|\,\mathbf{e}\,|$, with a given in (11) and not in the plane of $\mathbf{n}\,\mathbf{e}$ but rotated by the angle α given in (12). Thus, during rotations, the magnetic part lags behind the electric part by a constant angle.

The reader may check that the energy-momentum cliffor for wave (9) is

$$\frac{1}{2} f f^\dagger = e^{-2\gamma\, \mathbf{n}\cdot\mathbf{r}} k(k + k_0\, \mathbf{n})|\, \mathbf{c}_2\,|^2 \; .$$

Thus the ratio $|\,\mathbf{S}\,|/w = u\, k_0/k$ is not the maximal possible as for the travelling wave in a nonconducting medium.

If one starts with the opposite angular velocity in the exponent of (3), the whole procedure presented leads to the following solution

$$f' = e^{\gamma\, \mathbf{n}\cdot\mathbf{r}}\, e^{\widehat{\mathbf{n}}\,(\mathbf{k}\cdot\mathbf{r}+\omega\, t)}\,(k_0 + \gamma\, I - \mathbf{k})\,\mathbf{c}_1' + e^{-\gamma\, \mathbf{n}\cdot\mathbf{r}}\, e^{-\widehat{\mathbf{n}}\,(\mathbf{k}\cdot\mathbf{r}-\omega\, t)}(k_0 - \gamma\, I + \mathbf{k})\,\mathbf{c}_2' \; ,$$

replacing (8). The wave propagating only in the direction $+\mathbf{n}$ is obtained by choosing $\mathbf{c}_1' = 0$:

$$f' = e^{-\gamma\,\mathbf{n}\cdot\mathbf{r}}\,e^{-\widehat{\mathbf{n}}\,(\mathbf{k}\cdot\mathbf{r}-\omega\,t)}(k_0 - \gamma\,I + \mathbf{k})\,\mathbf{c}_2' \,.$$

This can be interpreted as a right-handed circularly polarized damped wave. An appropriate combination of this expression with (9), namely for $\mathbf{c}_2' = \mathbf{c}_2$ yields the wave with linear polarization

$$\begin{aligned} f'' &= e^{-\gamma\,\mathbf{n}\cdot\mathbf{r}}\big[e^{\widehat{\mathbf{n}}\,(\mathbf{k}\cdot\mathbf{r}-\omega\,t)}k_0(1 + a\,e^{\widehat{\mathbf{n}}\,\alpha}\mathbf{n})\,\mathbf{c}_2 \\ &\quad + e^{-\widehat{\mathbf{n}}\,(\mathbf{k}\cdot\mathbf{r}-\omega\,t)}k_0(1 + a\,e^{-\widehat{\mathbf{n}}\,\alpha}\mathbf{n})\,\mathbf{c}_2\big] \\ &= e^{-\gamma\,\mathbf{n}\cdot\mathbf{r}}2k_0[\cos(\mathbf{k}\cdot\mathbf{r} - \omega\,t)\,\mathbf{c}_2 + \cos(\mathbf{k}\cdot\mathbf{r} - \omega\,t + \alpha)a\mathbf{n}\,\mathbf{c}_2] \,. \end{aligned}$$

We observe again that the magnetic part has an amplitude a times greater and is retarded in phase by α with respect to the electric part.

§2. The Plane Harmonic Wave in a Nonuniform Medium

Let us consider the case when the constants ε and/or μ are functions of position — this is the case of a *nonuniform medium.* In practice, there is no need to consider the μ variable, so we assume that μ is constant. Since we are looking for electromagnetic waves, we start with the Maxwell equation (1.22) with $\tilde{j} = 0$ and with the third term missing:

$$\mathcal{D} f + \mathbf{e}\,\mathcal{D}\log\sqrt{\varepsilon} = 0 \,. \qquad (13)$$

This is very difficult to solve for an arbitrary function $\varepsilon(\mathbf{r})$. So we make the simplifying assumption that the medium is *plane-stratified* which means that its properties are constant on each plane perpendicular to a fixed direction $\mathbf{n} : \varepsilon(\mathbf{r}) = \varepsilon(\mathbf{n}\cdot\mathbf{r})$. In this case, $\mathcal{D}\log\sqrt{\varepsilon} = \mathbf{n}\sqrt{\varepsilon}'/\sqrt{\varepsilon}$ where prime means the differentiation with respect to the scalar variable $\mathbf{n}\cdot\mathbf{r}$. Thus Eq. (13) takes the form

$$\mathcal{D} f + \frac{\sqrt{\varepsilon}'}{\sqrt{\varepsilon}}\,\mathbf{e}\,\mathbf{n} = 0 \,. \qquad (14)$$

We now assume that f is a plane harmonic wave with round polarization also propagating along $\mathbf{n}$ (the case with a more general direction of propagation is hard to solve):

$$f(\mathbf{r}, t) = g(\mathbf{n}\cdot\mathbf{r})e^{\widehat{\mathbf{n}}\,\omega\,t} = [\mathbf{g}_v(\mathbf{n}\cdot\mathbf{r}) + \widehat{\mathbf{g}}_b(\mathbf{n}\cdot\mathbf{r})]e^{\widehat{\mathbf{n}}\,\omega\,t} \,,$$

where $\mathbf{g}_v \perp \mathbf{n}$, $\widehat{\mathbf{g}}_b \| \mathbf{n}$, which is equivalent to $\mathbf{n}\, g = -g\, \mathbf{n}$. We also denote $\mathbf{g}_v = \mathbf{g}_1, \widehat{\mathbf{g}}_b = \mathbf{n}\, \mathbf{g}_2$, where $\mathbf{g}_1, \mathbf{g}_2 \perp \mathbf{n}$:

$$f = [\mathbf{g}_1(\mathbf{n}\cdot\mathbf{r}) + \mathbf{n}\,\mathbf{g}_2(\mathbf{n}\cdot\mathbf{r})]e^{\widehat{\mathbf{n}}\omega t} \,. \tag{15}$$

In this case,

$$\begin{aligned} \mathcal{D} f = \nabla f + \sqrt{\varepsilon\mu}\frac{\partial f}{\partial t} &= (\mathbf{n}\, g' + \sqrt{\varepsilon\mu}\, g\, \widehat{\mathbf{n}}\,\omega)e^{\widehat{\mathbf{n}}\omega t} \\ &= (\mathbf{n}\, g' - \sqrt{\varepsilon\mu}\,\omega\, \widehat{\mathbf{n}}\, g)e^{\widehat{\mathbf{n}}\omega t} \,. \end{aligned}$$

We substitute this in (14) and divide by the exponential:

$$\mathbf{n}\, g' - \sqrt{\varepsilon\mu}\,\omega\, \mathbf{n}\, I g - \frac{\sqrt{\varepsilon}'}{\sqrt{\varepsilon}}\mathbf{n}\,\mathbf{g}_1 = 0 \,.$$

The common factor $\mathbf{n}$ can also be divided out:

$$g' = \sqrt{\varepsilon\mu}\,\omega\, I g + \frac{\sqrt{\varepsilon}'}{\sqrt{\varepsilon}}\mathbf{g}_1 \,.$$

This can be written as

$$\mathbf{g}_1' + \mathbf{n}\,\mathbf{g}_2' = \sqrt{\varepsilon\mu}\,\omega(I\,\mathbf{g}_1 + I\mathbf{n}\,\mathbf{g}_2) + \frac{\sqrt{\varepsilon}'}{\sqrt{\varepsilon}}\mathbf{g}_1 \,.$$

We equate separately the vector and bivector part:

$$\mathbf{g}_1' = \sqrt{\varepsilon\mu}\,\omega\, I\mathbf{n}\,\mathbf{g}_2 + \frac{\sqrt{\varepsilon}'}{\sqrt{\varepsilon}}\mathbf{g}_1 \,, \tag{16a}$$

$$\mathbf{n}\,\mathbf{g}_2' = \sqrt{\varepsilon\mu}\,\omega\, I\,\mathbf{g}_1 \,. \tag{16b}$$

It is useful to introduce a new function $\mathbf{h}$ through

$$\mathbf{g}_1(\mathbf{n}\cdot\mathbf{r}) = \sqrt{\varepsilon(\mathbf{n}\cdot\mathbf{r})}\,\mathbf{h}(\mathbf{n}\cdot\mathbf{r}) \,. \tag{17}$$

Then Eq. (16a) takes the form

$$\sqrt{\varepsilon}'\,\mathbf{h} + \sqrt{\varepsilon}\,\mathbf{h}' = \sqrt{\varepsilon\mu}\,\omega\,\widehat{\mathbf{n}}\,\mathbf{g}_2 + \sqrt{\varepsilon}'\,\mathbf{h} \,,$$

and yields

$$\mathbf{g}_2 = -\frac{1}{\sqrt{\mu}\,\omega}\widehat{\mathbf{n}}\,\mathbf{h}' \,. \tag{18}$$

We substitute this and (17) in (16b) such that

$$-\frac{1}{\sqrt{\mu}\omega} I\mathbf{h}'' = \varepsilon\sqrt{\mu}\omega\, I\mathbf{h}$$

which gives the equation

$$\mathbf{h}'' + \varepsilon\mu\omega^2\,\mathbf{h} = 0\ . \tag{19}$$

By introducing a new variable $\varsigma = \omega\,\mathbf{n}\cdot\mathbf{r}$, we get $\mathbf{h}' = d\mathbf{h}/d(\mathbf{n}\cdot\mathbf{r}) = \omega\, d\mathbf{h}/d\varsigma = \omega\,\dot{\mathbf{h}}$ where the dot means differentiation with respect to ς. After this substitution, Eq. (19) reduces to

$$\ddot{\mathbf{h}} + \varepsilon\mu\,\mathbf{h} = 0\ .$$

A linear second-order homogeneous differential equation is obtained. In spite of its simple form, the solutions for arbitrary coefficient functions $\varepsilon\mu$ are not known to mathematicians. We choose the specific dependence

$$\mu\varepsilon(\varsigma) = \frac{\alpha^2}{(\varsigma + \alpha c)^2}\ , \tag{20}$$

where α is an arbitrary positive dimensionless constant and $c = 1/\sqrt{\varepsilon_0\mu_0}$ is the velocity of light in the vacuum. Then we get the differential equation

$$\frac{d^2\mathbf{h}}{d\varsigma^2} + \frac{\alpha^2}{(\varsigma + \alpha c)^2}\mathbf{h} = 0\ .$$

According to handbooks on differential equations [19,20], the solutions have two different forms depending on α. Namely, for $\alpha > \frac{1}{2}$,

$$\mathbf{h}(\varsigma) = \left(\frac{\varsigma + \alpha c}{\alpha c}\right)^{\frac{1}{2}} \left[\mathbf{c}_1 \cos\left(s \log\frac{\varsigma + \alpha c}{\alpha c}\right) + \mathbf{c}_2 \sin\left(s \log\frac{\varsigma + \alpha c}{\alpha c}\right)\right]\ , \tag{21}$$

where $s = \sqrt{\alpha^2 - \frac{1}{4}}$, or for $\alpha < \frac{1}{2}$,

$$\mathbf{h}(\varsigma) = \mathbf{c}_1\left(\frac{\varsigma + \alpha c}{\alpha c}\right)^{\frac{1}{2}+s} + \mathbf{c}_2\left(\frac{\varsigma + \alpha c}{\alpha c}\right)^{\frac{1}{2}-s}\ , \tag{22}$$

where $s = \sqrt{\frac{1}{4} - \alpha^2}$, $\mathbf{c}_1, \mathbf{c}_2$ — arbitrary constant vectors (we add the assumption $\mathbf{c}_1, \mathbf{c}_2 \perp \mathbf{n}$).

We have chosen the additive constant αc in the denominator of (20) deliberately such that for $\varsigma = 0$, expression (20) yields $\mu\varepsilon = 1/c^2$, that is, the value $\mu_0\varepsilon_0$ for the vacuum. Equation (20) is illustrated in Fig. 99. The relation $\mu\varepsilon \geq \mu_0\varepsilon_0$ is satisfied for ordinary media, so the physically interesting region for ς is $-\alpha c < \varsigma \leq 0$.

The slope of the curve at the point $\varsigma = 0$ is $d(\varepsilon\mu)/d\varsigma = -2/\alpha c^3$. This implies that the greater the α, the flatter is the curve (see Fig. 100). Thus the limit of the uniform medium is $\alpha \to \infty$. So we choose first solution (21) for discussion.

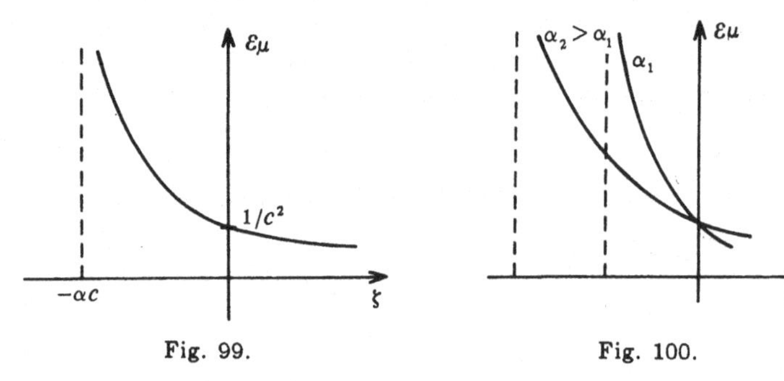

Fig. 99. Fig. 100.

Denote $\varphi = s \log \frac{\varsigma+\alpha c}{\alpha c}$ and substitute (21) in (17):

$$\mathbf{g}_1 = \frac{1}{\sqrt{\mu}} \frac{\alpha}{\varsigma + \alpha c} \mathbf{h} = \frac{1}{c\sqrt{\mu}} \left(\frac{\varsigma + \alpha c}{\alpha c} \right)^{-\frac{1}{2}} (\mathbf{c}_1 \cos\varphi + \mathbf{c}_2 \sin\varphi) \,,$$

and in (18),

$$\mathbf{g}_2 = \frac{-1}{\sqrt{\mu}} \hat{\mathbf{n}} \dot{\mathbf{h}} = \frac{-\hat{\mathbf{n}}}{\sqrt{\mu}\alpha c} \left(\frac{\varsigma + \alpha c}{\alpha c} \right)^{-\frac{1}{2}} \left[\left(\frac{1}{2} \mathbf{c}_1 + s\, \mathbf{c}_2 \right) \cos\varphi + \left(\frac{1}{2} \mathbf{c}_2 - s\, \mathbf{c}_1 \right) \sin\varphi \right] .$$

Thus we can write f according to (15):

$$f = \frac{1}{c\alpha\sqrt{\mu}} \left(\frac{\varsigma + \alpha c}{\alpha c} \right)^{-\frac{1}{2}} \left\{ \left[\alpha \mathbf{c}_1 - I\left(\frac{1}{2} \mathbf{c}_1 + s\, \mathbf{c}_2 \right) \right] \cos\varphi \right.$$
$$\left. + \left[\alpha \mathbf{c}_2 - I\left(\frac{1}{2} \mathbf{c}_2 - s\, \mathbf{c}_1 \right) \right] \sin\varphi \right\} e^{\hat{\mathbf{n}} \omega t} \,. \tag{23}$$

This is the general plane harmonic wave with round polarization passing through the plane stratified medium in the direction perpendicular to the layers when the medium characteristics depend on position as in (20).

One may introduce the refraction index of the medium with respect to the vacuum. According to (4.36),

$$n(\mathbf{n} \cdot \mathbf{r}) = \sqrt{\frac{\varepsilon(\mathbf{n} \cdot \mathbf{r})\mu}{\varepsilon_0 \mu_o}} = c\sqrt{\varepsilon\mu} \,,$$

and — as we notice — depends on the position. By dint of (20), one may write

$$n(\mathbf{n}\cdot\mathbf{r}) = \frac{\alpha c}{\varsigma + \alpha c}\,, \tag{24}$$

and substitute this in (23):

$$f = \frac{1}{c\alpha\sqrt{\mu}}\, n^{\frac{1}{2}}\Big\{\Big[I\Big(\frac{1}{2}\mathbf{c}_1 + s\,\mathbf{c}_2\Big) - \alpha\mathbf{c}_1\Big]\cos(s\,\log n) + \Big[I\Big(\frac{1}{2}\mathbf{c}_2 - s\,\mathbf{c}_1\Big) - \alpha\mathbf{c}_2\Big]\sin(s\,\log n)\Big\}e^{\widehat{\mathbf{n}}\,\omega\,t}\,.$$

It is interesting to notice that this electromagnetic wave depends on position only through the refraction index.

Let us assume that the medium is uniform for $\varsigma > 0$ and has the properties of a vacuum, i.e. $\varepsilon\mu = 1/c^2$. We now look for the particular wave (23), which at $\varsigma = 0$, is equal to the travelling wave outgoing from the plane $\mathbf{n}\cdot\mathbf{r} = 0$ in the direction $+\mathbf{n}$. We know from the discussion in §4.2 that this can be expressed as the condition

$$\mathbf{n}\,f = f \tag{25}$$

for all $\varsigma > 0$ and, by continuity, also for (23) at $\varsigma = 0$. The variable $\varphi = 0$ for $\varsigma = 0$, so the condition (25) applied to (23) yields

$$\mathbf{n}\Big[\alpha\mathbf{c}_1 - I\Big(\frac{1}{2}\mathbf{c}_1 + s\,\mathbf{c}_2\Big)\Big] = \alpha\mathbf{c}_1 - I\Big(\frac{1}{2}\mathbf{c}_1 + s\,\mathbf{c}_2\Big)\,.$$

After separately equating the vector and bivector parts, we obtain twice the same condition

$$\mathbf{c}_2 = \frac{\alpha\widehat{\mathbf{n}} - \frac{1}{2}}{s}\,\mathbf{c}_1\,.$$

Inserting this into (23) yields

$$f = \frac{1}{c\sqrt{\mu}}\Big(\frac{\alpha c}{\varsigma + \alpha c}\Big)^{\frac{1}{2}}\Big[(1{+}\mathbf{n})\Big(\cos\varphi + \frac{\alpha}{s}\,\widehat{\mathbf{n}}\sin\varphi\Big) + \frac{\mathbf{n}-1}{2s}\sin\varphi\Big]\,\mathbf{c}_1\,e^{\widehat{\mathbf{n}}\,\omega\,t}\,. \tag{26}$$

It is interesting to notice that this solution is not periodic in the space variable $\mathbf{n}\cdot\mathbf{r}$. It has some elements of periodicity in the logarithmic variable $\varphi = s\,\log\frac{\omega\mathbf{n}\cdot\mathbf{r}+\alpha c}{\alpha c}$, but this periodicity is difficult to test when n changes too fast. The change of φ from 0 to -2π would demand changing $n = \frac{\alpha c}{\varsigma+\alpha c}$ from 1 to $e^{2\pi/s}$ which could be too large for small s. (Smaller s corresponds to faster changes in n.) Even $s = 1$ must be considered small, because then $e^{2\pi/s} = e^{2\pi} \cong 534$ which is too large to happen in real media.

We omit tedious calculations leading to the following expression for the energy-momentum cliffor

$$\frac{1}{2} f f^\dagger = \frac{\alpha |\mathbf{c}_1|^2}{c\mu(\varsigma + \alpha c)} \left(\cos^2 \varphi + \frac{4\alpha^2 + 1}{4\alpha^2 - 1} \sin^2 \varphi + \mathbf{n} \right) .$$

Its vector part gives the energy flux

$$\mathbf{S} = \frac{1}{\sqrt{\varepsilon\mu}} \frac{\alpha |\mathbf{c}_1|^2}{c\mu(\varsigma + \alpha c)} \mathbf{n} ,$$

which, due to (20), is

$$\mathbf{S} = \frac{|\mathbf{c}_1|^2}{c\mu} \mathbf{n} .$$

We notice that this expression is independent of position. This result is physically correct because it confirms the conservation of energy — the energy passing the same area of different layers is the same. We are allowed to state that the *intensity* of the wave is constant along any straight line parallel to $\mathbf{n}$.

It is worth considering the limiting transition to the uniform medium, that is, $\alpha \to \infty$. Then $\frac{\alpha c}{\varsigma + \alpha c} \to 1$, $\frac{\alpha}{s} \to 1$, $\frac{1}{2s} \to 0$ and

$$\varphi = s \log \frac{\varsigma + \alpha c}{\alpha c} = s \log \left(1 + \frac{\varsigma}{\alpha c} \right) \to s \frac{\varsigma}{\alpha c} \to \frac{\varsigma}{c} = \frac{\omega}{c} \mathbf{n} \cdot \mathbf{r} .$$

So, after denoting $\mathbf{n} \frac{\omega}{c} = \mathbf{k}$, we obtain from (26)

$$f \to \frac{1}{c\sqrt{\mu}} (1 + \mathbf{n}) \left(\cos \frac{\varsigma}{c} + \widehat{\mathbf{n}} \sin \frac{\varsigma}{c} \right) \mathbf{c}_1 e^{\widehat{\mathbf{n}} \omega t}$$
$$= (1 + \mathbf{n}) e^{\widehat{\mathbf{n}} (\mathbf{k} \cdot \mathbf{r} - \omega t)} \frac{\mathbf{c}_1}{c\sqrt{\mu}} ,$$

which is the ordinary circularly polarized wave travelling in the direction $+\mathbf{n}$.

Now we are interested in the other limit $\alpha \to 0$, hence we choose solution (22) for discussion. We insert it into (17):

$$\mathbf{g}_1 = \frac{1}{\sqrt{\mu}} \frac{\alpha}{\varsigma + \alpha c} \mathbf{h} = \frac{1}{c\sqrt{\mu}} \left[\left(\frac{\varsigma + \alpha c}{\alpha c} \right)^{s - \frac{1}{2}} \mathbf{c}_1 + \left(\frac{\varsigma + \alpha c}{\alpha c} \right)^{-s - \frac{1}{2}} \mathbf{c}_2 \right] ,$$

and into (18):

$$\mathbf{g}_2 = \frac{-\widehat{\mathbf{n}}}{\sqrt{\mu} \alpha c} \left[\left(\frac{1}{2} + s \right) \left(\frac{\varsigma + \alpha c}{\alpha c} \right)^{s - \frac{1}{2}} \mathbf{c}_1 + \left(\frac{1}{2} - s \right) \left(\frac{\varsigma + \alpha c}{\alpha c} \right)^{-s - \frac{1}{2}} \mathbf{c}_2 \right] .$$

Finally we write f according to (15)

$$f = \frac{1}{c\alpha\sqrt{\mu}}\left\{\left[\alpha - I\left(\frac{1}{2}+s\right)\right]\left(\frac{\varsigma+\alpha c}{\alpha c}\right)^{s-\frac{1}{2}}\mathbf{c}_1 + \left[\alpha - I\left(\frac{1}{2}-s\right)\right]\left(\frac{\varsigma+\alpha c}{\alpha c}\right)^{-s-\frac{1}{2}}\mathbf{c}_2\right\}e^{\widehat{\mathbf{n}}\omega t}\,.$$

The initial condition (25) for $\varsigma = 0$ implies

$$\mathbf{c}_2 = -\frac{1+2s}{\alpha}(\alpha + s\,\widehat{\mathbf{n}})\,\mathbf{c}_1\,,$$

thus

$$f = \frac{1}{c\alpha\sqrt{\mu}}\left(\frac{\alpha c}{\varsigma+\alpha c}\right)^{\frac{1}{2}}\left\{\left[\alpha - I\left(\frac{1}{2}+s\right)\right]\left(\frac{\varsigma+\alpha c}{\alpha c}\right)^{s} + \frac{1+2s}{\alpha}\left[I\left(\frac{1}{2}-s\right)-\alpha\right](\alpha+s\,\widehat{\mathbf{n}})\left(\frac{\varsigma+\alpha c}{\alpha c}\right)^{-s}\right\}\mathbf{c}_1\,e^{\widehat{\mathbf{n}}\omega t}\,. \tag{27}$$

This solution is completely aperiodic as distinct from (26). We remind the reader that this occurs for $\alpha < \frac{1}{2}$, that is, for a sufficiently steep function (20). This is similar to the well-known phenomenon in mechanics: when the damping is too large, the damped oscillations change into aperiodic motion.

The energy-momentum cliffor for the field (27) is

$$\frac{1}{2}ff^{\dagger} = \frac{(\frac{1}{2}+s)|\mathbf{c}_1|^2}{2c\alpha^2\mu}\,\frac{\alpha}{\varsigma+\alpha c}\left[\left(\frac{\varsigma+\alpha c}{\alpha c}\right)^{2s} + \left(\frac{\varsigma+\alpha c}{\alpha c}\right)^{-2s} + 4(s^2\mathbf{n}-\alpha^2)\right]\,.$$

It contains the energy flux

$$\mathbf{S} = \frac{1}{\sqrt{\varepsilon\mu}}\,\frac{(1+2s)|\mathbf{c}_1|^2}{c\alpha^2\mu}\,\frac{\alpha}{\varsigma+\alpha c}\,s^2\mathbf{n}\,,$$

which, by (20), can be written as

$$\mathbf{S} = \frac{(1+2s)|\mathbf{c}_1|^2}{c\alpha^2\mu}\,s^2\,\mathbf{n}\,.$$

We again find that it is a constant quantity. This expression is singular under the limit $\alpha \to 0$, so we introduce a new constant $\mathbf{A} = \frac{1}{\alpha}\,\mathbf{c}_1$ and insert refraction index (24) into (27):

$$f = \frac{1}{c\sqrt{\mu}}\,n^{\frac{1}{2}}\left\{\left[\alpha - I\left(\frac{1}{2}+s\right)\right]n^{-s} + \frac{1+2s}{\alpha}\left[I\left(\frac{1}{2}-s\right)-\alpha\right](\alpha+s\,\widehat{\mathbf{n}})n^{s}\right\}\mathbf{A}\,e^{\widehat{\mathbf{n}}\omega t}\,. \tag{28}$$

Let us consider the medium characterized by our function (20) in the interval $\varsigma_1 \le \varsigma \le 0$ and constant outside it (see Fig. 101) with the values n_1 for $\varsigma \le \varsigma_1$ and 1 for $\varsigma \ge 0$. With the value n_1 kept constant, the α tending to 0 means a steeper and steeper function (20) and therefore, implies $\varsigma_1 \to 0$. In this manner, the limit $\alpha \to 0$ denotes transition to the step-like behaviour of the refraction index on the interface $\mathbf{n} \cdot \mathbf{r} = 0$ between two media.

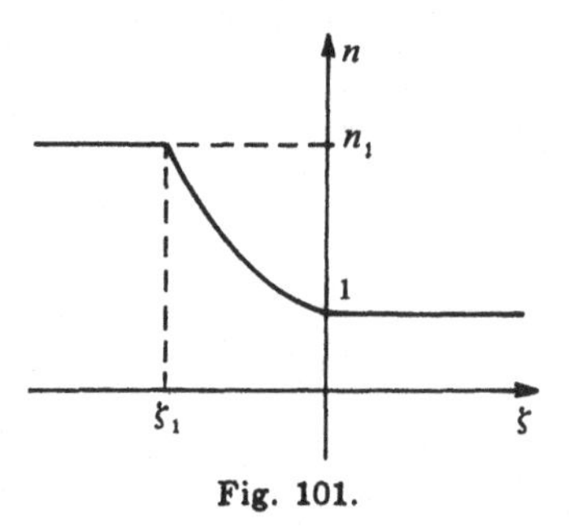

Fig. 101.

We discuss the behaviour of (28) under this limit. By virtue of $s = \sqrt{\frac{1}{4} - \alpha^2}$, we have $s \to \frac{1}{2}$ and $\frac{1}{\alpha}\left(\frac{1}{2} - s\right) \to 0$ for $\alpha \to 0$, hence

$$f \to \frac{1}{c\sqrt{\mu}} n_1^{\frac{1}{2}} \left(- I n_1^{-\frac{1}{2}} - \widehat{\mathbf{n}}\, n_1^{\frac{1}{2}} \right) \mathbf{A}\, e^{\widehat{\mathbf{n}}\,\omega t} = \frac{-1}{c\sqrt{\mu}} (\mathbf{n} + n_1)\, \widehat{\mathbf{n}}\, \mathbf{A}\, e^{\widehat{\mathbf{n}}\,\omega t} .$$

Having denoted $\mathbf{A}' = \frac{-1}{c\sqrt{\mu}} \widehat{\mathbf{n}}\, \mathbf{A}$, we may write

$$f \to (n_1 + \mathbf{n})\, \mathbf{A}'\, e^{\widehat{\mathbf{n}}\,\omega t} = e^{-\widehat{\mathbf{n}}\,\omega t} (n_1 + \mathbf{n}) \mathbf{A}' . \tag{29}$$

This should be compared with the Fresnel formula (4.45) for the normal incidence of the plane wave on an interface (in our case, from the side of negative ς). We put $\varepsilon_2 = \varepsilon_0$ and $\mu_2 = \mu_1 = \mu_0$ in (4.45) and obtain

$$\begin{aligned} f + f^{(r)}\Big|_{\mathbf{n}\cdot\mathbf{r}=0} &= 2g(\omega\, t) \frac{1 + \sqrt{\varepsilon_0/\varepsilon_1}\, \mathbf{n}}{1 + \sqrt{\varepsilon_0/\varepsilon_1}}\, \mathbf{c} \\ &= 2g(\omega\, t) \frac{\sqrt{\varepsilon_1/\varepsilon_0} + \mathbf{n}}{\sqrt{\varepsilon_1/\varepsilon_0} + 1}\, \mathbf{c} = 2g(\omega\, t) \frac{n_1 + \mathbf{n}}{n_1 + 1}\, \mathbf{c} . \end{aligned}$$

This coincides with (29) after identifications $g(\omega\, t) = e^{-\widehat{\mathbf{n}}\,\omega t}$, $[2/(n_1 + 1)]\, \mathbf{c} = \mathbf{A}'$. Thus we are allowed to claim that solution (27) (in the limit of the harp change of the refraction index) reproduces the Fresnel formulae on an interface between two media.

§3. Spherical Waves

As we established in Chap. 4, the travelling plane electromagnetic fields transport energy with the flux density described by the Poynting vector **S**. Up to now, various properties of such fields have been discussed without touching the question of its sources. We now intend to discuss the problem of generating energy transported by the electromagnetic fields.

At this stage, we should solve the Maxwell equation (1.23) for non-static sources. Uniting Lemmas 1 and 4 of §0.12, we obtain a solution to (1.23) by substituting the fundamental solution (0.159) into (0.152):

$$f(\mathbf{r},t) = \frac{1}{4\pi}\int dt'\int dv'\left[\frac{\mathbf{r}-\mathbf{r}'}{|\mathbf{r}-\mathbf{r}'|^3}\delta\left(t-t'-\frac{|\mathbf{r}-\mathbf{r}'|}{u}\right)\right.$$
$$\left.+\frac{1}{u|\mathbf{r}-\mathbf{r}'|}\left(1+\frac{\mathbf{r}-\mathbf{r}'}{|\mathbf{r}-\mathbf{r}'|}\right)\delta'\left(t-t'-\frac{|\mathbf{r}-\mathbf{r}'|}{u}\right)\right]\tilde{\jmath}(\mathbf{r}',t') .$$

Of the two solutions contained in Eq. (0.159), we have chosen the one with the negative sign since it corresponds to the retarded field fulfilling the *causality principle* (the cause $\tilde{\jmath}$ outstrips the effect f). Intergration with the Dirac delta and its derivative yields

$$\begin{aligned} f(\mathbf{r},t) &= \frac{1}{4\pi}\int dv'\frac{\mathbf{r}-\mathbf{r}'}{|\mathbf{r}-\mathbf{r}'|^3}\tilde{\jmath}(\mathbf{r}',t-|\mathbf{r}-\mathbf{r}'|/u) \\ &+\frac{1}{4\pi u}\int dv'\frac{1}{|\mathbf{r}-\mathbf{r}'|}\left(1+\frac{\mathbf{r}-\mathbf{r}'}{|\mathbf{r}-\mathbf{r}'|}\right)\frac{\partial}{\partial t'}\tilde{\jmath}(r',t')\Big|_{t'=t-|r-r'|/u} . \end{aligned} \tag{30}$$

The first integral corresponds to solution (1.33) taken in an appropriate later time. The second term is completely new and does not appear for static sources.

We now consider the field due to a time-varying electric dipole. We apply Eqs. (2.37) and (2.38):

$$\rho(\mathbf{r},t) = -(\mathbf{p}(t)\cdot\nabla)\delta^3(\mathbf{r}) , \qquad \mathbf{j}(\mathbf{r},t) = \frac{d\mathbf{p}(t)}{dt}\delta^3(\mathbf{r}) ,$$

where $\mathbf{p}$ is the electric dipole moment. Thus we should insert

$$\tilde{\jmath}(\mathbf{r}',t') = \frac{1}{\sqrt{\varepsilon}}\rho - \sqrt{\mu}\mathbf{j} = -\frac{1}{\sqrt{\varepsilon}}(\mathbf{p}(t')\cdot\nabla')\delta^3(\mathbf{r}') - \sqrt{\mu}\frac{d\mathbf{p}(t')}{dt'}\delta^3(\mathbf{r}')$$

into (30):

$$-4\pi f = \int dv' \frac{\mathbf{r}-\mathbf{r}'}{|\mathbf{r}-\mathbf{r}'|^3}\left[\frac{1}{\sqrt{\varepsilon}}(\mathbf{p}(t')\cdot\nabla')\,\delta^3(\mathbf{r}) + \sqrt{\mu}\,\frac{d\mathbf{p}(t')}{dt'}\,\delta^3(\mathbf{r}')\right.$$

$$+\frac{1}{u}\int dv' \frac{1}{|\mathbf{r}-\mathbf{r}'|}\left(1+\frac{\mathbf{r}-\mathbf{r}'}{|\mathbf{r}-\mathbf{r}'|}\right)\Bigg]\left[\frac{1}{\sqrt{\varepsilon}}\left(\frac{d\mathbf{p}(t')}{dt'}\cdot\nabla'\right)\delta^3(\mathbf{r}')\right.$$

$$\left.+\sqrt{\mu}\frac{d^2\mathbf{p}(t')}{dt'^2}\delta^3(\mathbf{r}')\right],$$

where $t' = t - |\mathbf{r}-\mathbf{r}'|/u$. Before the integration over $\mathbf{r}'$, the derivative ∇' should be transferred (with the sign reversed) from the Dirac delta onto the other functions under the integral. Since the variables $\mathbf{r}, \mathbf{r}'$ occur only in the difference $\mathbf{r}-\mathbf{r}'$, one can replace $-\nabla'$ by ∇. All these manipulations yield

$$-4\pi f = \frac{1}{\sqrt{\varepsilon}}(p(t')\cdot\nabla)\frac{\mathbf{r}}{r^3} + \frac{\mathbf{r}}{r^3}\left[\frac{1}{\sqrt{\varepsilon}}\nabla\cdot\mathbf{p}(t') + \sqrt{\mu}\dot{\mathbf{p}}(t')\right]$$

$$+\frac{1}{u}\left[\frac{1}{\sqrt{\varepsilon}}(\mathbf{p}(t')\cdot\nabla)\frac{1}{r}\left(1+\frac{\mathbf{r}}{r}\right) + \frac{1}{r}\left(1+\frac{\mathbf{r}}{r}\right)\left[\frac{1}{\sqrt{\varepsilon}}\nabla\cdot\dot{\mathbf{p}}(t') + \sqrt{\mu}\ddot{\mathbf{p}}(t')\right]\right],$$

where this time $t' = t - r/u$. We calculate separately

$$\nabla\cdot\mathbf{p}(t') = \frac{d\mathbf{p}(t')}{dt'}\cdot\nabla t' = \dot{\mathbf{p}}(t')\cdot\left(-\frac{1}{u}\frac{\mathbf{r}}{r}\right) = -\frac{1}{ur}\mathbf{r}\cdot\dot{\mathbf{p}}(t'),$$

and similarly, $\nabla\cdot\dot{\mathbf{p}}(t') = -\frac{1}{ur}\mathbf{r}\cdot\ddot{\mathbf{p}}(t')$. Hence after the substitution $\sqrt{\mu} = 1/u\sqrt{\varepsilon}$, we obtain

$$-4\pi f = (\mathbf{p}\cdot\nabla)\frac{\mathbf{r}}{r^3} + \frac{\mathbf{r}}{ur^3}\left(-\frac{\mathbf{r}}{r}\cdot\dot{\mathbf{p}}+\dot{\mathbf{p}}\right) + \frac{1}{u}(\dot{\mathbf{p}}\cdot\nabla)\frac{1}{r}\left(1+\frac{\mathbf{r}}{r}\right)$$

$$+\frac{1}{u^2r}\left(1+\frac{\mathbf{r}}{r}\right)\left(-\frac{\mathbf{r}}{r}\cdot\ddot{\mathbf{p}}+\ddot{\mathbf{p}}\right).$$

After the directional derivatives $\mathbf{p}\cdot\nabla$ and $\dot{\mathbf{p}}\cdot\nabla$ are calculated, we get

$$4\pi\sqrt{\varepsilon}\,f = \frac{3\,\mathbf{n}(\mathbf{n}\cdot\mathbf{p})-\mathbf{p}}{r^3} + \frac{1}{ur^2}\mathbf{n}(\mathbf{n}\cdot\dot{\mathbf{p}}) - \frac{1}{ur^2}\mathbf{n}\dot{\mathbf{p}} + \frac{2\mathbf{n}(\mathbf{n}\cdot\dot{\mathbf{p}})-\dot{\mathbf{p}}}{ur^2}$$

$$+\frac{1}{ur^2}(\mathbf{n}\cdot\dot{\mathbf{p}}) + \frac{1}{u^2r}(1+\mathbf{n})(\mathbf{n}\cdot\ddot{\mathbf{p}}) - \frac{1}{u^2r}(1+\mathbf{n})\ddot{\mathbf{p}},$$

where $\mathbf{n} = \mathbf{r}/r$. We group in separate brackets multivectors of the same rank

$$4\pi\sqrt{\varepsilon}\,f = \left[-\frac{1}{ur^2}(\mathbf{n}\cdot\dot{\mathbf{p}}) + \frac{1}{ur^2}(\mathbf{n}\cdot\dot{\mathbf{p}}) + \frac{1}{u^2r}(\mathbf{n}\cdot\ddot{\mathbf{p}}) - \frac{1}{u^2r}(\mathbf{n}\cdot\ddot{\mathbf{p}})\right]$$

$$+\left[\frac{3\,\mathbf{n}(\mathbf{n}\cdot\mathbf{p})-\mathbf{p}}{r^3} + \frac{1}{ur^2}\mathbf{n}(\mathbf{n}\cdot\dot{\mathbf{p}}) + \frac{2\mathbf{n}(\mathbf{n}\cdot\dot{\mathbf{p}})-\dot{\mathbf{p}}}{ur^2} + \frac{\mathbf{n}(\mathbf{n}\cdot\ddot{\mathbf{p}})-\ddot{\mathbf{p}}}{u^2r}\right]$$

$$-\left[\frac{1}{ur^2}\mathbf{n}\wedge\dot{\mathbf{p}} + \frac{1}{u^2r}\mathbf{n}\wedge\ddot{\mathbf{p}}\right].$$

The first bracket, containing only scalars, is zero. In the two others, we have vectors and volutors, as is required for the electromagnetic cliffor. In this way, we obtain the expression

$$f(\mathbf{r},t) = \frac{1}{4\pi\sqrt{\varepsilon}} \frac{3\mathbf{n}(\mathbf{n}\cdot\mathbf{p}) - \mathbf{p}}{r^3} + \frac{3\mathbf{n}(\mathbf{n}\cdot\dot{\mathbf{p}}) - \dot{\mathbf{p}}}{ur^2} - \frac{\mathbf{n}\cdot(\mathbf{n}\wedge\ddot{\mathbf{p}})}{u^2 r} + \frac{1}{4\pi\sqrt{\varepsilon}\,u}\left(\frac{\dot{\mathbf{p}}}{r^2} + \frac{\ddot{\mathbf{p}}}{ur}\right)\wedge\mathbf{n}\,, \tag{31}$$

in which the dipole moment $\mathbf{p}$ and its derivatives are taken in time $t' = t - r/u$. We should now discuss the result.

The term are ordered in the brackets according to the descending powers of r in the denominators. The last terms are therefore significant at large distances from the dipole. On the other hand, the first term in the first bracket becomes important for small distances — this term has no counterpart in the second bracket. Thus it describes a purely electric field similar to (2.39), the only difference being the retarded time argument.

Let us write out the field corresponding to the last terms in the brackets:

$$f^{(f)} = \mathbf{e}^{(f)} + \widehat{\mathbf{b}}^{(f)} = \frac{1}{4\pi\sqrt{\varepsilon}\,u^2 r}(1+\mathbf{n})[\ddot{\mathbf{p}}(t - r/u)\wedge\mathbf{n}]\,. \tag{32}$$

The superscript "*f*" denotes here *far fields*. We notice that the magnetic volutor lies in the plane determined by $\ddot{\mathbf{p}}$ and the radius vector $\mathbf{r}$, whereas the electric vector $\mathbf{e}^{(f)}$ is parallel to $\widehat{\mathbf{b}}^{(f)}$ and orthogonal to $\mathbf{r}$. We show their arrangement in Fig. 102. The relations

$$\mathbf{e}^{(f)} = \mathbf{n}\cdot\widehat{\mathbf{b}}^{(f)}\,, \qquad \widehat{\mathbf{b}}^{(f)} = \mathbf{n}\wedge\mathbf{e}^{(f)}\,, \tag{33}$$

take place between the separate parts of (32), similar to the ones for the plane field travelling in the direction $\mathbf{n}$ (compare with §4.2).

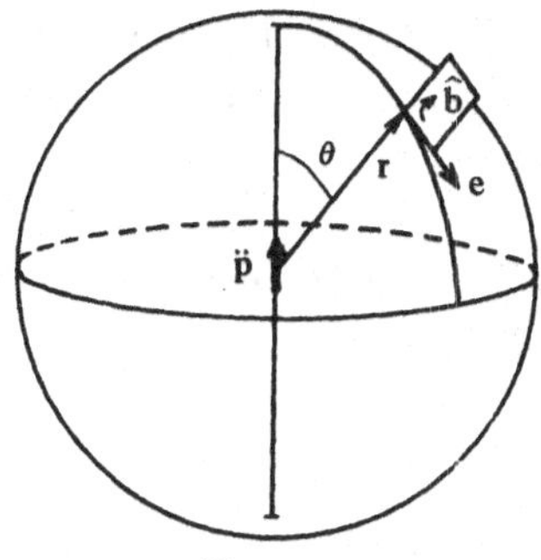

Fig. 102.

The Poynting vector

$$\mathbf{S}^{(f)} = u\,\widehat{\mathbf{b}}^{(f)} \cdot \mathbf{e}^{(f)} = \frac{|\ddot{\mathbf{p}} \wedge \mathbf{n}|^2}{(4\pi)^2 \varepsilon u^3 r^2}\,\mathbf{n}\,,$$

corresponds to field (32). We observe that $\mathbf{S}$ has the direction of $\mathbf{r}$ which means that the energy propagates radially out from the dipole. But the propagation is not isotropic, since after introducing the angle θ between $\mathbf{r}$ and $\ddot{\mathbf{p}}$ we get

$$|\mathbf{S}(\mathbf{r},\theta)| = \frac{|\ddot{\mathbf{p}}|^2 \sin^2\theta}{(4\pi)^2 \varepsilon u^3 r^2}\,. \tag{34}$$

The energy is not sent at all parallelly to $\ddot{\mathbf{p}}$, whereas the largest flux density is for $\theta = \pi/2$, that is, on the equator of the sphere in Fig. 102. The power r^2 in the denominator denotes that the same energy passes at each unit of time through each sphere with its center in the dipole position:

$$W = \int_{r=\text{const}} d\mathbf{s} \cdot \mathbf{S} = \frac{1}{(4\pi)^2 \varepsilon u^3} \int_0^{\pi} d\theta r^2 \sin^2\theta \int_0^{2\pi} d\phi \frac{\ddot{p}^2 \sin^2\theta}{r^2}\,,$$

$$W = \frac{\ddot{p}^2}{6\pi\varepsilon u^3}\,. \tag{35}$$

This energy flux reaches each sphere of arbitrarily large radius. This means that the dipole *radiates* electromagnetic energy or *emits electromagnetic radiation.* Expression (35) is treated as the *power of radiation.* This is the reason why the far field (32), formed of the last terms in the brackets of (31), is also called the *radiation field.* The other terms of (31) describe phenomena occurring in regions close to the dipole and do not transport the energy to an arbitrary distance because the surface integral of the Poynting vector, formed of these terms, over a sphere tends to zero when its radius tends to infinity. The terms contain lower derivatives of the dipole moment, therefore it is concluded that the electromagnetic radiation is determined by the second time derivative of the dipole moment of the system of charges. Field (32) is called *electric dipole radiation.*

Equation (30) can also be applied to the time-varying magnetic dipole. The charge and current densities

$$\rho(\mathbf{r},t) = 0\,, \quad \mathbf{j}(\mathbf{r},t) = (\widehat{\mathbf{M}}(t) \cdot \boldsymbol{\nabla})\delta^3(\mathbf{r})\,,$$

found in §2.4 should be inserted in it. Calculations similar to the previous ones give

$$f(\mathbf{r},t) = \frac{\sqrt{\mu}}{4\pi}\left[\frac{2\widehat{\mathbf{M}} - 3\mathbf{n}(\mathbf{n}\cdot\widehat{\mathbf{M}})}{r^3} + \frac{2\dot{\widehat{\mathbf{M}}} - 3\mathbf{n}(\mathbf{n}\cdot\dot{\widehat{\mathbf{M}}})}{ur^2} - \frac{\mathbf{n}(\mathbf{n}\cdot\ddot{\widehat{\mathbf{M}}})}{u^2 r}\right]$$

$$- \frac{\sqrt{\mu}}{4\pi u}\mathbf{n}\cdot\left(\frac{\dot{\widehat{\mathbf{M}}}}{r^2} + \frac{\ddot{\widehat{\mathbf{M}}}}{ur}\right),$$

where the magnetic moment $\widehat{\mathbf{M}}$ and its derivatives depend on $t' = t - r/u$. The following far field

$$f^{(f)} = -\frac{\sqrt{\mu}}{4\pi u^2 r}(1+\mathbf{n})[\mathbf{n}\cdot\ddot{\widehat{\mathbf{M}}}(t - r/u)] \tag{36}$$

corresponds to this. Field (36) also satisfies relations (33); the electric part $\mathbf{e}^{(f)} \sim \mathbf{n}\cdot\ddot{\widehat{\mathbf{M}}}$ lies in the plane of $\ddot{\widehat{\mathbf{M}}}$ and is orthogonal to the radius vector $\mathbf{r}$, whereas the magnetic part $\widehat{\mathbf{b}}^{(f)} = \mathbf{n}\wedge\mathbf{e}^{(f)}$ lies in the plane determined by $\mathbf{e}^{(f)}$ and $\mathbf{r}$. We show the arrangement of these quantities in Fig. 103. Notice that the whole electromagnetic field at a fixed point $\mathbf{r}$ is perpendicular to the field of Fig. 102 at the same point. The reason is that we have chosen $\ddot{\widehat{\mathbf{M}}}$ of Fig. 103 orthogonal to $\ddot{\mathbf{p}}$ of Fig. 102. For the points lying in the plane of $\ddot{\widehat{\mathbf{M}}}$, the field is parallel to this plane.

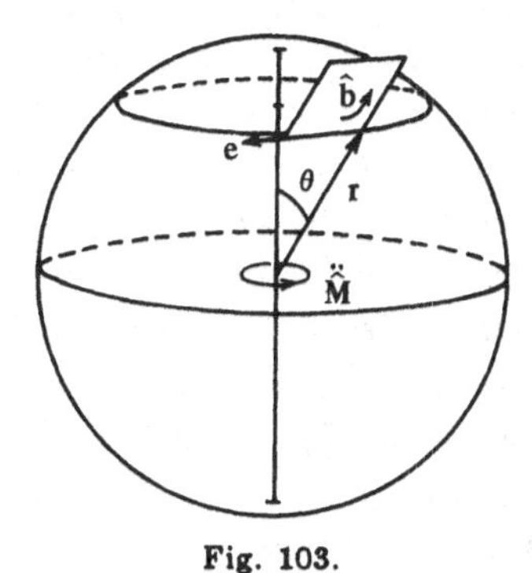

Fig. 103.

The electromagnetic field described by (36) is called *magnetic dipole radiation*. It is determined by the second derivative of the magnetic moment. The Poynting vector in this case is

$$\mathbf{S}^{(f)} = \frac{\mu|\mathbf{n}\cdot\ddot{\widehat{\mathbf{M}}}|}{(4\pi)^2 u^3 r^2}\mathbf{n},$$

and its magnitude is the following function of the angle θ between $\mathbf{n}$ and $\ddot{\mathbf{M}}$:

$$\mathbf{S}^{(f)}(r,\theta) = \frac{\mu\,|\ddot{\mathbf{M}}|^2 \sin^2\theta}{(4\pi)^2 u^3 r^2} \,. \tag{37}$$

The energy flux distribution — as we see — is the same as for the electric dipole radiation, hence its power is

$$W = \frac{\mu\,|\ddot{\mathbf{M}}|^2}{6\pi u^3} \,.$$

We now assume specific time dependence of the electric dipole moment. Let it be the harmonic oscillation

$$\mathbf{p}\,(t) = \mathbf{p}_0 \cos\omega\, t \,. \tag{38}$$

This kind of source was anticipated theoretically and realized experimentally by Heinrich Hertz in 1888. Therefore we call it the *Hertz dipole*. Substitution of (38) into (32) yields

$$f^{(f)} = \frac{\omega^2}{4\pi\sqrt{\varepsilon}\,u^2 r}\,(1+\mathbf{n})\,\mathbf{n}\wedge\mathbf{p}_0\cos(\omega\, t - kr) \,, \tag{39}$$

where $k = \omega/u$. We see that the solution is a harmonic periodic function of time, therefore it can be called the *harmonic electromagnetic wave*. If $\omega\, t - kr$ is treated as the *phase*, the loci of points with constant phase, known as the *wave fronts*, are spheres, in this case, therefore (39) is referred to as the *spherical wave*. With the flow of time, the radii of spheres of constant phase grow. In this way, the direction of the phase velocity is $\mathbf{n}$ for this wave and coincides with the direction of the Poynting vector. The electric vector at a fixed point changes only its magnitude and orientation — hence this is the *linearly polarized wave*. The plane of light oscillations is determined by $\mathbf{r}$ and $\mathbf{p}_0$. Figure 102 illustrates it well.

The power of the radiation field (39) is

$$W = \frac{\omega^4 p_0^2}{6\pi\varepsilon\, u^3}\cos^2(\omega\, t - kr) \,,$$

and after averaging over the period,

$$\langle W\rangle = \frac{\omega^4 p_0^2}{12\pi\varepsilon\, u^3} \,.$$

It is visible from this that the radiation has small power for low frequency. Therefore the Hertz experiments were successful only when the dipole was realized in the form of a spark gap with sufficiently high frequency.

Similarly, the harmonic oscillation of the magnetic moment

$$\widehat{\mathbf{M}}(t) = \widehat{\mathbf{M}}_0 \cos(\omega t + \alpha) , \tag{40}$$

substituted into (36) yields

$$f^{(f)} = \frac{\sqrt{\mu}\omega^2}{4\pi u^2 r}(1 + \mathbf{n})\mathbf{n} \cdot \widehat{\mathbf{M}}_0 \cos(\omega t - kr + \alpha) . \tag{41}$$

This is also a spherical harmonic electromagnetic wave with linear polarization. The plane of light oscillations is perpendicular to that of wave (39) if $\widehat{\mathbf{M}}_0 \perp \mathbf{p}_0$.

One may also consider the two sources (38) and (40) existing simultaneously at the same point and oscillating with the same circular frequency ω. In this case, the far field is a superposition of (39) with (41). If we assume the phase shift $\alpha = \pi/2$ between the sources, the same phase shift occurs for the radiation fields. Since the fields (39) and (41) are mutually orthogonal, the electric part of the superposed field traces an ellipse (this is a general property for the composition of nonparallel harmonic oscillations with the same frequency), so the net wave has *elliptic polarization.* The ellipses at each point $\mathbf{r}$ have the same ratio of major to minor semiaxes. If $\sqrt{\mu}\,|\mathbf{M}_0| = \frac{1}{\sqrt{\varepsilon}}\mathbf{p}_0$, the wave has circular polarization.

Another time dependence of the electric dipole moment is worth discussing. Let it be a rotation with circular frequency $\widehat{\boldsymbol{\omega}}$ and the initial moment $\mathbf{p}_0 \parallel \widehat{\boldsymbol{\omega}}$:

$$\mathbf{p}(t) = e^{-\frac{1}{2}\widehat{\boldsymbol{\omega}} t}\,\mathbf{p}_0\, e^{\frac{1}{2}\widehat{\boldsymbol{\omega}} t} = \mathbf{p}_0\, e^{\widehat{\boldsymbol{\omega}} t} . \tag{42}$$

The condition $\mathbf{p}_0 \parallel \widehat{\boldsymbol{\omega}}$ ensures that vector $\mathbf{p}(t)$ always lies in the plane of $\widehat{\boldsymbol{\omega}}$ and $\mathbf{p}_0\widehat{\boldsymbol{\omega}} = -\widehat{\boldsymbol{\omega}}\mathbf{p}_0$ which was used in the second equality (42). We insert the derivatives of (42)

$$\dot{\mathbf{p}}(t) = \mathbf{p}(t)\widehat{\boldsymbol{\omega}}, \quad \ddot{\mathbf{p}}(t) = \mathbf{p}(t)\widehat{\boldsymbol{\omega}}^2 = -\omega^2\mathbf{p}(t)$$

into (32):

$$f^{(f)}(\mathbf{r}, t) = \frac{k^2}{4\pi\sqrt{\varepsilon}\, r}(1 + \mathbf{n})\{\mathbf{n} \wedge [\mathbf{p}_0\, e^{\widehat{\boldsymbol{\omega}}(t - \frac{r}{u})}]\} . \tag{43}$$

A comparison of this with (4.69) leads to the observation that this is the spherical harmonic wave with *elliptic polarization.* At the points $\mathbf{r} \perp \widehat{\boldsymbol{\omega}}$, it reduces to circular polarization and at the points $\mathbf{r} \parallel \widehat{\omega}$, to linear polarization. In fact, the ellipse of electric oscillations has the same shape as the image of the circle $t \to \mathbf{p}_0 e^{\widehat{\omega} t}$ seen from the point $\mathbf{r}$ at which the electromagnetic field is considered. We illustrate the ellipses in Fig. 104. Points on the sphere correspond to various vectors $\mathbf{n}$. The equator is in the plane of $\widehat{\omega}$. This figure looks similar to Fig. 98 illustrating the sphere of possible elliptic polarizations, however the main distinction is that the ellipses of Fig. 104 are tangent to the sphere, whereas the ellipses of Fig. 98 are all parallel to the plane of the equator.

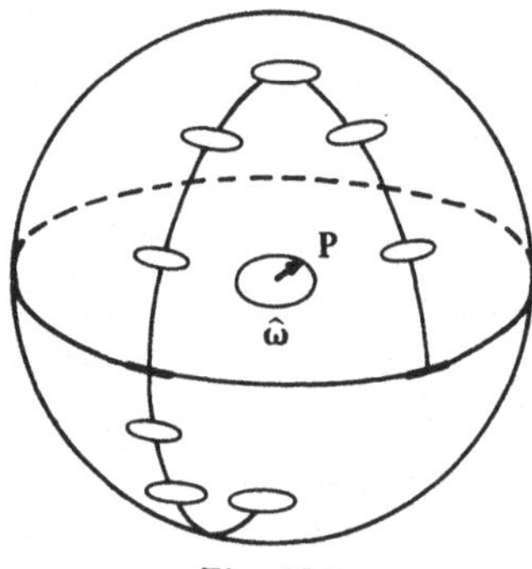

Fig. 104.

For distances r far from the radiating dipoles, in regions small in comparison with r, the spherical waves can be approximately treated as the plane waves (see Problem 1). This observation justifies the notion that the plane waves are also emitted by the radiating dipoles, but such waves are only approximations of the spherical waves.

Problems

1. Show that at far distances from the radiating dipole, in sufficiently small regions, the spherical waves (39), (41) and (43) can be conceded as plane waves.
2. Let the magnetic moment depend on time according to $\widehat{\mathbf{M}}(t) = \widehat{\mathbf{M}}_0 e^{\widehat{\omega} t}$ where $\widehat{\mathbf{M}}_0 \perp \widehat{\omega}$. Find the polarization of the wave obtained by substitution of this function into (36).

Chapter 6

SPECIAL RELATIVITY

§1. Lorentz Transformations

We shall not consider the physical premises leading to the creation of special relativity. We start by accepting Einstein's postulate called the *relativity principle*: all inertial reference frames are physically equivalent. In this case, the whole of electrodynamics and, in particular, the Maxwell equations have the same form in all inertial frames — this also concerns the conclusions following from the equations. The phase velocity of light in the vacuum is one such conclusion: according to (4.27) it is $c = 1/\sqrt{\varepsilon_0 \mu_0} = 299792458$ m/s. Thus we arrive at the following conclusion from the relativity principle: the velocity of light in the vacuum is the same in all inertial frames.

If a space is filled by a medium (dielectric or magnetic substance), a family of frames can be distinguished, namely, those in which the medium rests. The light has the velocity $u = 1/\sqrt{\varepsilon\mu}$ only in those frames. In other frames, the light may have other velocities.

We shall now derive the transformation formulae relating coordinates of events which are to be used in various frames if the postulate of equal velocity of light is to be satisfied. We remind the reader of the *Galilei* (that is, pre-relativistic) *transformations* in one space dimension:

direct	inverse
$x' = x - Vt$	$x = x' + Vt'$
$t' = t$	$t = t'$,

relating the space (x, x') and time (t, t') coordinates used in two frames moving with relative velocity V. The equations on the left show that the origin of the second frame $x' = 0$ is characterized in the first frame by the equation $x = Vt$ which describes the uniform motion with velocity V.

The simplest generalization of the transformation formulae consists of admitting a constant coefficient γ:

$$x' = \gamma(x - Vt) \, , \tag{1a}$$

$$x = \gamma(x' + Vt') \, . \tag{1b}$$

The adjective "constant" means here: independent of x and t; γ may, however, depend on the relative velocity V. We assume that γ is the same in both equations (1) by virtue of the prerequisite that the inverse transformation should be identical to the original one except for a change of V to $-V$. We admit also the possibility of transforming the time, which is already implicit in (1), as we shall see in a moment. Equations (1) are the simplest modifications of the Galilei transformations which preserve the implication $x' = 0 \Longrightarrow x = Vt$.

Let us calculate γ. For this purpose, we consider a special set of events matching the movement of a light signal $x = ct$. By virtue of the Einstein postulate, the similar condition $x' = ct'$ should correspond to it in the second frame. Thus Eqs. (1) take the form

$$x' = \gamma(ct - Vt) \, , \qquad x = \gamma(ct' + Vt') \, ,$$

or

$$ct' = \gamma(c - V)t \, , \qquad ct = \gamma(c + V)t' \, ,$$

for this set. We divide the two equations

$$\frac{c}{\gamma(c+V)} = \frac{\gamma(c-V)}{c} \, ,$$

hence

$$\gamma = \gamma(V) = \frac{1}{\sqrt{1 - V^2/c^2}} \, . \tag{2}$$

This is the expression which we were looking for. The coefficient γ is known as the *Lorentz factor*.

We now insert (1a) into (1b):

$$x = \gamma^2(x - Vt) + Vt' .$$

After using the explicit form (2) of γ, we solve this equation for t':

$$t' = \gamma\left(t - \frac{V}{c^2}x\right) ,$$

which is the transformation law for the time coordinate of the events. One similarly obtains

$$t = \gamma(t' + \frac{V}{c^2}x') .$$

The fact that time is also transformed is referred to as the *relativity of time*.

Summarizing, we have the following transformation laws for the time and space coordinates:

$$\begin{array}{ll} \underline{\text{direct}} & \underline{\text{inverse}} \\ x' = \dfrac{x - Vt}{\sqrt{1 - V^2/c^2}} & x = \dfrac{x' + Vt'}{\sqrt{1 - V^2/c^2}} \\ t' = \dfrac{t - Vx/c^2}{\sqrt{1 - V^2/c^2}} & t = \dfrac{t' + Vx'/c^2}{\sqrt{1 - V^2/c^2}} , \end{array}$$

known as the *Lorentz transformations.* Notice that for $x \ll ct$, when the ratio V/c tends to zero, the formulae tend towards the Galilei transformation formulae.

The Lorentz and Galilei transformations change a set of events matching a resting particle into a set of events matching the particle moving with velocity V or $-V$. Thus they transform the state of rest into movement. Therefore they are also called *boosts*. The boost should not be confused with acceleration because the change of velocity is not gradual in time, it is rather an abstract change of motion which is the same for all times.

The Lorentz transformations assume more symmetrical form in the time and space coordinates if a new time variable is introduced: $x_0 = ct$. This is a "time" measured in meters, equal to the distance travelled by the light in time interval t. For such a choice of units, the parameter $\beta = V/c$, as the quotient $\Delta x/\Delta x_0$, corresponds to the velocity. Then the Lorentz formulae assume the form

$$x' = \frac{x - \beta x_0}{\sqrt{1-\beta^2}} \qquad x = \frac{x' + \beta x_0'}{\sqrt{1-\beta^2}}$$

$$x_0' = \frac{x_0 - \beta x}{\sqrt{1-\beta^2}} \qquad x_0 = \frac{x_0' + \beta x'}{\sqrt{1-\beta^2}} \, .$$

Diagrams on a plane of the variables x_0, x are called *Minkowski diagrams.* Let us find, as an instance, a set of events matching the X'-axis determined by the condition $x_0' = 0$ which, by virtue of the above formula, denotes $x_0 = \beta x$. So the X'-axis is the straight line inclined to the X-axis by the angle φ such that $\tan\varphi = \beta$. Similarly, the events of the X_0'-axis determined by the condition $x' = 0$ fulfil the equation $x = \beta x_0$, so the X_0'-axis is the straight line inclined to the X_0-axis by the same angle φ. If the initial coordinate frame (X_0, X) is rectangular, the new frame (X_0', X') cannot be rectangular after the Lorentz transformation — we show both coordinate frames in Fig. 105. Similarly, it can be verified that when the frame (X_0', X') is rectangular, the axes X_0 and X are inclined, respectively, also by the angle φ but at the other side, as shown in Fig. 106. These observations lead to the conclusion that the naive notion of right angle (as we know it from the geometry of a sheet of paper) between the time and space axes is not invariant under the Lorentz transformations.

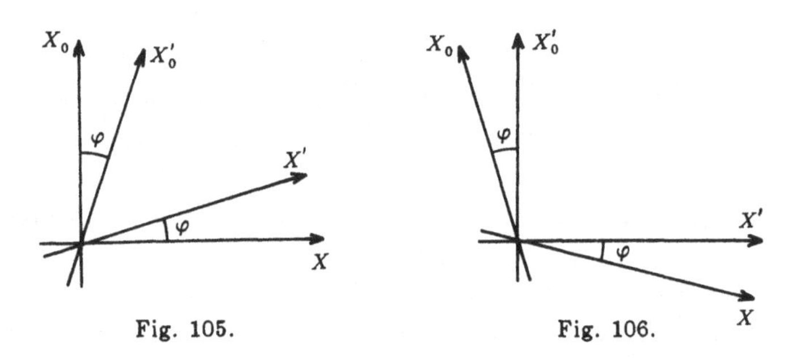

Fig. 105. Fig. 106.

The time coordinate x_0 and space coordinate x are assigned to an arbitrary event A by drawing lines parallel to the axes of the frame and finding their intersection points with the axes (see Fig. 107). The lines parallel to the X-axis collect events with the same x_0 coordinate, that is, events simultaneous for the first observer, and the lines parallel to X' — the events simultaneous for the second observer. The lines of one family are oblique to that of the other and this means that the events simultaneous for one observer are not so for the other. In this way, the ***relativity of simultaneity*** is manifested. We show in Fig. 108 two events A, B simultaneous for the first observer, but B is earlier than A for the second observer.

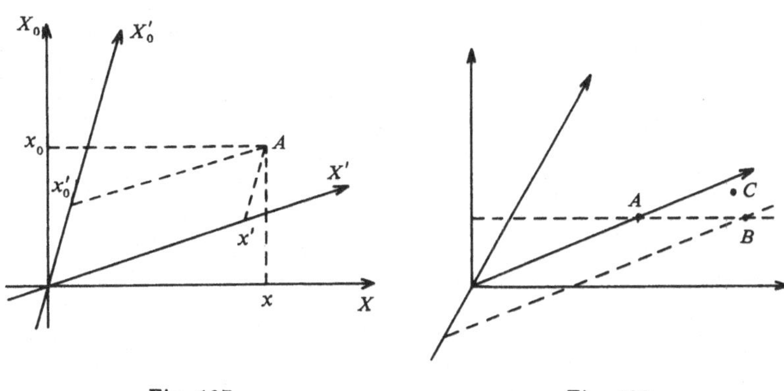

Fig. 107. Fig. 108.

Also, a reversion of the time order of events may take place — for instance event C depicted in Fig. 108 is later than A for the first observer, but is earlier than A for the second one. Such a reversion does not occur for all pairs of events (see Problem 8).

It was this relativity of simultaneity, along with the abovementioned relativity of time, which was the most striking conclusion at the time when the theory was being created. This was the reason why the Einstein postulate was given the name: the relativity principle and the theory based on it — the ***relativity theory***. In order to distinguish it from the later theory comprising also frames in nonuniform motion and gravitational phenomena, it acquired the name: special relativity theory or, in short: ***special relativity***.

Up to now, we have discussed events occurring only on one space axis parallel to the direction of the relative motion of the reference frames. So the transformation formulae must be supplemented by other ones for the two other dimensions. It is usual to assume that the two other Cartesian coordinates are invariant, so that the complete set of Lorentz transformation formulae is

$$x_0' = \frac{x_0 - \beta x}{\sqrt{1-\beta^2}}\,, \quad x' = \frac{x - \beta x_0}{\sqrt{1-\beta^2}}\,, \quad y' = y\,, \quad z' = z\,.$$

Henceforth they will be written as follows

$$x^{0'} = \frac{x^0 - \beta x^1}{\sqrt{1-\beta^2}}, \tag{3a}$$

$$x^{1'} = \frac{x^1 - \beta x^0}{\sqrt{1-\beta^2}}, \tag{3b}$$

$$x^{2'} = x^2, \qquad x^{3'} = x^3, \tag{3c}$$

understanding that the space radius vector has three coordinates $\mathbf{r} = (x^1, x^2, x^3)$. Introducing the columns

$$x = \begin{pmatrix} x^0 \\ x^1 \\ x^2 \\ x^3 \end{pmatrix}, \qquad x' = \begin{pmatrix} x^{0'} \\ x^{1'} \\ x^{2'} \\ x^{3'} \end{pmatrix},$$

we write (3) in the matrix form: $x' = \Lambda x$, where $\Lambda = \Lambda(\beta)$ is the following 4×4 matrix:

$$\Lambda = \begin{pmatrix} \gamma & -\beta\gamma & 0 & 0 \\ -\beta\gamma & \gamma & 0 & 0 \\ 0 & 0 & 1 & 0 \\ 0 & 0 & 0 & 1 \end{pmatrix}. \tag{4}$$

If after the Lorentz transformation corresponding to the velocity V_1, a further Lorentz transformation is performed corresponding to V_2 in the same direction, the velocity of the last frame with respect to the first one is

$$V = \frac{V_1 + V_2}{1 + \frac{V_1 V_2}{c^2}}, \tag{5}$$

which in terms of parameters β can be written as

$$\beta = \frac{\beta_1 + \beta_2}{1 + \beta_1\beta_2}.$$

The Eq. (5) is a particular case of the so-called *relativistic velocity addition law* (see Problem 1). In terms of the matrices (4), this denotes the equality

$$\Lambda(\beta_2)\Lambda(\beta_1) = \Lambda\left(\frac{\beta_2 + \beta_1}{1 + \beta_1\beta_2}\right).$$

Hence the velocity V (and the parameter β corresponding to it) is not an additive quantity under superposition of the Lorentz transformations.

It is convenient to introduce a new parameter ϑ through the relation

$$\beta = \frac{V}{c} = \tanh\vartheta \,, \tag{6}$$

which implies

$$\gamma = \frac{1}{\sqrt{1-\beta^2}} = \cosh\vartheta \,, \qquad \beta\gamma = \sinh\vartheta \,. \tag{7}$$

We insert this in the matrix (4):

$$\Lambda(\vartheta) = \begin{pmatrix} \cosh\vartheta & -\sinh\vartheta & 0 & 0 \\ -\sinh\vartheta & \cosh\vartheta & 0 & 0 \\ 0 & 0 & 1 & 0 \\ 0 & 0 & 0 & 1 \end{pmatrix} . \tag{8}$$

The reader may check (Problem 4) that the new parameter is additive, that is, $\Lambda(\vartheta_2)\Lambda(\vartheta_1) = \Lambda(\vartheta_2+\vartheta_1)$. The similarity of (8) to the rotation matrix with the trigonometric functions replaced by hyperbolic ones is the reason why (8) is called the *hyperbolic rotation* and ϑ — the *hyperbolic angle.* Another term, *rapidity,* is also used for ϑ. The relation

$$\vartheta = \frac{1}{2}\log\frac{1+\beta}{1-\beta} \,, \tag{9}$$

inverse to (6) can be derived (see Problem 5).

Ordinary rotations are said to be performed *around an axis.* The hyperbolic rotations (8) are made in four-dimensional linear space with coordinates (x^0, x^1, x^2, x^3), so two axes X^2 and X^3 can be found perpendicular to each other, which are invariant under (8). Therefore, the transformations (4) and (8) do not distinguish any single invariant axis. If they distinguish any axis, it is rather the X^1-axis, along which the two reference frames move. Thus the transformation (4) or (8) is referred to as the *boost along the X^1-axis.*

The Lorentz transformation (3) has an interesting property:

$$\begin{aligned}(x^{0'})^2 - (x^{1'})^2 &= \gamma^2[(x^0)^2 - 2\beta x^0x^1 + \beta^2(x^1)^2 - (x^1)^2 + 2\beta x^1x^0 - \beta^2(x^0)^2] \\ &= \gamma^2[(1-\beta^2)(x^0)^2 - (1-\beta^2)(x^1)^2] = (x^0)^2 - (x^1)^2 \,.\end{aligned}$$

We used (2) for γ in this derivation. We ascertain that (3) preserves the following quadratic form formed of the radius vector $\mathbf{r}$ and the time coordinate x^0:

$$(x^{0'})^2 - \mathbf{r}'^2 = (x^0)^2 - \mathbf{r}^2 \,. \tag{10}$$

In this respect, the Lorentz transformation has one more similarity to rotations. The rotations around the origin of coordinates preserve the square of length of the vectors. Expression (10) cannot, however, be treated as a square of any length since it can be both positive and negative, i.e. the form (10) is not *positive definite.*

We shall now find the boost in an arbitrary direction $\mathbf{n}$ with the velocity $\mathbf{V} = V\mathbf{n}$. The transformation of time (3a) can easily be generalized by introducing the vector parameter $\boldsymbol{\beta} = \mathbf{V}/c$ and noticing that $\boldsymbol{\beta} \cdot \mathbf{r}$ reduces to βx^1 when $\mathbf{V}$ is parallel to the X^1-axis. In this way, we obtain

$$x^{0'} = \frac{x^0 - \boldsymbol{\beta} \cdot \mathbf{r}}{\sqrt{1-\beta^2}} = \gamma(x^0 - \boldsymbol{\beta} \cdot \mathbf{r}) \,. \tag{11}$$

The component $\mathbf{r}_{\|}$ of $\mathbf{r}$ parallel to $\mathbf{n}$ should transform similarly to (3b)

$$\mathbf{r}'_{\|} = \frac{\mathbf{r}_{\|} - \boldsymbol{\beta} x^0}{\sqrt{1-\beta^2}} = \gamma(\mathbf{r}_{\|} - \boldsymbol{\beta} x^0) \,. \tag{12}$$

The perpendicular component $\mathbf{r}_{\perp}$ is not transformed: $\mathbf{r}'_{\perp} = \mathbf{r}_{\perp}$, so, adding this to (12), we get

$$\mathbf{r}' = \gamma(\mathbf{r}_{\|} - \boldsymbol{\beta} x^0) + \mathbf{r}_{\perp} \,,$$

which, after the substitution of $\mathbf{r}_{\|} = \mathbf{n}\,(\mathbf{n} \cdot \mathbf{r}), \mathbf{r}_{\perp} = \mathbf{r} - \mathbf{n}\,(\mathbf{n} \cdot \mathbf{r})$ can be written as

$$\mathbf{r}' = (\gamma - 1)\,\mathbf{n}\,(\mathbf{n} \cdot \mathbf{r}) + \mathbf{r} - \gamma\boldsymbol{\beta} x^0 \,. \tag{13a}$$

We add to this (11)

$$x^{0'} = \gamma(x^0 - \boldsymbol{\beta} \cdot \mathbf{r}) \tag{13b}$$

and ascertain that (13a,b) are the formulae for a Lorentz boost with the arbitrary velocity $\mathbf{V} = \boldsymbol{\beta} c = \beta c\,\mathbf{n}$. ("Arbitrary" means here: of arbitrary direction, not magnitude, because $|\mathbf{V}| = V$ must be less than c, otherwise (2) makes no sense.) Of course, this transformation also fulfils condition (10).

For $\mathbf{n} \cdot \mathbf{r} \ll x^0$, when $\boldsymbol{\beta}$ tends to zero, the formulae (13) become

$$x^{0'} = x^0 \,, \qquad \mathbf{r}' = \mathbf{r} - \boldsymbol{\beta} x^0 \,,$$

i.e.

$$t' = t \,, \qquad \mathbf{r}' = \mathbf{r} - \mathbf{V} t \,,$$

which is the Galilei boost with the arbitrary velocity $\mathbf{V}$.

§2. Minkowski Space

Property (10) of the Lorentz transformations suggests introducing a four-dimensional vector space containing our three-dimensional space together with time — it is called the *space-time*. Its elements are four-coordinate vectors called *four-vectors* or *space-time vectors* $x = (x^0, \mathbf{x}) = (x^0, x^1, x^2, x^3)$[1)] and correspond to the events (since one such vector determines a moment x^0 on the time axis and a position $\mathbf{x}$ in the three-dimensional space). We shall refer to x^0 as the *time part* of x and to $\mathbf{x} = (x^1, x^2, x^3)$ as the *space part* of x.

We define the following *scalar product*

$$x \cdot y = x^0 y^0 - \mathbf{x} \cdot \mathbf{y} = x^0 y^0 - x^1 y^1 - x^2 y^2 - x^3 y^3 \tag{14}$$

of two vectors, which for twice the same factor reproduces the quadratic form (10):

$$x^2 = x \cdot x = (x^0)^2 - \mathbf{x}^2 , \tag{15}$$

which could be named the *scalar square* of a vector.

The four-dimensional vector space endowed with the scalar product (14) is called the *Minkowski space* — this is a particular case of a *pseudo-Euclidean space*. One may check (Problem 6) that the Lorentz transformation (13) preserves (14) and this is the main reason for introducing the Minkowski space.

One may compare the scalar product (14) with the Euclidean scalar product:

$$(x, y) = x^0 y^0 + x^1 y^1 + x^2 y^2 + x^3 y^3 , \tag{16}$$

which can also be introduced in the four-dimensional vector space. There is a relation $x \cdot y = (x, Gy)$ between them, where G is the matrix

$$G = \begin{pmatrix} 1 & 0 & 0 & 0 \\ 0 & -1 & 0 & 0 \\ 0 & 0 & -1 & 0 \\ 0 & 0 & 0 & -1 \end{pmatrix} , \tag{17}$$

with the elements $g_{\mu\nu}, \mu, \nu \in \{0, 1, 2, 3\}$. Thus we may write (14) as

$$x \cdot y = \sum_{\mu,\nu=0}^{3} x^\mu g_{\mu\nu} y^\nu . \tag{18}$$

[1)] The four-vectors x will not be denoted by boldface type. This will only be reserved for the three-vectors of the Euclidean space.

The collection of elements $\{g_{\mu\nu}\}$ is traditionally called the *metric tensor.*

The mapping $y \to Gy$ reverses all the space coordinates y^1, y^2, y^3 and leaves the time coordinate y^0 invariant, hence it is inversion in the time axis, which we present in Fig. 109. In such a case, the vectors x and y yield $x \cdot y = 0$ with respect to (14) if, and only if, x and Gy are perpendicular with respect to (16). We illustrate this in Fig. 110. The relation $x \cdot y = 0$ is still called *orthogonality* of the vectors x and y. The axes X_0' and X' are orthogonal in Fig. 105, similarly X_0 and X in Fig. 106. This is not strange, since the Lorentz transformation preserves the scalar product (14), and therefore, also preserves the orthogonality of the axes. In particular, it may happen that $x \cdot x = 0$ for $x \neq 0$ — of course when x and Gx are perpendicular in the Euclidean sense. This occurs for all vectors inclined by the angle $\pi/4$ to the X^0-axis.

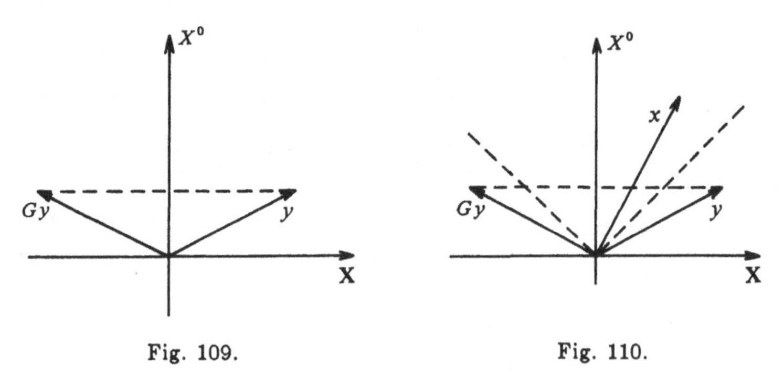

Fig. 109.　　　　Fig. 110.

Let us find a physical meaning of the expression (15). Let x be a vector connecting two events concerning a single particle moving uniformly with velocity $\mathbf{v}: x = (c\Delta t, \Delta\mathbf{r})$. Then

$$x^2 = c^2(\Delta t)^2 - (\Delta\mathbf{r})^2 .$$

Dividing both sides by $(\Delta t)^2$,

$$\frac{x^2}{(\Delta t)^2} = c^2 - \left(\frac{\Delta\mathbf{r}}{\Delta t}\right)^2 = c^2 - \mathbf{v}^2 .$$

This expression is nonnegative when $|\mathbf{v}|$ does not exceed c. The value zero occurs only if the particle moves with the velocity of light in the vacuum. When the particle rests, $x = (c\Delta\tau, 0)$ and then $x^2 = c^2(\Delta\tau)^2$. The time is here denoted by the Greek letter τ in order to display the fact that it describes a resting particle.

For a particle moving with velocity $v < c$, a reference frame exists in which it is resting (it suffices to perform the Lorentz transformation (13) with $\mathbf{V} = \mathbf{v}$), hence for each pair of events, from the history of the particle we may write $x' = (c\Delta\tau, 0)$. On the other hand, in a different frame, the same pair of events is connected by vector $x = (c\Delta t, \Delta\mathbf{r})$. We have $(x')^2 = x^2$ from (10), so

$$c^2(\Delta\tau)^2 = c^2(\Delta t)^2 - (\Delta\mathbf{r})^2 \,. \tag{19}$$

It is clear from this equality that $(\Delta\tau)^2$ and hence $\Delta\tau$ is invariant under the Lorentz transformations. Equation (19) displays, moreover, that the scalar square of a vector distance between two events concerning a single particle is proportional to the square of the time distance $\Delta\tau$ in the frame, where the particle is at rest. We find from (19) the expression for $\Delta\tau$, referred to as the *proper time* of the particle: $\Delta\tau = \Delta t\sqrt{1 - v^2/c^2}$ or

$$\Delta t = \gamma(\mathbf{v})\Delta\tau \,, \tag{20}$$

where velocity $\mathbf{v}$ is inserted, instead of $\mathbf{V}$ into γ given by (2). In this way, an interpretation is given to the Lorentz factor.

Equation (20) is called the *time dilation formula*. It relates the time separation Δt obtained from resting clocks (there must be at least two clocks in the positions of the two considered events) with the time separation $\Delta\tau$ shown on a clock in uniform motion. Since $\gamma > 1$ for $\mathbf{v} \neq 0$, then $\Delta t > \Delta\tau$ which is just the *time dilation*. This is another manifestation of the relativity of time.

If the vector $x = (c\Delta t, \Delta\mathbf{r})$ has a negative square, then

$$c^2 - \left(\frac{\Delta\mathbf{r}}{\Delta t}\right)^2 < 0 \,,$$

which means $|\Delta\mathbf{r}/\Delta t| > c$. If two events connected by such a vector x were joined by some signal, the signal should travel faster than light in the vacuum. Up to now, no particles have been discovered which achieve such velocities, so a pair of events connected by a vector x with $x^2 < 0$ cannot concern a single moving physical object. Two simultaneous events are separated by the vector $x = (0, \Delta\mathbf{r})$ with $x^2 < 0$, so the pure space vector has the negative scalar square.

The examples considered justify introducing the following terminology. Vectors x from the Minkowski space are called:

(i) *time-like* if $x^2 > 0$, i.e. $|x^0| > |\mathbf{x}|$;

(ii) *light-like* if $x^2 = 0$, i.e. $|x^0| = |\mathbf{x}|$;

(iii) *space-like* if $x^2 < 0$, i.e. $|x^0| < |\mathbf{x}|$.

Thus the Minkowski space can be divided into the sets shown in Fig. 111 (in which one space dimension is absent). The light-like vectors form a three-dimensional hypersurface called the *light cone* (its part with $x^0 > 0$ is called the *future light cone*, the part with $x^0 < 0$ — the *past light cone*), the time-like vectors lie inside this cone and are directed upwards or downwards, the space-like vectors lie outside the cone.

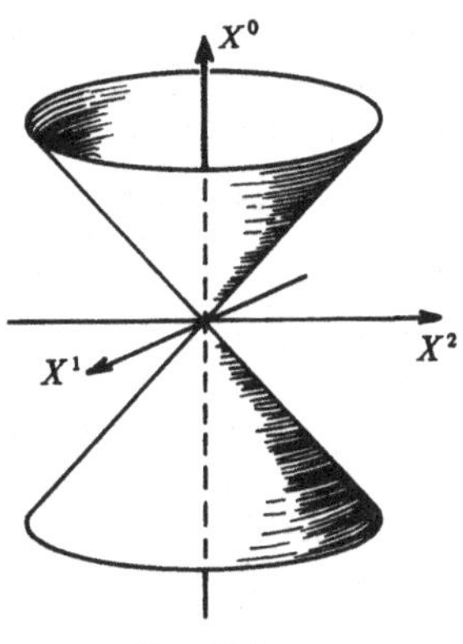

Fig. 111.

The Lorentz transformations preserve x^2, so they preserve the sets defined by the equations $x^2 = \text{const}$. According to the sign, they are:

(i) for $x^2 > 0$ hyperboloids of two sheets (if the time is not reversed, each sheet is preserved separately);

(ii) for $x^2 = 0$ the light cone;

(iii) for $x^2 < 0$ hyperboloids of one sheet.

Examples of sets of type (i) and (ii) are shown in Figs. 112 and 113, respectively.

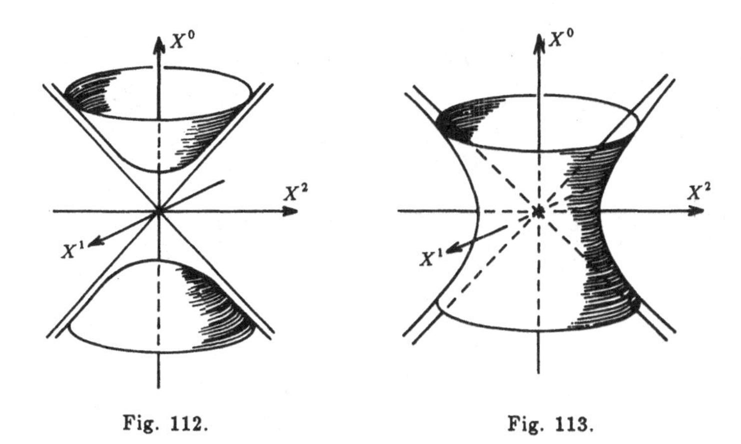

Fig. 112. Fig. 113.

It can be seen from the above considerations that if one is accustomed to a vector as a quantity having a direction and a magnitude (interpreted as the length), one is forced to resign from one feature, namely, the magnitude, or at least from the Lorentz invariant magnitude. The reason is that the only Lorentz invariant quantity built of coordinates of a single vector x is the expression (15) which is not positive definite, so one cannot take the real root of it. In this manner, we get a generalization of the notion of vector from Euclidean geometry. The vector is now a quantity to which direction can be ascribed, but not length (serving to compare nonparallel vectors). Some remnant, however, of length remains in comparing parallel vectors. We may, namely, state that for real λ, the vector λx is $|\lambda|$ times *longer* than x. Therefore, we say that λx has the *relative length* $|\lambda|$ with respect to x.

One may also introduce a *pseudolength* of x as in the expression $|x| = \sqrt{|x^2|}$, but it does not fulfil the triangle inequality: $|x+y| \leq |x| + |y|$. Nevertheless, we call a vector the *unit vector* if its pseudolength is one.

The invariance properties, discussed previously, suggest the following generalization of the terminology: each linear transformation of the Minkowski space, which preserves the scalar product (14), is named the *Lorentz transformation.* This definition is so broad that it includes, besides the boosts (13), the ordinary rotations, changing only the space part $\mathbf{x}$ of vectors $x = (x^0, \mathbf{x})$ since they preserve $\mathbf{x}^2$ and hence also x^2.

The preservation of (14) under a Lorentz transformation Λ is expressed by the equality

$$\Lambda x \cdot \Lambda y = x \cdot y \,, \tag{21}$$

that is,

$$(\Lambda x, G\,\Lambda y) = (x, Gy) \ .$$

By virtue of the well known properties of the Euclidean scalar product, we may write this as

$$(x, \Lambda^T G\,\Lambda y) = (x, Gy) \ ,$$

where Λ^T denotes the transpose of matrix Λ. Since x and y are arbitrary vectors, this implies the matrix equality

$$\Lambda^T G \Lambda = G \ . \tag{22}$$

It follows from the derivation that this condition is equivalent to (21) with arbitrary x and y, so (22) may be treated as the condition for Λ to determine the Lorentz transformation.

Let us calculate the determinant of (22):

$$|\Lambda^T||G||\Lambda| = |G| \ ,$$
$$|\Lambda^T||\Lambda| = 1 \ .$$

By dint of the identity $|\Lambda^T| = |\Lambda|$, we obtain $|\Lambda|^2 = 1$, so only two possibilities remain: $|\Lambda| = \pm 1$ as for the orthogonal matrices. A Lorentz transformation is called *proper* if it does not change sign of the time variable x^0 and has determinant $+1$. The condition $|\Lambda| = 1$ means that the Jacobian of the transformation is one, hence the four-dimensional "volume" element $d^4x = dx^0 dx^1 dx^2 dx^3$ is invariant for the proper Lorentz transformations. This is one more similarity to the rotations.

Condition (22) can be used to find the inverse matrix to Λ. We multiply (22) by G^{-1} from the left:

$$G^{-1}\Lambda^T G \Lambda = 1 \ ,$$

and by Λ^{-1} from the right (this matrix exists because it is a nonsingular matrix):

$$G^{-1}\Lambda^T G = \Lambda^{-1} \ . \tag{23}$$

The transposition of this equality yields

$$(\Lambda^{-1})^T = G^T \Lambda (G^{-1})^T = G \Lambda G^{-1} \ , \tag{24}$$

since G and G^{-1} are symmetric matrices.

The previous considerations allow us to introduce more general definitions of boosts and rotations as some particular Lorentz transformations. A proper Lorentz transformation Λ is called the *boost* if two time-like vectors x and y exist such that $y = x\Lambda$ and Λ leaves invariant all vectors orthogonal to the plane spanned by x and y. A proper Lorentz transformation is called the *space-like rotation* if it leaves invariant a time-like vector x. In this case, it is the rotation in the three-dimensional space orthogonal to x.

A basis in the Minkowski space must consist of four vectors. If the basis is *orthogonal* which means that the basis vectors are orthogonal to each other, only one of its vectors can be time-like (see Problem 9) — we denote it as e_0. We choose other vectors e_1, e_2, e_3 as space-like. We assume further that all the basis vectors have the unit pseudolength, i.e. $|e_\mu^2| = 1, \mu \in \{0,1,2,3\}$, which implies

$$e_0^2 = 1, \quad e_1^2 = e_2^2 = e_3^2 = -1 .$$

Such a basis is called *orthonormal.* All the above assumptions may be summarized in the formula

$$e_\mu \cdot e_\nu = g_{\mu\nu} \quad \text{for} \quad \mu,\nu \in \{0,1,2,3\} , \tag{25}$$

where $g_{\mu\nu}$ are elements of matrix (17), that is, the coordinates of the metric tensor.

One may introduce also the *dual basis* $\{e^\nu\}$ to $\{e_\mu\}$ defined as the basis which satisfies the condition $e^\nu \cdot e_\mu = \delta_\mu^\nu$ where δ_μ^ν is the Kronecker delta. One can check (Problem 10) that this basis has the properties

$$e^\mu \cdot e^\nu = g^{\mu\nu} , \quad e^\nu = \sum_{\kappa=0}^{3} g^{\nu\kappa} e_\kappa , \tag{26}$$

where $g^{\nu\kappa}$ are elements of the matrix inverse to G.

It is useful here to adopt two conventions; the first is that the Greek indices always run from 0 to 3. The second, known as the *Einstein convention,* states that when the repeated index (one an upper index, the other a lower index) occurs, the summation is understood over this index from 0 to 3. With these conventions the second formula (26) may be written as $e^\nu = g^{\nu\kappa} e_\kappa$.

An arbitrary vector x from the Minkowski space can be expressed by the basis $\{e_\mu\}$ as $x = x^\mu e_\mu$ or due to (26) as $x = x_\nu e^\nu$, where

$$x_\nu = g_{\nu\lambda} x^\lambda . \tag{27}$$

The coefficients x^μ are known as the *contravariant coordinates* and x_ν as the *covariant coordinates* of x. The relation

$$x^\lambda = g^{\lambda\nu} x_\nu , \tag{28}$$

is the inverse of (27). The scalar product of vectors can also be expressed by the covariant coordinates:

$$x \cdot y = x_0 y_0 - x_1 y_1 - x_2 y_2 - x_3 y_3 . \tag{29}$$

When the contravariant coordinates are subjected to the Lorentz transformation Λ,

$$x^\mu \to x'^\mu = \Lambda^\mu{}_\nu \, x^\nu ,$$

we find the transformation law for x_ν as follows:

$$x'_\mu = g_{\mu\lambda} \, x'^\lambda = g_{\mu\lambda} \, \Lambda^\lambda{}_\kappa \, x^\kappa = g_{\mu\lambda} \, \Lambda^\lambda{}_\kappa \, g^{\kappa\nu} \, x_\nu ,$$

where relations (27) and (28) were used. Thus the matrix of the transformation $x_\mu \to x'_\mu$ can be written as $G\Lambda G^{-1}$ and, due to (24), as $(\Lambda^{-1})^T$. The transformation determined by $(\Lambda^{-1})^T$ is called *contragradient* to Λ, so we state that the covariant coordinates of a vector transform under contragradient transformation to the Lorentz transformation of the contravariant coordinates.

Let us check how the partial derivatives transform under the Lorentz transformation:

$$\frac{\partial}{\partial x'^\mu} f(x) = \frac{\partial}{\partial x'^\mu} f(\Lambda^{-1}x') = \frac{\partial f(x)}{\partial x^\nu} (\Lambda^{-1})^\nu{}_\mu = (\Lambda^{-1})^\nu{}_\mu \frac{\partial}{\partial x^\nu} f(x) .$$

If we treat $\frac{\partial}{\partial x^0}, \frac{\partial}{\partial x^1}, \frac{\partial}{\partial x^2}, \frac{\partial}{\partial x^3}$ as coordinates of some vector, we see that they transform under the matrix $(\Lambda^{-1})^T$, so we ascertain that they should be considered as covariant coordinates ∂_ν. Thus we introduce the *gradient vector* ∂ in the Minkowski space through the formula

$$\partial = e^\nu \frac{\partial}{\partial x^\nu} = e^\nu \partial_\nu .$$

The operator-vector ∂ is the Minkowski space counterpart of the nabla operator.

The d'Alembert operator (0.129) for the vacuum assumes the form

$$\Box = \Delta - \frac{1}{c^2}\frac{\partial^2}{\partial t^2} = \left(\frac{\partial}{\partial x^1}\right)^2 + \left(\frac{\partial}{\partial x^2}\right)^2 + \left(\frac{\partial}{\partial x^3}\right)^2 - \frac{1}{c^2}\left(\frac{\partial}{\partial t}\right)^2 ,$$

or

$$\Box = \partial_1^2 + \partial_2^2 + \partial_3^2 - \partial_0^2 \,.$$

Due to (29), we may write $\Box = -\partial \cdot \partial$.

§3. Clifford Algebra of the Minkowski Space

In the Minkowski space, multivectors can also be introduced and a Clifford algebra built out of them. Essential changes appear, however, in comparison with the notions introduced in Chap. 0. They are caused firstly, by the four dimensions of the space and secondly, by the nonpositive definite scalar product. The first reason implies that non-simple bivectors occur, the second one — that the magnitude of a multivector cannot be introduced.

The notions serving to introduce the *two-dimensional direction* can be introduced for the Minkowski space in the same manner as in §0.1, because no appeal is made there to any magnitudes and the planes may be embedded in space of any dimension. Yet some way of identifying or distinguishing two parallel flat figures with compatible orientations is needed. We introduce this as follows.

Two parallel flat figures can be placed on one plane by parallel translations — let it be the plane passing through the origin and determined by two nonparallel vectors u_1 and u_2. Then for any vector $x = \alpha^1 u_1 + \alpha^2 u_2$ from this plane, we introduce its *Euclidean length* $|x| = \sqrt{(\alpha^1)^2 + (\alpha^2)^2}$ — this length determines the *Euclidean metric* on the plane: $\rho(x, y) = |x - y|$. This metric can be used to measure areas of figures in the plane: $d\alpha^1 d\alpha^2$ is an element of the area. In this way, the parallelogram U determined by vectors u_1 and u_2 has unit area and the area of all other figures can be named the *relative area* with respect to U.

Now we are ready to introduce our definition. A *volutor* is the equivalence class of parallel flat figures with compatible orientations of their boundaries and with equal areas in any Euclidean metric of the plane determined by their two-dimensional direction.[1] Each volutor D can be represented as the *outer product* of two vectors in the same manner as described in §0.1:

$$D = a \wedge b \,,$$

that is, as the equivalence class of the directed parallelogram based on vectors a and b.

---

[1] The reader should convince himself that if two figures have equal areas in one Euclidean metric, they would also have equal areas in any other one.

We define *bivector* as the abstract finite sum of volutors, assuming that the addition of bivectors has all the customary properties, including the distributive law of the outer multiplication:

$$a \wedge (b + c) = a \wedge b + a \wedge c\ .$$

The volutor will also be called the *simple bivector*. The set of all bivectors forms the linear space over real numbers. If $\{e_0, e_1, e_2, e_3\}$ is a basis in the Minkowski space, the six volutors

$$e_0 \wedge e_1, \quad e_0 \wedge e_2, \quad e_0 \wedge e_3, \quad e_1 \wedge e_2, \quad e_2 \wedge e_3, \quad e_3 \wedge e_1\ ,$$

form a basis in the linear space of bivectors, so it is six-dimensional. Each bivector D can be written

$$D = \frac{1}{2} D^{\mu\nu}\, e_\mu \wedge e_\nu\ , \tag{30}$$

where the coefficients $D^{\mu\nu}$ are antisymmetric: $D^{\mu\nu} = -D^{\nu\mu}$, which implies that, among 16 coefficients $D^{\mu\nu}$, only six are independent. The same bivector D can be written as

$$D = \frac{1}{2}\, D_{\mu\nu}\, e^\mu \wedge e^\nu\ .$$

$D^{\mu\nu}$ are called *contravariant* and $D_{\mu\nu}$ — *covariant coordinates* of D. One may derive the following relations between them:

$$D_{\mu\nu} = g_{\mu\lambda}\, g_{\nu\kappa}\, D^{\lambda\kappa}\ , \qquad D^{\mu\nu} = g^{\mu\lambda}\, g^{\nu\kappa}\, D_{\lambda\kappa}\ . \tag{31}$$

Expression (30) can be written

$$D = e_0 \wedge D^{0j}\, e_j + \frac{1}{2} D^{jk} e_j \wedge e_k\ , \tag{32}$$

with the summation over j and k from 1 to 3. The first term on the right-hand side is a volutor as the other product of two vectors e_0 and $D^{0j}\, e_j$. The second term is also a volutor, because it is composed of volutors from a three-dimensional space generated by three vectors e_1, e_2, e_3. In this manner, we come to the conclusion that arbitrary bivector D (from the Minkowski space) can be represented as a sum of two volutors:

$$D = a \wedge b + c \wedge d = D_1 + D_2 \tag{33}$$

where vector a is orthogonal to b, c, d and c is orthogonal to d (we shall use the notation $c \perp d$). The decomposition (33), of course, is not unique.

The inner products of a bivector with a vector or bivector are reduced by the distributivity to the inner products of its constitutent volutors with the same factor, for example,

$$D \cdot a = D_1 \cdot a + D_2 \cdot a \,.$$

The inner products of a volutor with a vector, as well as the inner and mingled products of two volutors are defined in the same manner as in Chap. 0, basing all the definitions on the scalar product (14) of vectors. Let us calculate, as an instance, the inner product of the bivector (33) with itself:

$$D \cdot D = (a \wedge b) \cdot (a \wedge b) + (a \wedge b) \cdot (c \wedge d) + (c \wedge d) \cdot (a \wedge b) + (c \wedge d) \cdot (c \wedge d) \,.$$

The two middle terms vanish since $a \perp c, d$. Thus, due to (0.25) and (0.17),

$$\begin{aligned} D \cdot D &= a \cdot [b \cdot (a \wedge b)] + c \cdot [d \cdot (c \wedge d)] = a \cdot [-a(b \cdot b)] + c \cdot [-c(d \cdot d)] \\ &= -a^2 b^2 - c^2 d^2 \,. \end{aligned}$$

If all four vectors a, b, c, d have nonzero squares, they need not be of the same sign. Therefore the inner square of a bivector may have two possible signs, like the scalar square of a vector.

The definition of trivector is the same as in §0.3. One can show that in the Minkowski space, all trivectors are simple (Problem 13). Four basic trivectors exist:

$$e_0 \wedge e_1 \wedge e_2, \quad e_0 \wedge e_2 \wedge e_3, \quad e_0 \wedge e_3 \wedge e_1, \quad e_1 \wedge e_2 \wedge e_3 \,,$$

hence the linear space of trivectors is four-dimensional. Each trivector T can be written as the sum

$$T = \frac{1}{6} T^{\mu\nu\lambda} \, e_\mu \wedge e_\nu \wedge e_\lambda \,,$$

or

$$T = \frac{1}{6} T_{\mu\nu\lambda} \, e^\mu \wedge e^\nu \wedge e^\lambda \,.$$

The coefficients $T^{\mu\nu\lambda}$ and $T_{\mu\nu\lambda}$ are antisymmetric under any interchange of the indices and we refer to them as the *contravariant* and *covariant coordinates* of T, respectively. The factor $\frac{1}{6} = \frac{1}{3!}$ appears since all permutations of the chosen values of the indices yield the same terms. From among 64 coefficients $T^{\mu\nu\lambda}$, only four are independent.

For the reader familiar with the tensor formalism, we add that $D^{\mu\nu}$ and $D_{\mu\nu}$ are second rank antisymmetric tensors, $T^{\mu\nu\lambda}$ and $T_{\mu\nu\lambda}$ are third rank antisymmetric tensors. The correspondence between k-vectors and k-th rank antisymmetric tensors is one-to-one.

The definitions of the inner products of trivectors with multivectors of the first, second and third rank can be easily repeated from Chap. 0 with the substitution of (14) as the scalar product. Of course, the inner squares of trivectors may have both signs.

There is room in the Minkowski space for multivectors of the fourth rank. A *quadrivector* is a geometric quantity characterized by a *magnitude* understood as the four-dimensional "volume" and an *orientation* as the composition of the helical motion with the translational motion in the fourth dimension. The set of all quadrivectors forms a one-dimensional linear space, since any quadrivector Q is proportional to the basic one:

$$Q = \alpha\, e_0 \wedge e_1 \wedge e_2 \wedge e_3 \ , \tag{34}$$

where α is a real number.

After introducing the four-dimensional *totally antisymmetric symbol*

$$\varepsilon_{\mu\nu\lambda\kappa} = \begin{cases} 1 & \text{if } (\mu,\nu,\lambda,\kappa) \text{ is an even permutation of } (0,1,2,3) \\ -1 & \text{if } (\mu,\nu,\lambda,\kappa) \text{ is an odd permutation of } (0,1,2,3) \\ 0 & \text{if any two indices coincide ,} \end{cases}$$

we express the anticommutativity of the basis vectors as follows:

$$e_\mu \wedge e_\nu \wedge e_\lambda \wedge e_\kappa = \varepsilon_{\mu\nu\lambda\kappa}\, e_0 \wedge e_1 \wedge e_2 \wedge e_3 \ . \tag{35}$$

We know that the proper Lorentz transformations preserve the four-dimensional "volume", therefore the quadrivectors are invariant under the proper Lorentz transformations.

The outer product of a trivector with a vector yields, obviously, a quadrivector. One may also obtain a quadrivector as the outer product of two bivectors. Let us calculate, as an instance, the outer square of the bivector (33):

$$D \wedge D = (a \wedge b) \wedge (a \wedge b) + (a \wedge b) \wedge (c \wedge d) + (c \wedge d) \wedge (a \wedge b) + (c \wedge d) \wedge (c \wedge d) \ .$$

The extreme terms vanish because they contain repeating factors. We use associativity and anticommutativity of the outer product of vectors:

$$D \wedge D = a \wedge b \wedge c \wedge d + c \wedge d \wedge a \wedge b = 2\, a \wedge b \wedge c \wedge d \ . \tag{36}$$

Lemma 1. A bivector D is simple if, and only if, $D \wedge D = 0$.

Proof. The necessary condition. Let D be simple, we take $c = 0$ in (33). This implies in (36) $D \wedge D = 0$.

The sufficient condition. We represent D in the form (33) with $c \perp b, c, d$ and $c \perp d$.

(i) If $a = 0$ then $D = c \wedge d$ and D is simple.

(ii) If $a \neq 0$ then, from $0 = D \wedge D = 2\, a \wedge b \wedge c \wedge d$, we conclude $b \wedge c \wedge d = 0$. If c and d are linearly dependent, $c \wedge d = 0$ and D is simple. If c and d are linearly independent, from $b \wedge c \wedge d = 0$, we obtain $b = \alpha c + \beta d$ with scalars α, β. Insert this in (33):

$$D = a \wedge (\alpha c + \beta d) + c \wedge d \,.$$

In this equation, D is composed of volutors from a three-dimensional space generated by three vectors a, c, d, hence D is simple. ∎

The definitions of the inner products of a quadrivector with any multivector are easy generalizations of Eqs. (0.54) and (0.55) based on the scalar product (14).

We introduce the *cliffor* as the sum of multivectors of arbitrary ranks. The set of all cliffors forms a linear space of the dimension $1+4+6+4+1 = 16 = 2^4$. We define in this set the *Clifford product* as in §0.6 with the scalar product (14) — in this manner, we obtain the *Clifford algebra of the Minkowski space*, known also as the *Dirac algebra*. The identities (0.78)–(0.80) are also valid here, hence the basis vectors satisfy the identity

$$e_\mu e_\nu + e_\nu e_\mu = 2\, e_\mu \cdot e_\nu = 2 g_{\mu\nu} \,, \tag{37}$$

and similarly for the dual basis vectors,

$$e^\mu e^\nu + e^\nu e^\mu = 2\, e^\mu \cdot e^\nu = 2\, g^{\mu\nu} \,.$$

Let us calculate the square of the general quadrivector (34):

$$\begin{aligned} Q^2 &= \alpha^2 e_0 e_1 e_2 e_3 e_0 e_1 e_2 e_3 = -\alpha^2 e_0 e_1 e_2 e_0 e_1 e_2 e_3^2 \\ &= \alpha^2 e_0 e_1 e_2 e_0 e_1 e_2 = \alpha^2 e_0 e_1 e_0 e_1 e_2^2 = -\alpha^2 e_0 e_1 e_0 e_1 \\ &= \alpha^2 e_0^2 e_1^2 = -\alpha^2 \leq 0 \,. \end{aligned}$$

We ascertain that all quadrivectors have nonpositive squares. We denote the unit basic quadrivector by $J = e_0 e_1 e_2 e_3$. Let us check whether J commutes with one of the basis vectors:

$$e_0 J = e_0^2 e_1 e_2 e_3 = -e_0 e_1 e_0 e_2 e_3 = e_0 e_1 e_2 e_0 e_3 = -e_0 e_1 e_2 e_3 e_0 = -J e_0 \,.$$

One may similarly ascertain the anticommutativity with other basis vectors: $e_\mu J = -Je_\mu$. It follows then that quadrivectors anticommute with all vectors, commute with all bivectors and anticommute with all trivectors.

One may define the *Hodge map* with the aid of J:

$$M \to JM ,$$

transforming a k-vector M into the $(4-k)$-vector JM. This mapping is linear and one-to-one. Now it is not strange that the linear space of vectors has the same dimension as that of trivectors.

Let b be a vector such that $b^2 \neq 0$. Then vector $c = b/b^2$ has the property $bc = cb = 1$, so we call c the *inverse* of b and denote $c = b^{-1}$. For any other vector a we have

$$a = (ab)b^{-1} = (a \cdot b)b^{-1} + (a \wedge b)b^{-1} .$$

Both terms at the right-hand side are vectors. Since $b \| b^{-1}$, the first term commutes with b, whereas the second one satisfies:

$$b[(a \wedge b)b^{-1}] = \frac{1}{2}b(ab - ba)b^{-1} = \frac{1}{2}ba - \frac{1}{2}b^2ab^{-1} .$$

Element b^2 is a scalar, so

$$b[(a \wedge b)b^{-1}] = \frac{1}{2}ba - \frac{1}{2}ab = -a \wedge b = -[(a \wedge b)b^{-1}]b .$$

We see that $(a \wedge b)b^{-1}$ anticommutes with b, therefore it is orthogonal to b. Thus we may treat $a_{\|} = (a \cdot b)b^{-1}$ as the component of a *parallel* to b whilst $a_\perp = (a \wedge b)b^{-1}$ as the component orthogonal to b.

Let a volutor be given $D = a \wedge b$. Can it be decomposed into orthogonal factors? If one of the factors has the property $b^2 \neq 0$, we write $a = a_{\|} + a_\perp$ for $a_{\|}, a_\perp$ found as in preceding paragraph, then $a \wedge b = a_{\|} \wedge b + a_\perp \wedge b = a_\perp \wedge b$ and the factorization $D = a_\perp \wedge b$ into orthogonal factors is obtained. If both factors a, b have zero square, this method does not work. Then we introduce two other vectors $c = \frac{1}{2}(a+b), d = a - b$ which yield

$$d \wedge c = \frac{1}{2}(a-b) \wedge (a+b) = \frac{1}{2}(a \wedge b - b \wedge a) = a \wedge b ,$$

and

$$d \cdot c = \frac{1}{2}(a-b) \cdot (a+b) = \frac{1}{2}(a^2 - b^2) = 0 .$$

In this manner the factorization $D = d \wedge c$ into orthogonal factors is obtained.

The Clifford product of two bivectors B and D is

$$BD = B \cdot D + B \dot{\wedge} D + B \wedge D ,$$

which can be shown in the same way as Eq. (0.70). The products $B \cdot D$ and $B \wedge D$ are commutative, whereas $B \dot{\wedge} D$ is anticommutative. An interesting question arises: when can two volutors be called orthogonal? The situation which we know from the three-dimensional space, namely, that the equality $B \cdot D = 0$ occurs for two volutors is not complete orthogonality, because a common vector factor a can be found such that $B = a \wedge b, D = a \wedge d$ for any pair of volutors embedded in a three-dimensional space (compare Fig. 14). In this case, for $b \perp d, B \dot{\wedge} D \neq 0$. Thus the volutors B, D such that $B \cdot D = 0$ but $B \dot{\wedge} D \neq 0$, will be called *partially orthogonal*. We call two volutors B, D *orthogonal* if $B \cdot D = 0$ and $B \dot{\wedge} D = 0$. In this case, $BD = B \wedge D$, so B commutes with D and BD is a quadrivector.

The Clifford square of a bivector is

$$D^2 = D \cdot D + D \wedge D , \tag{38}$$

which yields the sum of a scalar and a quadrivector, that is, two invariants of the proper Lorentz transformations. Let us express these invariants by the coordinates of the bivector as written in (30). The first invariant is

$$\begin{aligned} D \cdot D &= \left(\frac{1}{2} D^{\mu\nu} e_\mu \wedge e_\nu\right) \cdot \left(\frac{1}{2} D^{\lambda\kappa} e_\lambda \wedge e_\kappa\right) = \frac{1}{4} D^{\mu\nu} D^{\lambda\kappa} e_\mu \cdot [e_\nu \cdot (e_\lambda \wedge e_\kappa)] \\ &= \frac{1}{4} D^{\mu\nu} D^{\lambda\kappa} e_\mu \cdot [(e_\nu \cdot e_\lambda) e_\kappa - (e_\nu \cdot e_\kappa) e_\lambda] . \end{aligned}$$

We use (25)

$$\begin{aligned} D \cdot D &= \frac{1}{4} D^{\mu\nu} D^{\lambda\kappa} e_\mu \cdot (g_{\nu\lambda} e_\kappa - g_{\nu\kappa} e_\lambda) = \frac{1}{4} D^{\mu\nu} D^{\lambda\kappa} (g_{\mu\kappa} g_{\nu\lambda} - g_{\mu\lambda} g_{\nu\kappa}) \\ &= \frac{1}{4} D^{\mu\nu} (g_{\mu\kappa} g_{\nu\lambda} D^{\lambda\kappa} - g_{\mu\lambda} g_{\nu\kappa} D^{\lambda\kappa}) . \end{aligned}$$

Due to (31),

$$D \cdot D = \frac{1}{4} D^{\mu\nu} (D_{\nu\mu} - D_{\mu\nu}) = -\frac{1}{2} D^{\mu\nu} D_{\mu\nu} . \tag{39}$$

The second invariant is

$$D \wedge D = \left(\frac{1}{2}D^{\mu\nu}e_\mu \wedge e_\nu\right) \wedge \left(\frac{1}{2}D^{\lambda\kappa}e_\lambda \wedge e_\kappa\right) = \frac{1}{4}D^{\mu\nu}D^{\lambda\kappa}e_\mu \wedge e_\nu \wedge e_\lambda \wedge e_\kappa \,.$$

By virtue of (35),

$$D \wedge D = \frac{1}{4}D^{\mu\nu}D^{\lambda\kappa}\varepsilon_{\mu\nu\lambda\kappa}J \,. \tag{40}$$

We now consider the meaning of the equality $D^2 = 0$. First, the quadrivector part is zero, i.e. $D \wedge D = 0$ which, by Lemma 1, means that D is a volutor: $D = a \wedge b$. Second, the scalar part is zero, i.e. $D \cdot D = 0$, hence,

$$\begin{aligned}(a \wedge b) \cdot (a \wedge b) &= a \cdot [b \cdot (a \wedge b)] = a \cdot [(b \cdot a)b - b^2 a] \\ &= (a \cdot b)^2 - a^2 b^2 = 0 \,.\end{aligned}$$

If one chooses $a \perp b$, the equality means that $a^2 b^2 = 0$, hence one of the vectors a, b has zero square, so it is the light-like vector. In this way, we have shown that a bivector D with the property $D^2 = 0$ is a volutor tangent to the light cone — therefore we call it the *light-like volutor*. We show in Fig. 114 an example of such a volutor, in which a is the light-like vector.

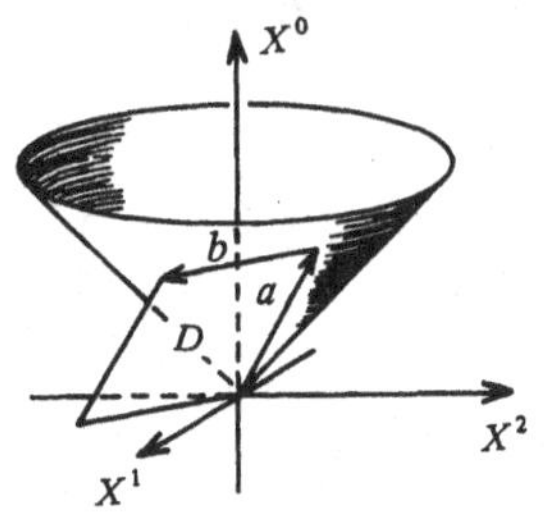

Fig. 114.

We may agree that the inequality $D^2 > 0$ will denote that $D \wedge D = 0$, i.e. D is a volutor, and that the inner product $D \cdot D$ is positive. We shall similarly understand the inequality $D^2 < 0$.

Let us take a bivector D with $D^2 > 0$. Then $D = a \wedge b$, and for $a \perp b, -a^2 b^2 > 0$ which means that a^2 and b^2 have opposite signs. In such a case, one of the vectors is time-like and we call D the *time-like volutor*.

Let $D^2 < 0$, then $D = a \wedge b$ and for $a \perp b, -a^2 b^2 < 0$, which implies that both a^2 and b^2 have the same sign. This can be satisfied only in one way: both have to be space-like (compare Problem 9). In this case, we call D a *space-like volutor*.

Notice that the division of volutors into time-like, space-like and light-like ones is invariant under the Lorentz transformations. We introduce a *pseudomagnitude* of a volutor as the expression $|D| = \sqrt{|D^2|}$ which does not satisfy the triangle inequality.

The Dirac algebra has a peculiarity in that not all nonzero multivectors are invertible. For instance, the light-like vectors and volutors do not have their inverses. On the other hand, if $b^2 \neq 0$ for a vector b, then $b^{-1} = b/b^2$.

A similar statement is valid for the bivectors. As we know from (38), $D^2 = \lambda + \alpha J$, where λ and α are scalars. The cliffor D^2 commutes with all bivectors and its inverse is

$$D^{-2} = \frac{\lambda - \alpha J}{\lambda^2 + \alpha^2},$$

as one can check by direct inspection. Now it is easy to verify that $DD^{-2} = D^{-2}D$ is the inverse of D, so we may write $D^{-1} = D/D^2$.

Let u be an arbitrary vector and B a bivector such that $B^2 \neq 0$. One may represent

$$u = (uB)B^{-1} = (u \cdot B + u \wedge B)B^{-1},$$

or

$$u = (u \cdot B)B^{-1} + (u \wedge B)B^{-1}. \tag{41}$$

Lemma 2. If B is a volutor and $B^2 \neq 0$, then the first term in (41) anticommutes with B and the second term commutes with B.

Proof. We calculate

$$B(u \cdot B)B^{-1} = \frac{1}{2}B(uB - Bu)B^{-1} = \frac{1}{2}Bu - \frac{1}{2}B^2uB^{-1}.$$

Due to Lemma 1, B^2 is a scalar, so it commutes with u:

$$B(u \cdot B)B^{-1} = \frac{1}{2}Bu - \frac{1}{2}uB = -(u \cdot B) = -[(u \cdot B)B^{-1}]B.$$

Similarly,

$$\begin{aligned} B(u \wedge B)B^{-1} &= \frac{1}{2}B(uB + Bu)B^{-1} = \frac{1}{2}Bu + \frac{1}{2}B^2uB^{-1} \\ &= \frac{1}{2}Bu + \frac{1}{2}uB = u \wedge B = [(u \wedge B)B^{-1}]B. \blacksquare \end{aligned}$$

When the assumptions of Lemma 2 are satisfied, the vector

$$u_{\|} = (u \cdot B)B^{-1}, \tag{42}$$

by (0.79) has the property $u_{\parallel} \wedge B = 0$, so we may call $u_{\parallel}$ the component of u *parallel* to B. Similarly for

$$u_{\perp} = (u \wedge B)B^{-1} , \tag{43}$$

we get $u_{\perp} \cdot B = 0$, so we may call $u_{\perp}$ the component of u *orthogonal* to B. Notice that one cannot speak about components parallel or orthogonal to a light-like volutor.

We now proceed to the question of whether an arbitrary bivector F from the Minkowski space can be decomposed on a sum of two orthogonal volutors. By virtue of (32), we write $F = e_0\, E + B$, where e_0 is a unit time-like vector, E — a space-like vector, B — a space-like volutor, and

$$e_0 \perp E , \qquad e_0 \perp B . \tag{44a}$$

Thus

$$F^2 = (e_0 E + B)(e_0 E + B) = e_0 E e_0 E + e_0 E B + B e_0 E + B^2 .$$

Relations (44a) ensure that e_0 commutes with B and anticommutes with E:

$$F^2 = -E^2 + e_0(EB + BE) + B^2 . \tag{44b}$$

E and B are space-like, hence $E^2 = -|E|^2$, $B^2 = -|B|^2$, so

$$F^2 = |E|^2 - |B|^2 + 2e_0(E \wedge B) .$$

Having denoted $\alpha = |E|^2 - |B|^2, \beta J = e_0(E \wedge B)$, we divide F^2 into its scalar and quadrivector parts:

$$F^2 = \alpha + 2\beta J . \tag{45}$$

The scalars α and β are invariant under the proper Lorentz transformations.

We seek orthogonal volutors F_1, F_2 such that

$$F = F_1 + F_2 . \tag{46}$$

Then

$$F^2 = F_1^2 + F_1 F_2 + F_2 F_1 + F_2^2 .$$

The condition $F_1 \perp F_2$ implies that

$$F_1 F_2 = F_2 F_1 = \gamma J , \tag{47}$$

with real γ. Moreover, each volutor has real square, hence

$$F_1^2 = \lambda_1 , \qquad F_2^2 = \lambda_2 , \tag{48}$$

with real λ_1, λ_2, so

$$F^2 = \lambda_1 + \lambda_2 + 2\gamma J .$$

Comparing this with (45) leads to $\lambda_1 + \lambda_2 = \alpha$, $\gamma = \beta$. Also $\lambda_1 \lambda_2 = F_1^2 F_2^2 = (F_1 F_2)^2 = (\gamma J)^2 = -\gamma^2$, hence we obtain the system of two equations

$$\lambda_1 + \lambda_2 = \alpha , \qquad \lambda_1 \lambda_2 = -\beta^2 .$$

Substitute $\lambda_2 = \alpha - \lambda_1$ in the second equation

$$\lambda_1(\alpha - \lambda_1) = -\beta^2 ,$$
$$\lambda_1^2 - \alpha\lambda_1 - \beta^2 = 0 .$$

The same equation is obtained for λ_2 after the substitution $\lambda_1 = \alpha - \lambda_2$. Thus we conclude that λ_1, λ_2 are two roots of the second degree equation:

$$\lambda^2 - \alpha\lambda - \beta^2 = 0 . \tag{49}$$

Let $\delta = \alpha^2 + 4\beta^2$ be its discriminant, then

$$\lambda_1 = \frac{\alpha + \sqrt{\delta}}{2} , \qquad \lambda_2 = \frac{\alpha - \sqrt{\delta}}{2} .$$

Of course, λ_1, λ_2 are built of proper Lorentz invariants and therefore are also invariants. Now it is easy to check that

$$F_1 = \frac{1}{\sqrt{\delta}}(\lambda_1 - \beta J)F , \qquad F_2 = \frac{1}{\sqrt{\delta}}(\beta J - \lambda_2)F , \tag{50}$$

satisfy the desired conditions (46)–(48). The coefficients standing in front of F are invariants, so the decomposition (46) of F into orthogonal volutors is invariant under the (proper) Lorentz transformations.

It is interesting to know whether the two roots λ_1, λ_2 of Eq. (49) may coincide. This may happen only when $\delta = 0$, that is, when $\alpha = \beta = 0$. The equality $\beta = 0$ denotes that $F \wedge F = 0$, i.e. that F is simple. In such a case, the decomposition (46) is immediate: $F_1 = F$, $F_2 = 0$. In this manner, we are convinced that for nonsimple F, the two volutors (50) are both nonzero and not proportional.

After a careful inspection of all possibilities $\alpha > 0, \alpha < 0, \alpha = 0$ with $\beta \neq 0$, the reader may ascertain that $\lambda_1 > 0, \lambda_2 < 0$, hence by (48), F_1 is time-like, F_2 is space-like.

We now look for a formula similar to (0.102) which allows us to express the Lorentz transformations as a multiplication by some cliffors. We remember from §0.8 that the volutorial angle $\widehat{\alpha}$ and the cliffor $e^{\frac{1}{2}\widehat{\alpha}}$ are important for a rotation. The Lorentz boost is for us a hyperbolic rotation in the plane of e_0e_1, so we employ the hyperbolic angle ϑ and form tha *volutorial rapidity* $\theta = \vartheta e_0 \wedge e_1 = \vartheta e_0 e_1$. Its square

$$\theta^2 = \vartheta^2 e_0 e_1 e_0 e_1 = -\vartheta^2 e_0^2 e_1^2 = \vartheta^2 \ ,$$

is positive, hence it is a time-like volutor. All even powers of θ are also scalars: $\theta^{2n} = \vartheta^{2n}$, whereas the odd powers are volutors: $\theta^{2n+1} = \vartheta^{2n+1} e_0 e_1$, thus

$$e^{\theta} = \sum_{n=0}^{\infty} \frac{\theta^n}{n!} = \cosh\vartheta + e_0 e_1 \sinh\vartheta \ . \tag{51}$$

Let $x = x^{\mu} e_{\mu}$ be an arbitrary vector from the Minkowski space. We verify what the mapping $x \to x' = U^{-1} x U$ is for

$$U = e^{-\frac{\theta}{2}} = \cosh\frac{\vartheta}{2} - e_0 e_1 \sin\frac{\vartheta}{2} \ .$$

Of course,

$$U^{-1} = U^{\dagger} = e^{\frac{\theta}{2}} = \cosh\frac{\vartheta}{2} + e_0 e_1 \sinh\frac{\vartheta}{2} \ .$$

The volutor e_0e_1 commutes with the vectors e_2, e_3 and anticommutes with e_0, e_1, so

$$U^{-1} e_0 U = e_0 U^2 \ , \quad U^{-1} e_1 U = e_1 U^2 \ , \quad U^{-1} e_2 U = e_2 \ , \quad U^{-1} e_3 U = e_3 \ ,$$

and

$$U^{-1} x U = (x^0 e_0 + x^1 e_1) U^2 + x^2 e_2 + x^3 e_3 \ .$$

We notice that the component of x orthogonal to the plane of e_0e_1 is invariant, whilst the parallel component is

$$\begin{aligned}(x^0 e_0 + x^1 e_1)(\cosh\vartheta - e_0 e_1 \sinh\vartheta) \\ = (x^0 \cosh\vartheta - x^1 \sinh\vartheta) e_0 + (-x^0 \sinh\vartheta + x^1 \cosh\vartheta) e_1 \ .\end{aligned}$$

Thus the coordinates transform as follows:

$$
\begin{aligned}
x'^0 &= x^0 \cosh\vartheta - x^1 \sinh\vartheta\,, \\
x'^1 &= -x^0 \sinh\vartheta + x^1 \cosh\vartheta\,, \\
x'^2 &= x^2\,, \\
x'^3 &= x^3\,,
\end{aligned}
$$

which is described by the matrix (8). In this manner, we have demonstrated that boost (8) may be written as

$$x \to x' = e^{\frac{\theta}{2}} x\, e^{-\frac{\theta}{2}}\,, \tag{52}$$

with $\theta = \vartheta e_0 e_1$.

We use (7) also to write e^θ differently. Due to (51),

$$e^\theta = \cosh\vartheta + e_0 e_1 \sinh\vartheta = \gamma(1 + e_0 e_1 \beta)\,.$$

If we introduce the volutor[1)] $W = c\beta e_0 e_1$ of relative velocity between the two reference frames, we get

$$e^\theta = \gamma\left(1 + \frac{1}{c}W\right)\,. \tag{53}$$

This expression is easy to generalize into boosts in arbitrary directions. It is sufficient to substitute $W = c\beta e_0 n$, where n is not the basis vector but any unit space-like vector orthogonal to e_0.

The cliffor $e^{\frac{\theta}{2}}$, that is the square root of (53), is needed in transformation (52). We leave to the reader the verification that

$$e^{\frac{\theta}{2}} = \frac{1 + \gamma + \gamma W/c}{\sqrt{2(1+\gamma)}}\,. \tag{54}$$

It is worth noticing that (53) can be written as

$$e^\theta = e_0' e_0\,. \tag{55}$$

1)The reader may ask: why a volutor, not a vector? The answer is based on the symmetry prerequisite: the relation of the prime frame $\{e'_\mu = e^{\theta/2}\, e_\mu\, e^{-\theta/2}\}$ to the initial frame $\{e_\mu\}$ should be identical with the inverse relation except for the change of its characteristic W into $-W$. If one chooses W as a vector $W = c\,\beta\, e_1$, one should, by the above prerequisite, take $W' = -\,c\,\beta\, e'_1$ for the inverse relation. But this is not equal to $-W$. The volutorial characteristic $W = c\,\beta\, e_0\, e_1$ fulfils the requirement $W' = -\,c\,\beta\, e'_0\, e'_1 = -\,W$ as the reader may inspect using the properties of hyperbolic functions.

Indeed, the volutor W anticommutes with e_0, hence $e_0 e^{-\frac{\theta}{2}} = e^{\frac{\theta}{2}} e_0$ and by (52),

$$e_0' = e^{\frac{\theta}{2}} e_0 e^{-\frac{\theta}{2}} = e^{\theta} e_0 \, .$$

After multiplying both sides by e_0 from the right, we get (55). In this way, Eq. (55) allows us to find cliffor (51) if only two time-like vectors are known: the initial vector e_0 and its image e_0' such that the boost leaves invariant all vectors orthogonal to the plane of $e_0' e_0$.

The space-like rotations can also be represented in the form (52). It is enough to put a space-like volutor Φ in place of θ in the exponents.

A Lorentz transformation of a new kind is obtained when a light-like volutor $\Psi = kn$ (where k, n are vectors, $k \cdot n = 0, k^2 = 0, n^2 = 1$) is placed in the exponent. Because of the identity $\Psi^2 = 0$, only two terms are present in the power series:

$$e^{\frac{\Psi}{2}} = \sum_{m=0}^{\infty} \frac{1}{m!} \left(\frac{\Psi}{2} \right)^m = 1 + \frac{\Psi}{2} \, .$$

In this case, $k\Psi = -\Psi k = 0$, hence

$$e^{\frac{\Psi}{2}} k e^{-\frac{\Psi}{2}} = (1 + \frac{\Psi}{2}) k (1 - \frac{\Psi}{2}) = k + \frac{\Psi}{2} k - k \frac{\Psi}{2} - \frac{\Psi}{2} k \frac{\Psi}{2} = k \, .$$

Thus for this particular Lorentz transformation, a light-like vector exists which is invariant under it. Another invariant vector is a space-like vector orthogonal to both k and n.

A non-simple bivector F can also put in the exponent. In such a case, we employ the previous discussion and represent F as the sum $F = \theta + \Phi$ where θ is a time-like volutor, Φ is a space-like volutor and $\theta \perp \Phi$. The orthogonality implies that $\theta \Phi = \Phi \theta$, hence by the Lemma of Sec. 0.7, we have

$$e^{\frac{F}{2}} = e^{\frac{\theta}{2} + \frac{\Phi}{2}} = e^{\frac{\theta}{2}} e^{\frac{\Phi}{2}} \, ,$$

and the transformation (52) can be written as

$$x \to x' = e^{\frac{\theta}{2}} \left(e^{\frac{\Phi}{2}} x e^{-\frac{\Phi}{2}} \right) e^{-\frac{\theta}{2}} = e^{\frac{\Phi}{2}} \left(e^{\frac{\theta}{2}} x e^{-\frac{\theta}{2}} \right) e^{-\frac{\Phi}{2}} \, ,$$

that is, as a composition of a space-like rotation and a boost in any order.

§4. Kinematics

Motion of a particle in the three-dimensional space is described by a function $t \to \mathbf{r}(t)$ ascribing to each time t a position $\mathbf{r}(t)$. A geometric image of the motion is a curve in the space $R \times R^3$, that is, in the space-time. Thus in the Minkowski space, this image is a set of points of the form $x(t) = (ct, \mathbf{r}(t))$ which is known as the *world line* of the particle. The vector tangent to this line is contained in the light cone (with its apex on the world line) because of the limitation $|\mathbf{v}| \leq c$ on the velocities of the particle (compare Fig. 115).

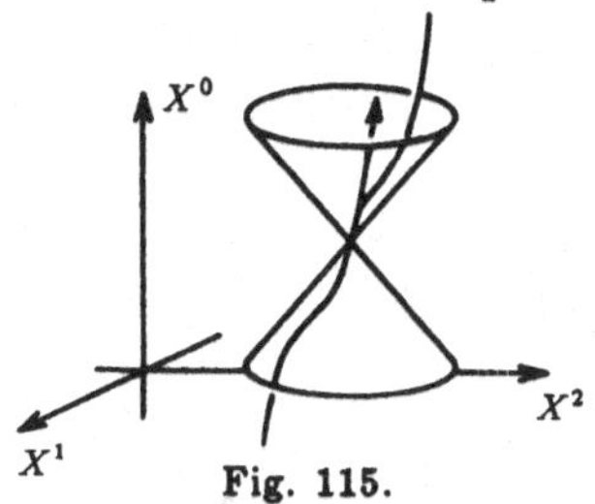

Fig. 115.

One may consider the derivative of the function $t \to x(t)$:

$$\frac{dx}{dt} = \left(c, \frac{d\mathbf{r}}{dt}\right) = (c, \mathbf{v}) .$$

The collection of four numbers $(c, \mathbf{v}) = (c, v^1, v^2, v^3)$ unfortunately, is not a space-time vector, since it does not behave properly under the Lorentz transformations. The reason is in the difference quotient $\Delta x/\Delta t$ (needed for calculating the limit $dx/dt = \lim_{\Delta t \to 0} \Delta x/\Delta t$), its denominator is not a Lorentz invariant.

Therefore, another vector is introduced with differentiation over the proper time τ:

$$u = \frac{dx}{d\tau} = \frac{dt}{d\tau}\frac{dx}{dt} = \gamma(c, \mathbf{v}) = \left(\frac{c}{\sqrt{1 - v^2/c^2}}, \frac{\mathbf{v}}{\sqrt{1 - v^2/c^2}}\right) , \qquad (56)$$

which is already the space-time vector. It is called the *four-velocity* or *celerity*. Its space part

$$\mathbf{u} = \gamma \mathbf{v} = \frac{\mathbf{v}}{\sqrt{1 - v^2/c^2}} , \qquad (57)$$

satisfies $|\mathbf{u}| > |\mathbf{v}|$ for $\mathbf{v} \neq 0$, since $\gamma > 1$. The scalar square of u is

$$u^2 = \gamma^2 c^2 - \gamma^2 v^2 = \gamma^2 c^2 (1 - v^2/c^2) = c^2 . \qquad (58)$$

We find that it is constant, i.e. independent of $\mathbf{v}$. This is a natural result, because each vector from the Minkowski space has the pseudolength invariant under the Lorentz transformations. Thus the pseudolength $|u|$ cannot depend on the velocity $\mathbf{v}$ which changes when passing to other reference frames. The physical quantities behaving properly under the Lorentz transformations will be called *covariant quantities.* Celerity is a covariant quantity. In the rest frame of the particle, the celerity is $u = (c, 0)$, so u has the direction of the time axis: $u = ce_0$. This relation of the celerity to the unit vector defining the time axis is obviously the same after any Lorentz transformation: $u' = ce_0'$, hence we shall often identify the direction of u with the time axis of the rest frame for the particle.

Notice that $\gamma \to \infty$ for $v \to c$, hence the four-velocity (56) makes no sense for particles travelling with the velocity of light. Notice also that the relation (20) demands $\Delta\tau \to 0$ which is expressed loosely in the words: "the proper time does not flow for the photon", whereas one should strictly say that the proper time makes no sense for the photon.

An important conserved quantity exists in classical mechanics, the *momentum* as the product of the mass and velocity of the particle: $\mathbf{p} = m\mathbf{v}$. The momentum conservation law is satisfied in all inertial frames connected by the Galilei boosts. Let us check whether the quantity $\Sigma_i m_i \mathbf{v}_i$ (with the summation over particles present) is conserved in various reference frames connected by the Lorentz boosts.

Consider the reference frames S and S_0, S moving relative to S_0 with velocity V along the X^1-axis. Assume that two identical particles move with velocities V and $-V$ along X^1 relative to S. Then by virtue of (5), the same two particles have velocities

$$U = \frac{2V}{1 + V^2/c^2} \tag{59}$$

and 0, respectively, relative to S_0 (this is the place where the Lorentz transformations are used). Now let an inelastic collision occur, consisting of gluing the particles together. This means that, after the collision, the particles have common velocity 0 in S and V in S_0. The momentum conservation law in S has the form

$$mV - mV = 2m0 \,,$$

where the momentum prior to the collision stands at the left-hand side, and after the collision, at the right-hand side. Instead, in S_0, the momentum prior to the collision is $2mV/(1 + V^2/c^2)$, while later it is $2mV$, so this quantity is not conserved.

In this situation, one comes to the conclusion that the very definition of momentum ought to be changed. A possible change consists in admitting that the mass m standing as the coefficient in front of $\mathbf{v}$ may depend on the velocity: $m = m(v)$. Let us apply this to our example of two colliding particles. In S_0, the particles have masses $m(U)$ and $m(0)$, respectively, before the collision, so the momentum is then $m(U)U$. Also assuming the conservation of the mass, we may state the total mass of the particles after the collision is $m(U)+m(0)$ and, since their common velocity is V, their momentum is $[m(U)+m(0)]V$. We write down the momentum conservation in S_0:

$$m(U)U = [m(U)+m(0)]V \ . \tag{60}$$

We insert $V = (c^2/U)(1-\sqrt{1-U^2/c^2})$, obtained from (59), into (60):

$$m(U)U^2/c^2 = [m(U)+m(0)](1-\sqrt{1-U^2/c^2}) \ ,$$

and get from this

$$m(U) = \frac{m(0)}{\sqrt{1-U^2/c^2}} \ .$$

In this manner, we have obtained the expression

$$\mathbf{p} = \frac{m_0}{\sqrt{1-v^2/c^2}}\mathbf{v} \ , \tag{61}$$

for the momentum of a particle, where $m_0 = m(0)$ is called the *rest mass* or *proper mass* of the particle — it is constant intrinsic property of the particle similar to the mass in Newtonian mechanics. On the other hand, we have the *relative mass*

$$m_r = \frac{m_0}{\sqrt{1-v^2/c^2}} \ , \tag{62}$$

which grows as the velocity increases.

Let us now find the energy of a particle as a function of the velocity. We start with the formula already used in §3.1: $d\mathcal{E} = \mathbf{v}\cdot d\mathbf{p}$. Due to (61), the vectors $\mathbf{v}$ and $\mathbf{p}$ are parallel, hence $d\mathcal{E} = vdp = \frac{dp}{dv}vdv$. We get, by differentiation of (61),

$$d\mathcal{E} = \left[\frac{m_0}{(1-v^2/c^2)^{1/2}} + \frac{m_0v^2}{c^2(1-v^2/c^2)^{3/2}}\right]v\,dv = \frac{m_0\,v\,dv}{(1-v^2/c^2)^{3/2}} \ .$$

The integral of this expression is

$$\mathcal{E}(v) = \frac{m_0c^2}{\sqrt{1-v^2/c^2}} + C \ , \tag{63}$$

where C is an integration constant. Employing the notation (62), we may write

$$\mathcal{E} = m_r\, c^2 + C\ .$$

This equality states that the energy is a linear function of the particle's mass, so the energy increases are proportional to the mass increases: $\Delta\mathcal{E} = c^2\,\Delta m$. This is a famous discovery of Einstein on the equivalence of mass and energy. After accepting this equivalence, we choose the integration constant C according to the following reasoning. Since the energy increases are proportional to the mass increases, let the whole energy be proportional to the relative mass. If the resting particle has a nonzero mass, let it also have nonzero energy given by $\mathcal{E}(0) = m_0 c^2$. This is called the ***rest energy*** or ***proper energy*** of the particle. In this way, $C = 0$ and Eq. (63) becomes

$$\mathcal{E}(v) = \frac{m_0\, c^2}{\sqrt{1 - v^2/c^2}}\ . \tag{64}$$

From now on, we shall omit the subscript 0 at the symbol of rest mass.

The difference between $\mathcal{E}(v)$ and rest energy is, of course, kinetic energy:

$$T = \frac{mc^2}{\sqrt{1 - v^2/c^2}} - mc^2\ .$$

If $v \ll c$, the expression can be developed in the power series in v^2/c^2, which yields $T = \frac{1}{2}\, mv^2$, that is, the well known formula for kinetic energy in Newtonian mechanics.

The quantities (61) and (64) can be united in a single four-component quantity:

$$p = \left(\frac{\mathcal{E}}{c}, \mathbf{p}\right) = \left(\frac{mc}{1 - v^2/c^2}, \frac{m\mathbf{v}}{1 - v^2/c^2}\right)\ . \tag{65}$$

The comparison with (56) allows us to notice the relation

$$p = m u\ . \tag{66}$$

The product of the four-vector u with the space-time scalar m stands at the right-hand side, so p is the four-vector. It is referred to as the ***energy-momentum*** or ***four-momentum vector***. Using (58), we find that its scalar square is

$$p^2 = m^2 c^2\ . \tag{67}$$

Equation (64) shows that the particle with nonzero rest mass can never achieve the velocity of light — this would demand an infinite energy.

The expressions (61) and (64) make no sense for particles moving with the velocity of light, hence the same must be true for (65). Nevertheless, the energy-momentum vector is also introduced for such particles by the following reasoning.

When a particle moves uniformly, its energy-momentum vector by virtue of (66) is proportional to the time rate of the space-time position:

$$p = m\frac{\Delta x}{\Delta \tau}\,. \tag{68}$$

Only p and Δx are four-vectors in this equality, so they have the same direction. The faster the particle moves, the farther Δx is deflected from the X^0-axis, which means that p is closer to the light cone. Thus in the limit, when we finally consider a particle moving with velocity c, for example a photon, we ascribe it the vector p lying on the light cone. We already know that such vectors have their scalar square zero, hence due to (67), $m = 0$, i.e. the rest mass of particles travelling with the velocity of light is zero — therefore they are called *massless particles.* Equation (68) also leads to $m = 0$ because for increasing velocity, $\Delta\tau \to 0$, so in order to keep (68) finite, we have to assume $m \to 0$. The limit of the quotient $m/\Delta\tau$ (for a fixed Δx) need not always be the same, thus various momenta and energies are possible for the photon which, after all, move with the same velocity c (compare Problem 16).

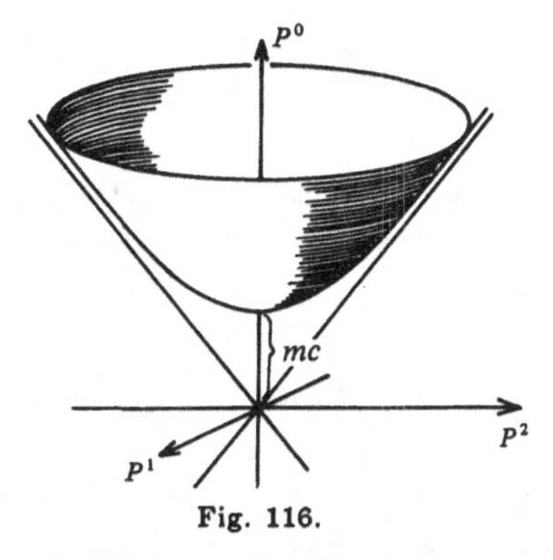

Fig. 116.

The set of all possible energy momentum vectors of a given particle may be represented on the Minkowski diagram as the upper sheet of the two-sheet hyperboloid (Fig. 116). The intersection point of the hyperboloid with the P^0-axis corresponds to the state of rest. The value of the p^0-coordinate of this point is mc. Different particles have different hyperboloids labelled by their rest masses, therefore such a hyperboloid is called the *mass shell* of a given particle. The mass shell for the photon is the upper half of the light cone, of course without its apex. This fact helps us to understand the impossibility of some decay processes (compare Problem 17).

We obtain from (65) the identity

$$\mathbf{p} = \frac{p_0}{c}\mathbf{v} = \frac{\mathcal{E}}{c^2}\mathbf{v} , \tag{69}$$

valid for particles with any rest mass. In particular, for the massless particles, that is, when $|\mathbf{v}| = c$, this yields $|\mathbf{p}| = p_0 = \mathcal{E}/c$.

§5. Accelerated Motion

We are looking for a relativistic counterpart of acceleration as a physical quantity. The traditional one $\mathbf{a} = d\mathbf{v}/dt = d^2\mathbf{r}/dt^2$ cannot serve as the space part of any four-vector, because it contains the differentiation over t, which is not an invariant. The derivative $d\mathbf{u}/dt$ has the same disadvantage. Only the expression $\mathbf{b} = d\mathbf{u}/d\tau = d^2\mathbf{r}/d\tau^2$ may be used to build the space-time vector:

$$b = \frac{du}{d\tau} = \frac{d}{d\tau}(\gamma, \gamma\mathbf{v}) , \tag{70}$$

which is known as the *four-acceleration.*

We get, by differentiation of (58),

$$\frac{d}{d\tau}u^2 = \frac{d}{d\tau}(u \cdot u) = 2u \cdot \frac{du}{d\tau} = 0 ,$$

or $u \cdot b = 0$, that is, the four-acceleration is orthogonal to the four-velocity. Since u is time-like, b is space-like. In an instantaneous rest frame of the particle, b has the form

$$b = (0, \mathbf{b}) = (0, \mathbf{a} + \mathbf{v}\frac{d\gamma}{d\tau}) = (0, \mathbf{a}) ,$$

because $\mathbf{v} = 0$ in the rest frame. In this case, b is called the ***rest acceleration.*** Of course, vector b cannot be defined for massless particles.

Let us consider whether an equivalent of the uniformly accelerated motion exists in relativity theory. The ordinary acceleration $\mathbf{a}$ obviously cannot be constant in time, bescause this would denote a uniform increase in the velocity $\mathbf{v}$, so, after some time, v would achieve the velocity of light, which is impossible for a particle with nonzero rest mass. On the other hand, if b is constant, then from (70), $u = b\tau + u_0$. This means that u cannot have a constant square, contrary to (58). In this manner, we have to reject the two simplest possibilities.

Another possibility is to assume that b^2 = constant which implies that the rest acceleration is the same at each time τ. Assume, moreover, that the motion is *rectilinear* and find functions describing the motion.

If one chooses the basis vector e_1 along the motion, the celerity can be written as

$$u(\tau) = c\, e_0 \cosh \varphi(\tau) + c\, e_1 \sinh \varphi(\tau) \ , \tag{71}$$

where φ is the unknown function of the proper time. Then the four-acceleration has the form

$$b(\tau) = c \, \frac{d\varphi}{d\tau} \left(e_0 \sinh \varphi(\tau) + e_1 \cosh \varphi(\tau) \right) \ .$$

Its square is $b(\tau)^2 = -c^2 \left(\frac{d\varphi}{d\tau} \right)^2$. This is assumed to be constant in time, so we get

$$\frac{d\varphi}{d\tau} = \alpha = \text{const} \ , \tag{72}$$

hence $\varphi(\tau) = \alpha\tau + \varphi_0$ and

$$u(\tau) = c\, e_0 \cosh(\alpha\tau + \varphi_0) + c\, e_1 \sinh(\alpha\tau + \varphi_0) \ .$$

We choose $\varphi_0 = 0$, so the particle rests in the initial moment $\tau = 0$:

$$u(\tau) = e_0 (\cosh \alpha\tau + e_0 e_1 \sinh \alpha\tau) = c\, e_0 e^{\alpha\tau e_0 e_1} \ . \tag{73}$$

By integrating the equation $dx/d\tau = u$, we get

$$x(\tau) = \frac{c}{\alpha} e_1 (\cosh \alpha\tau - e_1 e_0 \sinh \alpha\tau) + x_{(0)} = \frac{c}{\alpha} e_1 e^{\alpha\tau e_0 e_1} + x_{(0)} \ . \tag{74}$$

This yields the following parametric equations for the two relevant coordinates of the world line

$$t(\tau) = \frac{1}{\alpha} \sinh \alpha\tau + t_{(0)} \ , \qquad x^1(\tau) = \frac{c}{\alpha} \cosh \alpha\tau + x^1_{(0)} \ .$$

We choose $t_{(0)} = 0$ in order to adjust the initial moments of the two times t and τ:

$$t(\tau) = \frac{1}{\alpha} \sinh \alpha\tau \ , \qquad x^1(\tau) = \frac{c}{\alpha} \cosh \alpha\tau + x^1_{(0)} \ . \tag{75}$$

Such functions satisfy the equation

$$(x^1 - x^1_{(0)})^2 - c^2t^2 = \frac{c^2}{\alpha^2} ,$$

of the hyperbola shown in Fig. 117 — therefore, the motion described by (74) is called the *hyperbolic motion*. The asymptotes of the hyperbola are two light-like lines going out from the point $x^1_{(0)}e_1$, and the initial position of the particle is $\left(\frac{c}{\alpha} + x^1_{(0)}\right)e_1$.

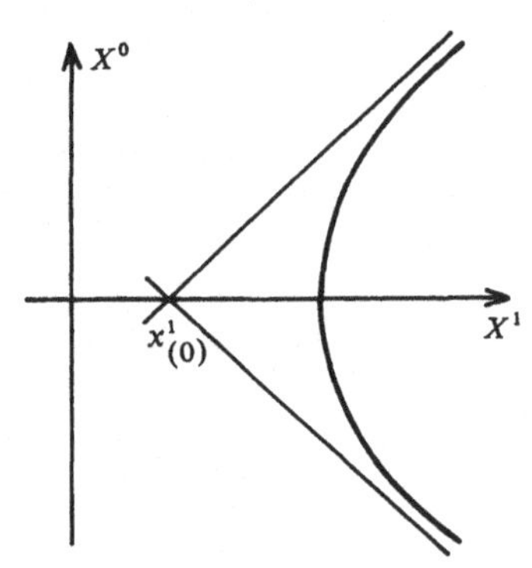

Fig. 117.

We see from expression (74) that the function $x(0) - x_{(0)} \to x(\tau) - x_{(0)}$ is a Lorentz boost with the rapidity proportional to the proper time: $\theta = \alpha\tau e_0 e_1$. This also shows that the world line of the particle is planar in the space-time: the world line lies in the plane passing through $x_{(0)}$ and parallel to e_0e_1.

Elimination of the proper time from Eqs. (75) gives

$$x^1(\tau) = c\sqrt{\frac{1}{\alpha^2} + t^2} + x^1_{(0)} .$$

We find, by differentiation, the ordinary velocity and acceleration:

$$v^1(t) = \frac{ct}{\sqrt{1/\alpha^2 + t^2}} = \frac{c\alpha t}{\sqrt{1 + (\alpha t)^2}} , \qquad (76)$$

$$a^1(t) = \frac{c/\alpha^2}{(1/\alpha^2 + t^2)^{3/2}} = \frac{c\alpha}{[1 + (\alpha t)^2]^{3/2}} .$$

Notice that the acceleration a^1 in the initial moment of rest is αc, so the parameter α introduced in (72) is proportional to the rest acceleration. The above functions have the limits $v^1(t) \to \pm c, a^1(t) \to 0$ for $t \to \pm\infty$.

By comparing (73) with (75), we observe that $c\alpha t = u^1$ which, along with (76), means that $v^1 = u^1/\sqrt{1+(\alpha t)^2}$. This in turn implies by (57) that $\gamma(t) = \sqrt{1+(\alpha t)^2}$ and by (56) that $\gamma(t) = \cosh \alpha\tau$. Thus the celerity is

$$u(t) = c\sqrt{1+(\alpha t)^2}e_0 + \alpha c t e_1 \,.$$

We ascertain that the noncovariant quantity $d\mathbf{u}/dt = \alpha c\, \mathbf{e}_1$ is a constant of the motion.

The four-acceleration as the function of time,

$$b(t) = c\alpha^2 t e_0 + c\alpha\sqrt{1+(\alpha t)^2}\, e_1 \,,$$

has the nonzero components growing to infinity. We notice also the following relation of the two constants of the motion: $|d\mathbf{u}/dt|^2 = -b^2$.

If the initial condition $x^1_{(0)} = 0$ is chosen, the equation of the hyperbola transforms into

$$(x^1)^2 - c^2t^2 = \frac{c^2}{\alpha^2} \,,$$

and becomes invariant under Lorentz boosts along X^1. This is an interesting property similar to that of the uniformly accelerated motion in Newtonian mechanics: the world line looks the same in a class of different reference frames moving parallelly to the particle. (For instance, the parabola $x = \frac{1}{2}at^2 + d$ goes over into $x' = \frac{1}{2}at'^2 + d$ under the Galilei transformation $x' = x - Vt + V^2/2a, t' = t - V/a$ for arbitrary V.)

We transform motion (74) (with $x_{(0)} = 0$) according to (52) by the Lorentz boost along X^2 with the volutorial rapidity $\theta = \vartheta e_0 e_2$:

$$\begin{aligned} x'(\tau) = e^{\frac{\theta}{2}} x(\tau) e^{-\frac{\theta}{2}} &= \left(\cosh\frac{\vartheta}{2} + e_0e_2 \sinh\frac{\vartheta}{2}\right) \\ &\times \left(\frac{c}{\alpha}e_0 \sinh\alpha\tau + \frac{c}{\alpha}e_1\cosh\alpha\tau\right)\left(\cosh\frac{\vartheta}{2} - e_0e_2\sinh\frac{\vartheta}{2}\right) \\ &= \frac{c}{\alpha}e_0\cosh\vartheta\sinh\alpha\tau + \frac{c}{\alpha}e_1\cosh\alpha\tau - \frac{c}{\alpha}e_2\sinh\vartheta\sinh\alpha\tau \,. \end{aligned} \tag{77}$$

This means that

$$\begin{aligned} t'(\tau) &= \frac{1}{\alpha}\cosh\vartheta\sinh\alpha\tau \,, \\ x'^1(\tau) &= \frac{c}{\alpha}\cosh\alpha\tau \,, \\ x'^2(\tau) &= -\frac{c}{\alpha}\sinh\vartheta\sinh\alpha\tau \,, \\ x'^3(\tau) &= 0 \,. \end{aligned}$$

We get rid of the parameter τ:

$$x'^1(t') = c\sqrt{\frac{1}{\alpha^2} + \frac{t'^2}{\cosh^2\vartheta}}, \quad x'^2(t') = -ct'\tanh\vartheta = -Vt' , \qquad (78)$$

where $V = c\tanh\vartheta$. After eliminating the time t' from these equations, we obtain the following equation of the trajectory:

$$(x'^1)^2 - \frac{(x'^2)^2}{\sinh^2\vartheta} = \frac{c^2}{\alpha^2} . \qquad (79)$$

This is a hyperbola of distance c/α from the origin and with the asymptotes $x'^2 = \pm x'^1 \sinh\vartheta$ (see Fig. 118). The angle φ between the asymptote and X'^1 is $\tan\varphi = \sinh\vartheta = \frac{V/c}{\sqrt{1-V^2/c^2}}$. In particular, for $V = 0$ we get $\varphi = 0$ and the motion reduces to the previous, rectilinear one.

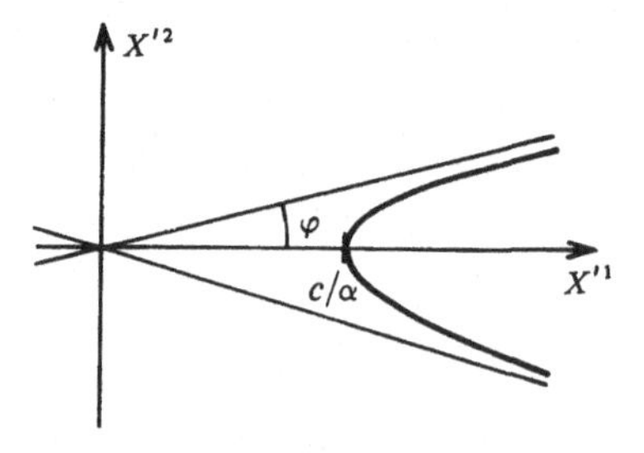

Fig. 118.

We may look differently at the boost (77). Let us use (74) and the commutativity of θ with e_1:

$$\begin{aligned} x'(\tau) &= e^{\frac{\theta}{2}}\left(\frac{c}{\alpha}e_1 e^{\alpha\tau e_0 e_1}\right)e^{-\frac{\theta}{2}} = \frac{c}{\alpha}e_1 e^{\frac{\theta}{2}} e^{\alpha\tau e_0 e_1} e^{-\frac{\theta}{2}} \\ &= \frac{c}{\alpha}e_1 \exp\left(e^{\frac{\theta}{2}}\alpha\tau e_0 e_1 e^{-\frac{\theta}{2}}\right) = \frac{c}{\alpha}e_1 \exp\left(\alpha\tau e^{\frac{\theta}{2}} e_0 e^{-\frac{\theta}{2}} e_1\right) \\ &= \frac{c}{\alpha}e_1 e^{\alpha\tau e'_0 e_1} , \end{aligned}$$

where we denoted $e'_0 = e^{\frac{\theta}{2}} e_0 e^{-\frac{\theta}{2}} = e_0\cosh\vartheta - e_2\sinh\vartheta$. This shows that the world line is still planar — it lies in the plane of $e'_0 e_1$.

Motion (78) is the relativistic counterpart of the oblique throw. Our construction shows that it is a composition of accelerated motion in the direction e_1 with uniform motion in the direction e_2. Again the nonrelativistic parabola is changed into the hyperbola — this time in the trajectory equation (79).

The velocity is the following function of time:

$$v'^1(t') = \frac{c\alpha t'}{\cosh\vartheta\sqrt{\cosh^2\vartheta + (\alpha t')^2}}, \quad v'^2(t') = -V, \tag{80}$$

for motion (78). This yields $\gamma(t') = \sqrt{\cosh^2\vartheta + (\alpha t')^2}$ and

$$u'^1(t') = \frac{c\alpha t'}{\cosh\vartheta}, \qquad u'^2(t') = -V\sqrt{\cosh^2\vartheta + (\alpha t')^2}.$$

Thus

$$\frac{du'^1}{dt'} = \frac{c\alpha}{\cosh\vartheta}, \qquad \frac{du'^2}{dt'} = \frac{-V\alpha^2 t'}{\sqrt{\cosh^2\vartheta + (\alpha t')^2}}.$$

We see that only the first coordinate of du'/dt' is a constant of the motion.

It follows from (80) that the set of possible velocity vectors, that is, *the hodograph* is a segment shown as AB in Fig. 119. The distances of A and B from the origin are c. For $t \to -\infty$, the velocity tends to A, for $t \to \infty$, it tends to B. Thus for large times, the velocity tends to the velocity of light.

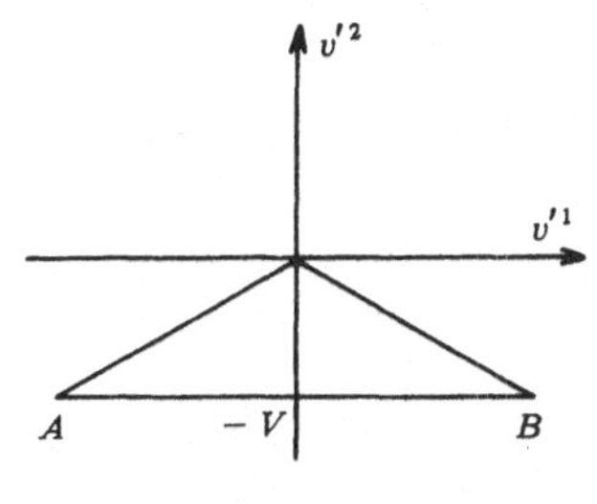

Fig. 119.

§6. Dynamics of a Single Particle

A principal goal posed by Einstein, when constructing his theory, was the demonstration that physical laws do not favour any particular reference frame. Thus the laws ought to have the same form in each reference frame. This is one more manifestation of the relativity principle already known to us.

This means mathematically that equations in which the physical laws are expressed have to be covariant and the fulfilment of an equation in one reference frame should guarantee its fulfilment in any other frame. This is ensured by assuming that all terms in the equation alter under the same linear transformation law.

Let us find a covariant form of the second Newton law. We start with the formula

$$\mathbf{f} = \frac{d\mathbf{p}}{dt}, \tag{81}$$

in which the relativistic momentum $\mathbf{p} = m\mathbf{u} = m\gamma\mathbf{v}$ stands at the right-hand side. This expression cannot be accepted as the space part of any space-time vector. One should rather take $d\mathbf{p}/d\tau = \gamma d\mathbf{p}/dt$, so we introduce $\boldsymbol{\mathcal{F}} = \gamma\mathbf{f}$ as the space part of the covariant vector of force. In this manner, we get the relativistic Newton equation

$$\mathcal{F} = \frac{dp}{d\tau},$$

where $\mathcal{F} = (\mathcal{F}^0, \boldsymbol{\mathcal{F}})$ is called the *four-force* or *Minkowski force*. Its time coordinate is

$$\mathcal{F}^0 = \frac{dp^0}{d\tau} = \frac{1}{c}\frac{d\mathcal{E}}{d\tau} = \frac{\gamma}{c}\frac{d\mathcal{E}}{dt}.$$

The expression $d\mathcal{E}/dt$ is the power used for accelerating the particle, hence $\mathcal{F}$ is also called the *power-force vector*. Because of the relation $p = mu$, the equality

$$\mathcal{F} = m\frac{du}{d\tau},$$

is satisfied, which shows that the four-force is proportional to the four-acceleration.

One should notice that the vector $\mathbf{f}$, in general, is not proportional to the ordinary acceleration, since

$$\mathbf{f} = \frac{d}{dt}(m_r\mathbf{v}) = \frac{dm_r}{dt}\mathbf{v} + m_r\frac{d\mathbf{v}}{dt} = \frac{dm_r}{dt}\mathbf{v} + m_r\mathbf{a},$$

where m_r is given by (62).

We shall solve the Newton equation (81) for a particle in a constant uniform field of force. We choose the following initial condition: in $t = 0$, the particle has the position $\mathbf{r} = 0$ and the velocity $\mathbf{v}_{(0)}$ perpendicular to $\mathbf{f}$. We choose the X^1-axis parallel to $\mathbf{f}$ and the X^2-axis to $\mathbf{v}_{(0)}$. The assumed force will not lead the particle out of the plane e_1e_2, so we may consider the planar problem. Then two coordinates of (81) are relevant:

$$\frac{dp^1}{dt} = f, \qquad \frac{dp^2}{dt} = 0, \tag{82}$$

which, under our conditions, have the solutions

$$p^1(t) = ft\,, \qquad p^2(t) = p_{(0)}\,. \tag{83}$$

Thus

$$\mathbf{p}^2 = (p^1)^2 + (p^2)^2 = (p_{(0)})^2 + (ft)^2\,,$$

and

$$\mathcal{E}(t) = c\sqrt{m^2/c^2 + p_{(0)}^2 + (ft)^2} = \sqrt{\mathcal{E}_{(0)}^2 + (cft)^2}\,.$$

We obtain from (69) $\mathbf{v} = \frac{c^2}{\mathcal{E}}\,\mathbf{p}$, so

$$\frac{dx^1}{dt} = v^1 = \frac{c^2 ft}{\sqrt{\mathcal{E}_{(0)}^2 + (cft)^2}}\,, \qquad \frac{dx^2}{dt} = v^2 = \frac{c^2 p_{(0)}}{\sqrt{\mathcal{E}_{(0)}^2 + (cft)^2}}\,. \tag{84}$$

We notice that for time increasing from zero, the coordinate v^1 grows to c, and v^2 decreases to zero. This time dependence is different from (80), where the second coordinate was constant. The coordinates (84) fulfil the equation of the ellipse:

$$(v^1)^2 + \left(\frac{\mathcal{E}_{(0)}}{p_{(0)}c}\,v^2\right)^2 = c^2\,.$$

We show the hodograph of this motion in Fig. 120 for $p_{(0)} < 0$. It is clearly different from that of Fig. 119.

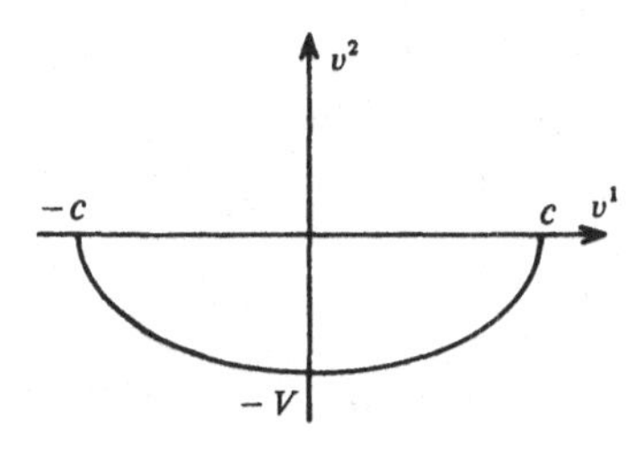

Fig. 120.

Integration of Eqs. (84), taking into account the initial conditions, yields

$$\left.\begin{aligned} x^1(t) &= \frac{1}{f}\sqrt{\mathcal{E}_{(0)}^2 + (cft)^2} - \frac{\mathcal{E}_{(0)}}{f}\,, \\ x^2(t) &= \frac{cp_{(0)}}{f}\,\mathrm{ar\,sinh}\,\frac{cft}{\mathcal{E}_{(0)}}\,. \end{aligned}\right\} \tag{85}$$

The inverse function to the hyperbolic sine is in the second formula. By eliminating the time t, we obtain the trajectory equation:

$$x^1 = \frac{\mathcal{E}_{(0)}}{f}\left(\cosh\frac{fx^2}{cp_0} - 1\right).$$

This is the so-called *catenary curve* of Fig. 121, differing from that of Fig. 118 mainly in that it has no asymptotes.

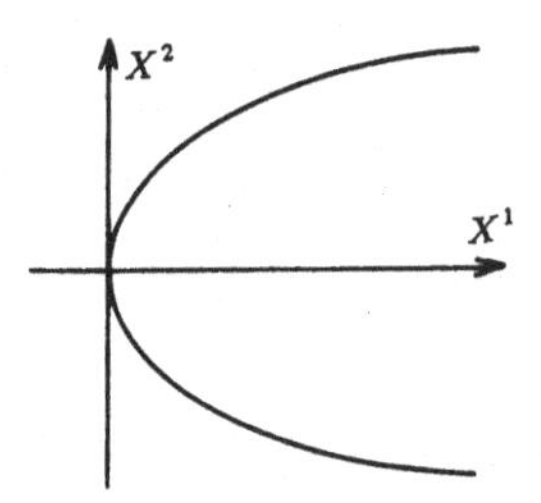

Fig. 121.

Since $d\mathbf{p}/dt = m\,d\mathbf{u}/dt$, it follows from (82) that $d\mathbf{u}/dt$ is constant in time:

$$\frac{d\mathbf{u}}{dt} = \frac{1}{m}\mathbf{f}.$$

By comparing the motion found in (85) with that considered previously (78), we observe that the motion of a particle in a uniform constant force field, with velocity nonparallel to the force, is not a composition of a one-dimensional hyperbolic motion parallel to the force with a uniform motion perpendicular to the force. Nevertheless, both motions are relativistic counterparts of the oblique throw known from Newtonian mechanics.

One may calculate from (85) $\gamma = \sqrt{\mathcal{E}_{(0)}^2 + (cft)^2}/mc^2$, hence

$$\frac{d\tau}{dt} = \frac{1}{\gamma} = \frac{mc^2}{\sqrt{\mathcal{E}_{(0)}^2 + (cft)^2}}.$$

This differential equation can be integrated:

$$\tau = \frac{mc}{f}\operatorname{ar\,sinh}\frac{cft}{\mathcal{E}_{(0)}},$$

with the initial condition $\tau(0) = 0$. The inverse relation

$$t = \frac{\mathcal{E}_{(0)}}{cf}\sinh\frac{f\tau}{mc},$$

can be substituted in (85) yielding

$$x^1(\tau) = \frac{\mathcal{E}_{(0)}}{f}\left(\cosh\frac{f\tau}{mc} - 1\right), \quad x^2(\tau) = \frac{p_{(0)}\tau}{m}.$$

All last three equations can be united into the single four-vector formula

$$\begin{aligned} x(\tau) &= \frac{\mathcal{E}_{(0)}}{f} e_0 \sinh\frac{f\tau}{mc} + \frac{\mathcal{E}_{(0)}}{f} e_1\left(\cosh\frac{f\tau}{mc} - 1\right) + \frac{p_{(0)}}{m}\tau e_2 \\ &= \frac{\mathcal{E}_{(0)}}{f} e_1 \exp\left(\frac{f\tau}{mc} e_0 e_1\right) + \frac{p_{(0)}}{m} e_2\tau - \frac{\mathcal{E}_{(0)}}{f} e_1 . \end{aligned} \tag{86}$$

This form of the solution resembles the motion (3.5) of a charged particle in a magnetic field, only the rotation is replaced by the boost. We shall discuss this point again in the next chapter.

Problems

1. A particle has the velocity v along the X^1-axis relative to a reference frame S. Another frame S' moves relative to S with velocity V also along X^1. Find the velocity v' of the particle relative to S'. The relation $v'(v, V)$ is called the *relativistic velocity addition law.*
2. Let the frame S' move relative to S with velocity V along the X^1-axis. Prove that if a particle has velocity c along X^2 relative to S, then it moves relative to S' in the plane of $e'_1 e'_2$ with a velocity of magnitude c.
3. Two reference frames situated relative to each other as in Fig. 122 are in relative motion with velocity V parallel to the X^1-axis. Find the Lorentz transformation connecting the frames and the inverse transformation. Are they proper Lorentz transformations?

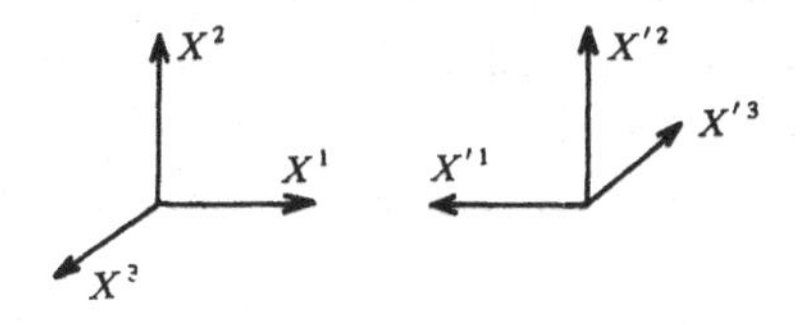

Fig. 122.

4. Find the Jacobian of transformation (8). Using Problem 1, show that $\Lambda(\vartheta_2)\Lambda(\vartheta_1) = \Lambda(\vartheta_2 + \vartheta_1)$.
5. Demonstrate that $\vartheta = \frac{1}{2}\log\frac{1+\beta}{1-\beta}$ is the inverse relation to (6).
6. Show explicitly that transformation (13) preserves the scalar product in the Minkowski space.
7. Illustrate on the Minkowski diagram the length contraction and time dilation.
8. Demonstrate that a reversion of time order of two events can happen only if the vectorial separation between the events is space-like. Discuss the relation of this fact to the causality principle.
9. Show that among two nonzero orthogonal vectors from the Minkowski space, only one can be time-like.
10. Prove the identities (26) for the dual basis.
11. Check that the matrix

$$\begin{pmatrix} 3 & 2 & 2 & 0 \\ 2 & 2 & 1 & 0 \\ 2 & 1 & 2 & 0 \\ 0 & 0 & 0 & 1 \end{pmatrix}$$

corresponds to a Lorentz transformation. Is this a boost? If so, along what direction?
12. Show that from among all time-like world lines connecting two events A and B, the longest separation of proper time is for the straight line of uniform motion. Hint: Perform the reasoning in the reference frame, in which the straight line is the time axis.
13. Show that in the four-dimensional space, all trivectors are simple.
14. Demonstrate that the Hodge map of a time-like volutor is a space-like volutor and vice versa, and that of a light-like volutor is a light-like one.
15. Show that Lorentz transformation (52) with a light-like exponent may be obtained as a composition of a boost with a space-like rotation.
16. Let a particle move along X^1 with the velocity of light. Show that a boost along X^1 does not alter this velocity. How does the same boost change the energy-momentum vector of the particle?
17. Use the notion of the mass shell to prove that an isolated photon cannot decay on two particles with nonzero rest masses. In what circumstances is the decay of photon into an electron-positron pair observed?
18. The expression

$$\mathbf{r} = \frac{\mathcal{E}_{(1)}\mathbf{r}_{(1)} + \mathcal{E}_{(2)}\mathbf{r}_{(2)}}{\mathcal{E}_{(1)} + \mathcal{E}_{(2)}} ,$$

is called the *center of energy* of a system of two particles where $\mathcal{E}_{(i)}$ are the energies and $\mathbf{r}_{(i)}$ are the position vectors of the particles. Let two identical particles move with opposite velocities parallelly to X^1 and have constant coordinates $x^2_{(1)} = d/2, x^2_{(2)} = -d/2$. How does the center of energy change in the function of time? How does it change in another reference frame? Are the energy centers, counted in different frames, related by a Lorentz transformation?

19. Find four-accelerations for the motions (78) and (85) in functions of time. Are their scalar squares constant?

Chapter 7

RELATIVITY AND ELECTRODYNAMICS

§1. The Covariance of Electrodynamics

In the historical development of physics, covariance of the Maxwell equations under the Lorentz transformations was first discovered (we owe this to Hendrik Antoon Lorentz and Henri Poincaré) and afterwards, special relativity was formulated (by Albert Einstein, as we have already mentioned). In this exposition, we have chosen a different order of presentation, so we now show only how electrodynamics can be adopted to special relativity.

The guiding line of our reasoning will be to express the electrodynamics quantities and equations by multivectors and cliffors over the Minkowski space. Then the fulfilment of the equations will be independent of the choice of basis in the Minkowski space. A quantity is a *bona fide* multivector if its coordinates behave properly under the Lorentz transformations.

We start with the electric charge Q. Our postulate is that it is an invariant for the proper Lorentz transformations. We consider two ways of expressing the charge: Firstly, by the space density of charge ρ

$$Q = \int_V dV \rho = \int_V \rho dx^1 dx^2 dx^3 \,, \tag{1}$$

and secondly, by the surface density of current $\mathbf{j}$

$$Q = \int_T dt \int_\Sigma d\mathbf{s} \cdot \mathbf{j} = \frac{1}{c} \iiint_{cT\times\Sigma} (j^1 dx^0 dx^2 dx^3 + j^2 dx^0 dx^3 dx^1 + j^3 dx^0 dx^1 dx^2) , \tag{2}$$

where Σ is a surface, $d\mathbf{s}$ — its vectorial element of area. Thus (2) is the charge flowing through Σ in the time interval T. Both expressions (1) and (2) are integrals over three-dimensional subsets — trivectors are appropriate quantities describing the directed measures of such sets. In (2), only three coordinates of a trivector are present because the region $cT \times \Sigma$ has a particular direction in the Minkowski space (direction parallel to the time axis) — hence the charge flows orthogonally to it. Expression (1), on the other hand, corresponds to a purely space region V; (1) can also be interpreted as a charge "flowing" in space-time, but parallelly to the time axis.

Therefore, we conclude that for a three-dimensional region Ω situated arbitrarily in the space-time, all four terms should be present in the integral giving Q as

$$Q = \frac{1}{c} \iiint_\Omega (c\rho dx^1 dx^2 dx^3 + j^1 dx^0 dx^2 dx^3 + j^2 dx^0 dx^3 dx^1 + j^3 dx^0 dx^1 dx^2) .$$

Multiplying this by the unit quadrivector $J = e_0 e_1 e_2 e_3$,

$$QJ = \frac{1}{c} \iiint_\Omega (c\rho dx^1 dx^2 dx^3 + \ldots + j^3 dx^0 dx^1 dx^2) e_0 \wedge e_1 \wedge e_2 \wedge e_3 .$$

If we introduce the space-time vector $j = c\rho e_0 + j^1 e_1 + j^2 e_2 + j^3 e_3$ and the space-time trivector

$$\begin{aligned} d\Omega &= dx^1 dx^2 dx^3 e_1 \wedge e_2 \wedge e_3 + dx^0 dx^1 dx^2 e_0 \wedge e_1 \wedge e_2 \\ &\quad + dx^0 dx^2 dx^3 e_0 \wedge e_2 \wedge e_3 + dx^0 dx^3 dx^1 e_0 \wedge e_3 \wedge e_1 \\ &= \frac{1}{6} dx^\mu dx^\nu dx^\lambda e_\mu \wedge e_\nu \wedge e_\lambda \end{aligned}$$

as the directed measure of the integration region, we recognize the outer product under the integral

$$QJ = \frac{1}{c} \int_\Omega j \wedge d\Omega . \tag{3}$$

This is a covariant equation having quadrivectors on both sides. This reasoning leads us to the following conclusion: The charge density ρ and the current density $\mathbf{j}$ should be united into the space-time vector $j = (c\rho, \mathbf{j})$ which is known as the *four-current* or the *charge-current density vector.*

When Ω is a closed manifold, i.e. when it is a boundary of a four-dimensional region Ξ, the integral (3) should vanish due to the principle of charge conservation (the same charge flows into Ξ and out of Ξ):

$$\int_{\partial\Xi} j \wedge d\Omega = 0 . \tag{4}$$

The four-dimensional Gauss law

$$\int_{\partial\Xi} j \wedge d\Omega = -\int_{\Xi} (\partial \cdot j) d\Xi$$

with the use of four-dimensional nabla ∂ at the right-hand side (we do not derive this law — it is a counterpart of Eq. (0.142) from three dimensions) allows us to write

$$\int_{\Xi} (\partial \cdot j) d\Xi = 0$$

for any four-dimensional region Ξ. Since Ξ is arbitrary, we obtain from this

$$\partial \cdot j = 0 , \tag{5}$$

which is the *local* or *differential law of charge conservation.* When written in terms of coordinates, this is $\partial(c\rho)/\partial(ct) + \partial j^k/\partial x^k$ or

$$\frac{\partial \rho}{\partial t} + \nabla \cdot \mathbf{j} = 0 .$$

We recognize (1.28) in this. In this manner, we have found that (5) is the covariant recording of the electric charge continuity equation.

We now pass to the electromagnetic potentials in the vacuum. We know from §2.1 that in an appropriate gauge, they satisfy the d'Alembert equations (2.7) and (2.8):

$$\Box \varphi = -\frac{1}{\varepsilon_0}\rho , \quad \Box \mathbf{A} = -\mu_0 \mathbf{j} .$$

We multiply the left equation by $\frac{1}{c} = \sqrt{\varepsilon_0 \mu_0}$ such that

$$\Box \frac{1}{c}\varphi = -\mu_0 c \rho , \quad \Box \mathbf{A} = -\mu_0 \mathbf{j} .$$

The right-hand sides of the equations form the space-time vector $-\mu_0 j$, the d'Alembert operator is a scalar (it is important that $\Box$ is considered in the vacuum, compare equality $\Box = -\partial \cdot \partial$), hence in order to maintain the covariance we have to accept that the electric potential furnishes the lacking fourth component of the vector potential: $A^0 = \frac{1}{c}\varphi$. In this notation, we may write out

$$\Box A^0 = -\mu_0 j^0 , \quad \Box \mathbf{A} = -\mu_0 \mathbf{j} ,$$

or jointly

$$\Box A = -\mu_0 j . \tag{6}$$

The space-time vector $A = (\frac{1}{c}\varphi, \mathbf{A})$ is called the *four-potential* or the *electromagnetic potential.*

The Lorentz condition should also be transformed to a covariant form. Applying the identity $\varepsilon_0\mu_0 = 1/c^2$, we may change $\nabla \cdot \mathbf{A} + \varepsilon_0\mu_0 \frac{\partial\varphi}{\partial t}$ into

$$\nabla \cdot \mathbf{A} + \frac{\partial}{\partial(ct)}\left(\frac{1}{c}\varphi\right) = 0$$

or, in terms of coordinates, $\partial_k A^k + \partial_0 A^0 = 0$, that is,

$$\partial \cdot A = 0 . \tag{7}$$

Thus, as a scalar equation, the Lorentz condition is satisfied in all inertial frames, if it is satisfied in one of them.

It is now time to consider the fields $\mathbf{E}$ and $\hat{\mathbf{B}}$. We write their relation to the potentials: $\mathbf{E} = -\nabla\varphi - \partial\,\mathbf{A}/\partial t$, $\hat{\mathbf{B}} = \nabla \wedge \mathbf{A}$ in terms of the coordinates

$$E^k = -\frac{\partial\varphi}{\partial x^k} - c\frac{\partial A^k}{\partial x^0} = c(-\partial_k A^0 - \partial_0 A^k) ,$$
$$B^{kl} = \frac{\partial A^l}{\partial x^k} - \frac{\partial A^k}{\partial x^l} = \partial_k A^l - \partial_l A^k \quad k,l \in \{1,2,3\} .$$

Some indices are written as lower indices at the right-hand side by virtue of the fact that $\partial_\mu = \frac{\partial}{\partial x^\mu}$ are covariant coordinates of the nabla operator. If we wish to make them upper indices, we should use the prescription $\partial^\mu = g^{\mu\nu}\partial_\nu$ which denotes the reversal of sign of the space components:

$$E^k = c(\partial^k A^0 - \partial^0 A^k) , \quad B^{kl} = \partial^l A^k - \partial^k A^l .$$

These six equations can be written in a single formula after introducing the two-indices symbol

$$F^{\mu\nu} = \partial^\mu A^\nu - \partial^\nu A^\mu , \qquad \mu,\nu \in \{0,1,2,3\} , \tag{8}$$

which has the following relation to the fields $\mathbf{E}$ and $\hat{\mathbf{B}}$:

$$F^{0k} = -\frac{1}{c}E^k \,, \qquad F^{kl} = -B^{kl} \,. \tag{9}$$

The symbol (8), known as the *electromagnetic field tensor*, is antisymmetric with respect to the interchange of the indices, so it contains only six independent quantities. We know that antisymmetric symbols can be identified with the coordinates of bivectors, so we introduce the space-time *electromagnetic field bivector*

$$F = \frac{1}{2}F^{\mu\nu}e_\mu \wedge e_\nu = F^{0k}e_0 \wedge e_k + \frac{1}{2}F^{kl}e_k \wedge e_l \,, \tag{10a}$$

$$F = -\frac{1}{c}e_0 \wedge E - B \,. \tag{10b}$$

This is a unification of the electric and magnetic fields into a single quantity, the unification following from relativity. The electromagnetic field is simply a bivector. Its division into the electric and magnetic parts depends on the reference frame or on the observer: the magnetic part is the component orthogonal to the time axis, the electric part — parallel to it.

In this way, the apparent discrepancy in the three-dimensional description of the electromagnetic field is removed. In previous chapters, the electric part was a vector, the magnetic part — a bivector. Now we find that in actual fact the electric field is also a bivector, or rather a component parallel to the time axis and this component cannot reveal its bivector character in three dimensions. We can perceive a vector as its "projection" on the three-dimensional subspace of constant time.

Relation (8) may be written in the bivectorial equation

$$F = \partial \wedge A \,, \tag{11}$$

which means that F is the outer derivative of A. The sum of Eqs. (7) and (11) yields $F = \partial \wedge A + \partial \cdot A$ or

$$F = \partial A \,, \tag{12}$$

meaning that F is the Clifford derivative of the electromagnetic potential. We should, however, emphasise that this is valid only when the potential satisfies the Lorentz condition.

We now study the Maxwell equations (1.17) in the vacuum. The pair of inhomogeneous equations

$$\nabla\cdot\widehat{\mathbf{B}}+\varepsilon_0\mu_0\frac{\partial\,\mathbf{E}}{\partial t}=-\mu_0\,\mathbf{j}\;,\qquad \nabla\cdot\mathbf{E}=\frac{1}{\varepsilon_0}\rho$$

can be written in terms of coordinates:

$$\partial_k B^{kl}+\frac{1}{c}\partial_0 E^l=-\mu_0 j^l\;,\qquad \partial_k E^k=\frac{1}{\varepsilon_0}\rho\;.$$

We insert here the coordinates of F:

$$-\partial_k F^{kl}-\partial_0 F^{0l}=-\mu_0 j^l\;,\qquad \partial_k F^{0k}=\frac{1}{c^2\varepsilon_0}j^0\;.$$

Both equations can be written jointly as $\partial_\mu F^{\mu\nu}=\mu_0 j^\nu$ and, with the use of bivector (10a)

$$\partial\cdot F=\mu_0 j\;. \tag{13}$$

This is the covariant version of the inhomogeneous equations and it states that the charge-current density vector is proportional to the inner derivative of F.

The pair of homogeneous equations

$$\nabla\wedge\mathbf{E}+\frac{\partial\,\widehat{\mathbf{B}}}{\partial t}=0\;,\qquad \nabla\wedge\widehat{\mathbf{B}}=0\;,$$

after being written in terms of coordinates

$$\partial_j E^k-\partial_k E^j+c\partial_0 B^{jk}=0\;,\qquad \partial_3 B^{12}+\partial_2 B^{31}+\partial_1 B^{23}=0$$

can be expressed by F:

$$-\partial_j F^{0k}+\partial_k F^{0j}-\partial_0 F^{jk}=0\;,\qquad -\partial_3 F^{12}-\partial_2 F^{31}-\partial_1 F^{23}=0\;.$$

We rearrange the indices and write them all as upper indices:

$$\partial^0 F^{kj}+\partial^j F^{0k}+\partial^k F^{j0}=0\;,\qquad \partial^3 F^{12}+\partial^2 F^{31}+\partial^1 F^{23}=0\;.$$

The left equation contains three nontrivial equalities for $(j,k)\in\{(1,2),(3,1),(2,3)\}$, the right one is single. We recognize in them the four coordinates of a trivector which, because of the right-hand side, is zero:

$$\partial\wedge F=0\;. \tag{14}$$

This is the covariant notation for the homogeneous Maxwell equations and it states that the outer derivative of F is zero.

The sum of (13) and (14) yields $\partial \cdot F + \partial \wedge F = \mu_0 \jmath$, or

$$\partial F = \mu_o \jmath \,. \tag{15}$$

In this manner, all the Maxwell equations are contained in a single cliffor equation. As distinct from the three-dimensional case, it contains only multivectors of two ranks: vectors and trivectors. These two multivector parts bring us back to the reconstruction of the separate Eqs. (13) and (14). The covariance of (15) under the Lorentz transformations is assured by the fact that all multivectors transform under the same law $M \to e^{-D/2} M e^{D/2}$ with the bivector exponents.

Equation (15) is the relativistic counterpart of (1.23). The operator $\partial = e^{\mu}\partial_{\mu}$ contains the time derivation like $\not{D} = \nabla + \sqrt{\varepsilon\mu}\partial/\partial t$ though in a different manner — that is with the unit vector factor e^0.

Finally, we consider the Lorentz force

$$\mathbf{f} = q\,\mathbf{E} - q\mathbf{v} \cdot \widehat{\mathbf{B}} \,.$$

We know from §6.6 that this should be multiplied by γ in order to obtain the space part of a space-time vector,

$$\mathcal{F} = q\gamma\,\mathbf{E} - q\gamma\,\mathbf{v} \cdot \widehat{\mathbf{B}} \,,$$

or, in terms of coordinates,

$$\mathcal{F}^k = q\gamma E^k - q\gamma v^j B^{jk} \,.$$

Inserting the electromagnetic field tensor,

$$\mathcal{F}^k = q\gamma F^{k0} + q\gamma v^j F^{jk} \,.$$

We remember that $(c\gamma, \gamma\mathbf{v}) = u$ is the space-time vector, so $\mathcal{F}^k = qu^0 F^{k0} + qu^j F^{jk} = q(F^{k0}u_0 + F^{kj}u_j)$ or

$$\mathcal{F}^k = qF^{k\nu}u_\nu \,. \tag{16}$$

We need a fourth coordinate to this. For this purpose, we use $d\mathcal{E} = vdp$ and the Lorentz force:

$$\frac{d\mathcal{E}}{dt} = \mathbf{v} \cdot \frac{d\mathbf{p}}{dt} = \mathbf{v} \cdot (q\,\mathbf{E} - q\,\mathbf{v} \cdot \widehat{\mathbf{B}}) = q\,\mathbf{v} \cdot \mathbf{E} - q\,\mathbf{v} \cdot (\mathbf{v} \cdot \mathbf{B}) = q\,\mathbf{E} \cdot \mathbf{v} \,.$$

We know from §6.6 that the left-hand side gives the zeroth coordinate of the power-force vector, hence

$$\mathcal{F}^0 = -qF^{0j}u^j = qF^{0j}u_j \ .$$

One may add zero, so

$$\mathcal{F}^0 = q(F^{0j}u_j + F^{00}u_0) = qF^{0\nu}u_\nu \ .$$

This equality along with the three versions of (16) for $k \in \{1,2,3\}$ furnish four coordinates of the vector equation

$$\mathcal{F} = qF \cdot u \ . \tag{17}$$

This is the relativistic counterpart of the Lorentz equation.

Summarizing: we have found covariant forms of the principal equations in electrodynamics: (5), (6), (12) (or only (11) if the Lorentz condition is not satisfied), (15) and (17). Their covariance is guaranteed by the fact that they contain cliffors transforming under the same law.

Some derivations are very easy in this formalism. For instance, by the substitution of (12) in (15), one obtains $\partial(\partial A) = \mu_0 j$. The associativity of the Clifford product allows us to write $\partial^2 A = -\Box A$, so we get

$$\Box A = -\mu_0 j \ ,$$

which was considered earlier as Eq. (6).

By taking the inner derivative of (13), one obtains $\partial \cdot (\partial \cdot F) = \mu_0 \partial \cdot j$. The double inner derivative is zero, so $\partial \cdot j = 0$ which is the continuity equation (5).

We established in §6.3 that the Clifford square of a bivector over the Minkowski space consists of two multivectors invariant under the proper Lorentz transformations: the scalar part and the quadrivector part. Equation (6.44b) expresses them in a notation almost matching the electromagnetic field:

$$F \cdot F = \frac{1}{c^2}|E|^2 - |B|^2 \ , \tag{18a}$$

$$F \wedge F = \frac{2}{c} e_0 (E \wedge B) \ . \tag{18b}$$

The invariance of the quadrivector part under the active Lorentz transformations (compare §6.3) denotes that the equality $e_0(E \wedge B) = e_0(E' \wedge B')$ holds. Multiplication by e_0 from the left yields

$$E \wedge B = E' \wedge B' \ .$$

When translated into the Clifford algebra of the three-dimensional Euclidean space, we may write this as

$$\mathbf{E} \wedge \widehat{\mathbf{B}} = \mathbf{E}' \wedge \widehat{\mathbf{B}}' \,. \tag{19}$$

When considering the behaviour of the electromagnetic field under Lorentz transformations, an interesting question arises: For a given field F, does a reference frame exist in which one of the parts: electric E' or magnetic B' vanishes? Equation (19) informs us that a necessary condition for this is $E \wedge B = 0$. By virtue of (18b), we have then $F \wedge F = 0$ which, by Lemma 1, §6.3, shows that F is a volutor. If this condition is satisfied, expression (18a) decides which of the parts: electric or magnetic can be zero. Namely, for $F \cdot F > 0$, the magnetic part can vanish and for $F \cdot F < 0$, the electric part can disappear in some reference frame.

Let us now find the explicit transformation formulae under a Lorentz boost for two parts of F separately. We start from $F' = e^{\theta/2} F e^{-\theta/2} = -e^{\theta/2}(e_0 \frac{E}{c} + B)e^{-\theta/2}$ and employ (6.54):

$$\begin{aligned}
-F' &= \frac{1}{2(1+\gamma)}(1+\gamma+\gamma e_0\,\beta)\Big(e_0\frac{E}{c}+B\Big)(1+\gamma-\gamma e_0\,\beta) \\
&= \frac{1}{2(1+\gamma)}\Big[(1+\gamma)^2\Big(e_0\frac{E}{c}+B\Big)+(1+\gamma)\gamma e_0\,\beta\Big(e_0\frac{E}{c}+B\Big) \\
&\quad -(1+\gamma)\gamma\Big(e_0\frac{E}{c}+B\Big)e_0\,\beta-\gamma^2 e_0\,\beta\Big(e_0\frac{E}{c}+B\Big)e_0\,\beta\Big] .
\end{aligned}$$

Vector e_0 commutes with B and anticommutes with space vectors E and β:

$$\begin{aligned}
-F' &= \frac{1}{2}\Big[(1+\gamma)\Big(e_0\frac{E}{c}+B\Big)+\gamma\Big(-e_0^2\,\beta\frac{E}{c}+e_0\,\beta B\Big) \\
&\quad -\gamma\Big(-e_0^2\frac{E}{c}\beta+e_0 B\beta\Big)-\frac{\gamma^2}{1+\gamma}\Big(e_0^2\,\beta\frac{E}{c}+e_0\,\beta B\Big)e_0\beta\Big] .
\end{aligned}$$

We know that $e_0^2 = 1$, so

$$\begin{aligned}
-F' &= \frac{1}{2}\Big[(1+\gamma)\Big(e_0\frac{E}{c}+B\Big)+\frac{\gamma}{c}(E\beta-\beta E)+\gamma e_0(\beta B-B\beta) \\
&\quad +\frac{\gamma^2}{1+\gamma}\Big(e_0\,\beta\frac{E}{c}\beta+\beta B\beta\Big)\Big] .
\end{aligned}$$

Substituting $-F' = e_0 \frac{E'}{c} + B'$,

$$\begin{aligned}
e_0 E' + cB' =& \frac{1+\gamma}{2}(e_0 E + cB) + \gamma E \wedge \beta + c\gamma e_0(\beta \cdot B) \\
&+ \frac{\gamma^2}{1+\gamma}(e_0\,\beta E\beta + c\beta B\beta) .
\end{aligned}$$

We introduce the unit space vector $n = \beta/|\beta|$ and equate separately the time and space parts of the bivectors:

$$e_0E' = \frac{1+\gamma}{2}e_0E + c\gamma e_0(\beta\cdot B) + \frac{\gamma-1}{2}e_0nEn\ ,$$
$$cB' = \frac{1+\gamma}{2}cB + \gamma E\wedge\beta + \frac{\gamma-1}{2}cnBn\ .$$

The identity $\gamma^2|\beta|^2/(1+\gamma) = \gamma-1$ was used here. We divide the first equality by e_0, the second one by c:

$$E' = \frac{\gamma+1}{2}E + c\gamma(\beta\cdot B) + \frac{\gamma-1}{2}nEn\ ,$$
$$B' = \frac{\gamma+1}{2}B + \frac{\gamma}{c}E\wedge\beta + \frac{\gamma-1}{2}nBn\ .$$

The components parallel and orthogonal to n behave as follows:

$$nE_{\|}n = E_{\|}n^2 = -E_{\|}\ , \qquad nE_{\perp}n = -E_{\perp}n^2 = E_{\perp}\ ,$$
$$nB_{\|}n = -B_{\|}n^2 = B_{\|}\ , \qquad nB_{\perp}n = B_{\perp}n^2 = -B_{\perp}\ ,$$

hence we obtain

$$E' = \frac{\gamma+1}{2}(E_{\|}+E_{\perp}) + \gamma(V\cdot B) + \frac{\gamma-1}{2}(E_{\perp}-E_{\|})\ ,$$
$$B' = \frac{\gamma+1}{2}(B_{\|}+B_{\perp}) - \frac{\gamma}{c^2}V\wedge E + \frac{\gamma-1}{2}(B_{\|}-B_{\perp})\ .$$

The quantities obtained are contained in the three-dimensional space orthogonal to the time axis, so we pass to the Clifford algebra of the three-dimensional Euclidean space (this passage demands the relation $V\cdot B = -\mathbf{V}\cdot\widehat{\mathbf{B}}$):

$$\mathbf{E}' = \mathbf{E}_{\|} + \gamma(\mathbf{E}_{\perp} - \mathbf{V}\cdot\widehat{\mathbf{B}})\ , \tag{20a}$$
$$\widehat{\mathbf{B}}' = \widehat{\mathbf{B}} + \gamma(\widehat{\mathbf{B}}_{\|} - \frac{1}{c^2}\mathbf{V}\wedge\mathbf{E})\ . \tag{20b}$$

The formulae show that if $\mathbf{V}\|\mathbf{E}$ and $\mathbf{V}\perp\widehat{\mathbf{B}}$, the electromagnetic field does not change for such a particular boost.

We now assume that F is space-like: $F^2 < 0$ and we look for a Lorentz boost which changes F into a purely magnetic field. We see from (20a) that the velocity $\mathbf{V}$ characterizing the boost must be orthogonal to $\mathbf{E}$ (the component $\mathbf{E}_{\|}$ does not change, so it must be zero from very beginning). From equating (20a) to zero, we obtain $\mathbf{E} = \mathbf{V}\cdot\widehat{\mathbf{B}}$. This implies $\mathbf{E}\wedge\widehat{\mathbf{B}} = 0$ (i.e. by (18.b), F

is simple which confirms our assumption $F^2 < 0$). We choose for convenience $\mathbf{V} \| \widehat{\mathbf{B}}$ which ensures $\mathbf{V} \cdot \widehat{\mathbf{B}} = \mathbf{V}\widehat{\mathbf{B}}$, so

$$\mathbf{E} = \mathbf{V}\widehat{\mathbf{B}} .$$

Thus, by employing the properties of the Clifford algebra, we obtain

$$\mathbf{V} = \mathbf{E}\widehat{\mathbf{B}}^{-1} = \mathbf{E}\Big(- \frac{\widehat{\mathbf{B}}}{|\widehat{\mathbf{B}}|^2}\Big) = -\frac{\mathbf{E}\cdot\widehat{\mathbf{B}}}{|\widehat{\mathbf{B}}|^2} = \frac{\widehat{\mathbf{B}}\cdot\mathbf{E}}{|\widehat{\mathbf{B}}|^2} . \tag{21}$$

The velocity $\mathbf{V}$ completely characterizes the boost (see Eq. (6.53)), so our problem is solved. The magnitude of $\mathbf{V}$ is $|\mathbf{V}| = |\mathbf{E}|/|\mathbf{B}|$. We now see the importance of the condition $F \cdot F < 0$: The inequality $|\mathbf{E}| < c|\widehat{\mathbf{B}}|$ assures that velocity (21) is less than that of light. Notice also that (21) is identical to $\mathbf{v}_d$, the electric drift velocity discussed in §3.1. This fact will be explained in §3.

Let us now assume that F is time-like: $F^2 > 0$, and find a boost which changes F into a purely electric field. It follows from (20b) that $\mathbf{V} \| \widehat{\mathbf{B}}$; we choose $\mathbf{V} \perp \mathbf{E}$, then $\mathbf{V} \wedge \mathbf{E} = \mathbf{V}\mathbf{E}$. By equating (20b) to zero we get

$$\widehat{\mathbf{B}} = \frac{1}{c^2}\mathbf{V}\wedge\mathbf{E} = \frac{1}{c^2}\mathbf{V}\mathbf{E} ,$$

and then the velocity

$$\mathbf{V} = c^2\widehat{\mathbf{B}}\mathbf{E}^{-1} = c^2\frac{\widehat{\mathbf{B}}\cdot\mathbf{E}}{|\mathbf{E}|^2} . \tag{22}$$

Its magnitude is $|\mathbf{V}| = c\frac{c|\widehat{\mathbf{B}}|}{|\mathbf{E}|}$. This time, the inequality $c|\widehat{\mathbf{B}}| < |\mathbf{E}|$ assures that $\mathbf{V}$ has its magnitude less than c.

The transformation formulae (20) permit other observations. If, for instance, the field F is purely magnetic, then after a boost with velocity $\mathbf{V}$, the electric field $\mathbf{E}' = -\mathbf{V}\cdot\widehat{\mathbf{B}}$ emerges. On the other hand, if F is purely electric, then a boost with velocity $\mathbf{V}$ "creates" the magnetic field $\widehat{\mathbf{B}}' = -\frac{1}{c^2}\mathbf{V}\wedge\mathbf{E}$.

§2. Examples of Electromagnetic Fields and Potentials

2.1. The Uniform Field

An electromagnetic field which is uniform in space and constant in time is clearly uniform in the space-time. The potentials found in Chap. 2,

$$\varphi(\mathbf{r}) = -\mathbf{r}\cdot\mathbf{E} , \qquad \mathbf{A}(\mathbf{r}) = \frac{1}{2}\mathbf{r}\cdot\widehat{\mathbf{B}} , \tag{23}$$

are constant in time, but nonuniform in space, therefore the four-potential $A = (\frac{1}{c}\varphi, \mathbf{A})$ is not uniform in the Minkowski space. This implies that a reference frame exists, in which A is not constant in time. The time is absent in formulae (23), so they are noncovariant.

A better candidate for a covariant potential is given in Problem 2.6:

$$\varphi = -\frac{1}{2}\mathbf{r}\cdot\mathbf{E}, \qquad \mathbf{A} = \frac{1}{2}\mathbf{r}\cdot\hat{\mathbf{B}} - \frac{1}{2}t\mathbf{E}\,. \tag{24}$$

We write them with the use of F:

$$A^0 = \frac{1}{2}x^j F^{0j}, \qquad A^k = -\frac{1}{2}x^j F^{jk} + \frac{1}{2}x^0 F^{0k},$$

and rearrange the indices

$$A^0 = \frac{1}{2}x_j F^{j0}, \qquad A^k = \frac{1}{2}x_j F^{jk} + x_0 F^{0k}\,.$$

Both expressions may be contained in the single formula $A^\nu = \frac{1}{2}x_\mu F^{\mu\nu}$ which can be represented by multivectors:

$$A = \frac{1}{2}x\cdot F\,.$$

This is the covariant formula expressing the electromagnetic potential as a function of the position four-vector x.

2.2. Field of the Electric Charge

Our goal is now to find a covariant notation for the Coulomb field produced by a single charge in uniform motion. We start with a charge Q resting at the origin of space coordinates — its field is purely electric,

$$E = \frac{Q}{4\pi\varepsilon_0}\frac{\mathbf{r}}{|\mathbf{r}|^3}$$

by virtue of (1.36). Thus the space-time bivector for this field is

$$F = -\frac{1}{c}e_0 E = -\frac{Q}{4\pi c\varepsilon_0}\frac{e_0 r}{|r|^3}\,.$$

After introducing the space-time radius vector $x = x^0 e_0 + r$, we notice $e_0 \wedge x = e_0 r$ and $|e_0\wedge x|^2 = (e_0 r)^2 = -r^2 = |r|^2$, hence

$$F = -\frac{Q}{4\pi c\varepsilon_0}\frac{e_0\wedge x}{|e_0\wedge x|^3}\,. \tag{25}$$

This expression is not yet covariant since it contains the distinguished vector e_0. But the presence of e_0 is connected with our assumption that the charge rests at the origin, so in this case, we can use the celerity $u = ce_0$ of the charge and substitute this in (25):

$$F = -\frac{Qc}{4\pi\varepsilon_0} \frac{u \wedge x}{|u \wedge x|^3} . \tag{26}$$

This is the covariant record of the Coulomb field of charge Q moving with the constant four-velocity u.

In the space-time formulation of electrodynamics, one may also introduce *force surfaces* if the electromagnetic bivector F is simple. Their definition is analogous to that given in §1.1: they are oriented surfaces to which the field F is tangent at each point and has compatible orientation. As can be seen from (17), the Minkowski forces are tangent to these surfaces. The Coulomb field of a single particle is, as we notice from (26), the volutor field, so we may look for the force surfaces for it.

When the charge rests at the origin, we see from (25) that the force surfaces are such as those in Fig. 123 (the orientation is adjusted to positive Q). This figure has the obvious defect of lacking a third space axis, but the reader can agree that the X^3-axis is on equal footing with X^1 and X^2. Thus the force surfaces are half-planes passing through the time axis and diverge radially in all three dimensions from it. Intersections of the surfaces with any three-space of constant time are half-straight lines diverging radially from the origin — they are the well known force lines of the electric field produced by a charge. Figure 123 is very similar to Fig. 34 — this is not strange, because a charge "sitting" in the origin is a kind of current flowing through the space-time along the time axis which is the world line for the charged particle.

When the charge moves uniformly, the picture of force surfaces is similar, but with the following modification: All the half-planes pass through the world line of the charged particle, which is now inclined to the time axis. We illustrate this in Fig. 124 for a charge moving in the direction e_2. We see that F in various points can be decomposed into two components: parallel and orthogonal to X^0. This means that both electric and magnetic fields are present — the plane of e_0e_2 is an exception — only the electric field exists on the X^2-axis in the three-space.

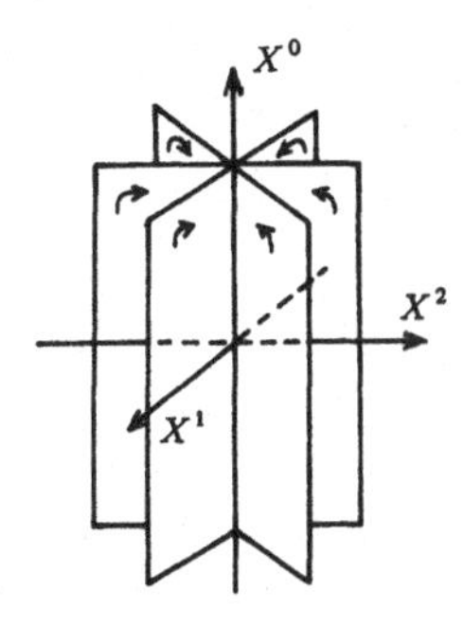

Fig. 123.

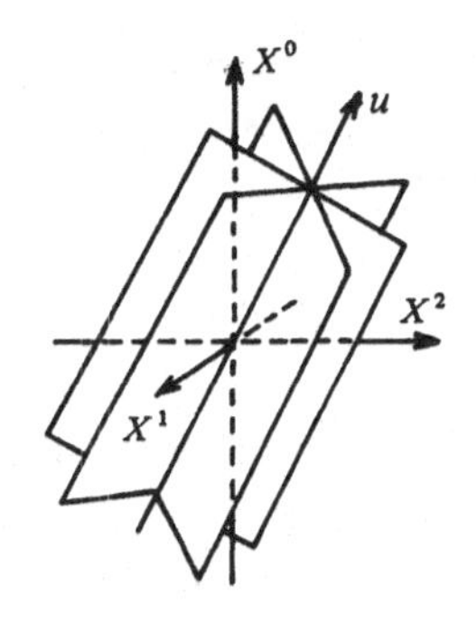

Fig. 124.

Let us find the two constituent fields from (26). Now

$$u \wedge x = (\gamma c e_0 + \gamma v) \wedge (x^0 e_0 + r) = e_0(\gamma c r - \gamma x^0 v) + \gamma v \wedge r$$

and

$$|u \wedge x|^2 = |\gamma c r - \gamma c t v|^2 - |\gamma v \wedge r|^2 = \gamma^2(c^2|r - vt|^2 - |v \wedge r|^2) .$$

Thus we may write (26) as

$$F = -\frac{Qc}{4\pi\varepsilon_0\gamma^2}\frac{c e_0(r - vt) + v \wedge r}{(c^2|r - vt|^2 - |v \wedge r|^2)^{\frac{3}{2}}} .$$

By referring to (10b), we obtain

$$E = \frac{Qc^2}{4\pi\varepsilon_0\gamma^2}\frac{c(r - vt)}{(c^2|r - vt|^2 - |v \wedge r|^2)^{\frac{3}{2}}} ,$$

$$B = \frac{Qc}{4\pi\varepsilon_0\gamma^2}\frac{v \wedge r}{(c^2|r - vt|^2 - |v \wedge r|^2)^{\frac{3}{2}}} ,$$

and in the notation of the Euclidean three-space,

$$\mathbf{E} = \frac{Q}{4\pi\varepsilon_0\gamma^2}\frac{\mathbf{r} - \mathbf{v}\,t}{(|\mathbf{r} - \mathbf{v}\,t|^2 - |\mathbf{v} \wedge \mathbf{r}\,|^2/c^2)^{\frac{3}{2}}} ,$$

$$\hat{\mathbf{B}} = \frac{Q}{4\pi\varepsilon_0 c^2\gamma^2}\frac{\mathbf{v} \wedge \mathbf{r}}{(|\mathbf{r} - \mathbf{v}\,t|^2 - |\mathbf{v} \wedge \mathbf{r}\,|^2/c^2)^{\frac{3}{2}}} .$$

The motion of charge is described by the function $\mathbf{r}_Q(t) = \mathbf{v}\,t$, moreover $\varepsilon_0 c^2 = 1/\mu_0$, hence

$$\mathbf{E} = \frac{Q}{4\pi\varepsilon_0\gamma^2}\frac{\mathbf{r} - \mathbf{r}_Q(t)}{[|\mathbf{r} - \mathbf{r}_Q(t)|^2 - |\mathbf{v} \wedge \mathbf{r}|^2/c^2]^{\frac{3}{2}}} , \tag{27a}$$

$$\hat{\mathbf{B}} = \frac{Q\mu_0}{4\pi\gamma^2}\frac{\mathbf{v} \wedge \mathbf{r}}{[|\mathbf{r} - \mathbf{r}_Q(t)|^2 - |\mathbf{v} \wedge \mathbf{r}|^2/c^2]^{\frac{3}{2}}} . \tag{27b}$$

The relation $\hat{\mathbf{B}} = \frac{1}{c^2}\mathbf{v} \wedge \mathbf{E}$ is satisfied at each point, so $\mathbf{E} \| \hat{\mathbf{B}}$ as should be expected for the simple bivector F. In particular, for the X^2-axis, i.e. on the trajectory of the charge,

$$\mathbf{E} = \frac{Q}{4\pi\varepsilon_0\gamma^2} \frac{\mathbf{r} - \mathbf{r}_Q(t)}{|\mathbf{r} - \mathbf{r}_Q(t)|^3}, \qquad \hat{\mathbf{B}} = 0 . \tag{28}$$

When comparing this $\mathbf{E}$ with (1.36), one could suppose that this field is γ^2 times smaller than the field of charge at rest. On the other hand, Eq. (20a) asserts that both electric fields ($\mathbf{E}'$ in the rest frame and $\mathbf{E}$ in the frame where the particle has velocity $\mathbf{v}$) should be equal. This apparent discrepancy can, nevertheless, be eliminated if one realizes that the directed distance $\mathbf{r} - \mathbf{r}_Q$ also changes by the factor γ, which yields

$$|\mathbf{E}| = \frac{Q}{4\pi\varepsilon_0\gamma^2|\mathbf{r} - \mathbf{r}_Q|^2} = \frac{Q}{4\pi\varepsilon_0|\mathbf{r}' - \mathbf{r}'_Q|^2} = |\mathbf{E}'| .$$

At other particular points $\mathbf{r}$ such that $\mathbf{r} - \mathbf{r}_Q \perp \mathbf{v}$ (we denote $\mathbf{r} - \mathbf{r}_Q = \mathbf{r}_\perp$ now), expressions (27) yield

$$\mathbf{E} = \frac{Q}{4\pi\varepsilon_0\gamma^2} \frac{\mathbf{r}_\perp}{(\mathbf{r}_\perp^2 - \mathbf{r}_\perp^2 v^2/c^2)^{\frac{3}{2}}} = \frac{Q\gamma}{4\pi\varepsilon_0} \frac{\mathbf{r}_\perp}{|\mathbf{r}_\perp|^3},$$
$$\hat{\mathbf{B}} = \frac{Q\mu_0\gamma}{4\pi} \frac{\mathbf{v} \wedge \mathbf{r}_\perp}{|\mathbf{r}_\perp|^3} .$$

The magnitude of $\mathbf{E}$ is γ^3 times larger than that of (28) for the same magnitude of $\mathbf{r} - \mathbf{r}_Q$.

The electromagnetic field (27) is illustrated in Fig. 125 on a plane passing through the (positive) charge Q and parallel to its velocity.

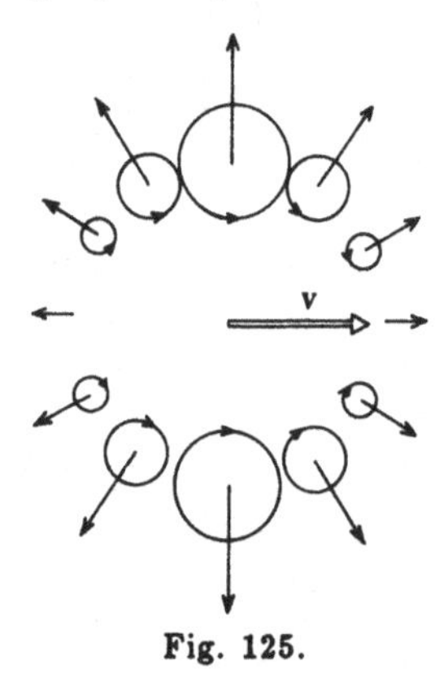

Fig. 125.

If we form the electromagnetic cliffor $f = \mathbf{e} + \widehat{\mathbf{b}}$ of the field (27), we obtain

$$f = \frac{Q}{4\pi\sqrt{\varepsilon_0}\gamma^2} \frac{\mathbf{r} - \mathbf{r}_Q + \boldsymbol{\beta} \wedge \mathbf{r}}{(|\mathbf{r} - \mathbf{r}_Q|^2 - |\boldsymbol{\beta} \wedge \mathbf{r}|^2)^{\frac{3}{2}}} .$$

The energy-momentum cliffor for this field is

$$\frac{1}{2} f f^{\dagger} = \frac{Q^2}{16\pi^2\varepsilon_0\gamma^4} \frac{|\mathbf{r} - \mathbf{r}_Q|^2 + |\boldsymbol{\beta} \wedge \mathbf{r}|^2 + 2(\boldsymbol{\beta} \wedge \mathbf{r}) \cdot (\mathbf{r} - \mathbf{r}_Q)}{(|\mathbf{r} - \mathbf{r}_Q|^2 - |\boldsymbol{\beta} \wedge \mathbf{r}|^2)^3} .$$

Its vector part determines the energy flux

$$S = \frac{Q^2}{8\pi^2\varepsilon_0\gamma^4} \frac{(\mathbf{v} \wedge \mathbf{r}) \cdot (\mathbf{r} - \mathbf{r}_Q)}{(|\mathbf{r} - \mathbf{r}_Q|^2 - |\boldsymbol{\beta} \wedge \mathbf{r}|^2)^3} .$$

Because of the properties of the inner product of a vector with a volutor, S is everywhere perpendicular to $\mathbf{r} - \mathbf{r}_Q$. We illustrate this flux in Fig. 126. We learn that the energy propagates principally along with the charge. Additional movements can be described as the accumulation of the energy on the X^2-axis before the charge and the diffusion of the energy from the X^2-axis behind the charge.

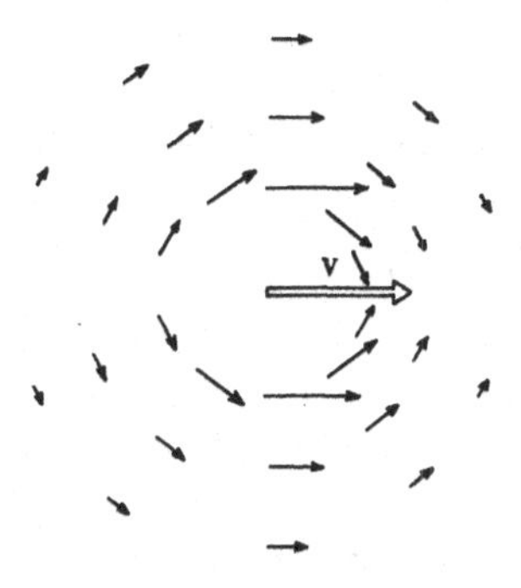

Fig. 126.

We also want to find a covariant potential for this situation. If the charge rests, we assume that the magnetic potential is zero, so the four-potential has to be parallel to the time axis

$$A \doteq \frac{\varphi}{c} e_0 = \frac{Q}{4\pi c\varepsilon_0} \frac{e_0}{|r|} = \frac{Q}{4\pi c\varepsilon_0} \frac{e_0}{|e_0 \wedge x|} .$$

Now the generalization on the uniformly moving charge stems from the substitution $u = ce_0$:

$$A = \frac{Q}{4\pi c\varepsilon_0} \frac{u}{|u \wedge x|} . \tag{29}$$

This potential, as the space-time vector, is parallel to the four-velocity of its source, therefore it is a counterpart of the vectorial potentials parallel to the currents, considered in §2.3.

One may check that the outer derivative of (29) yields field (26) but we leave this to the reader (Problem 4).

Equation (29) contains the following potentials of the three-dimensional Euclidean space:

$$\varphi = \frac{Q}{4\pi\varepsilon_0} \frac{1}{\sqrt{|\mathbf{r}-\mathbf{v}t|^2 - |\mathbf{v}\wedge\mathbf{r}|^2/c^2}} ,$$

$$\mathbf{A} = \frac{Q}{4\pi c\varepsilon_0} \frac{\mathbf{v}}{\sqrt{c^2|\mathbf{r}-\mathbf{v}t|^2 - |\mathbf{v}\wedge\mathbf{r}|^2}} .$$

2.3. Field of the Linear Current

We now present qualitative considerations demonstrating how the magnetic field of a current arises from the electric field of charges forming the current. One may look at the electric current as at a sequence of charges moving in a single file parallel to their common velocity. Hence at first, we have the purely electric field of a single positive charge shown in Fig. 123, secondly, the field of a moving charge as in Fig. 124. This field still has the electric part as the predominant part. Next, we should add many such fields produced by the charges of the sequence. In this way, by adding fields similar to that of Fig. 124, but shifted relative to each other in the direction e_2, we obtain the force surfaces for a file of identical positive charges moving along X^2 at the same velocity and with the distance between them disregarded. During such an addition, the components parallel to the plane of e_0e_2 are cancelled and the net force surfaces are half-planes partially perpendicular to e_0e_2 and inclined by the same angle to the X^2-axis. We exhibit them in the three-dimensional picture of Fig. 127 — they are parallel to X^1 and seem to be parallel to each other. But the same should be true for X^3 in place of X^1, since in the three-space of constant time, the field has the axial symmetry around X^2. The field obtained is not, therefore, uniform contrary to the impression given by Fig. 127 which, after all, is not complete. The field is singular on the plane of e_0e_2 which could be called a *world sheet* of the wire carrying the current and which separates two half-planes with opposite orientations. We notice, as in Fig. 124, that both electric and magnetic parts are nonzero. The volutor F is time-like, so the electric part still predominates.

In this example, the conducting wire is not neutral but positively charged — we consider only the sequence of positive charges. If we wish to obtain the field of a neutral wire, we ought to introduce another sequence of negative charges moving in the opposite direction. We depict in Fig. 128 the field produced by such charges. The net field corresponding to the neutral wire is the sum $F_1 + F_2 = F$ shown in Fig. 129. F_1 corresponds to the positive charges moving in direction e_2, F_2 to the negative charges in direction $-e_2$. F is the volutor orthogonal to the time axis, i.e. purely spatial. By the decomposition (10b), this signifies that only the magnetic field is present. The force surfaces of this field are shown in Fig. 130. Of course the field is not homogeneous, contrary to the impression suggested by the figure.

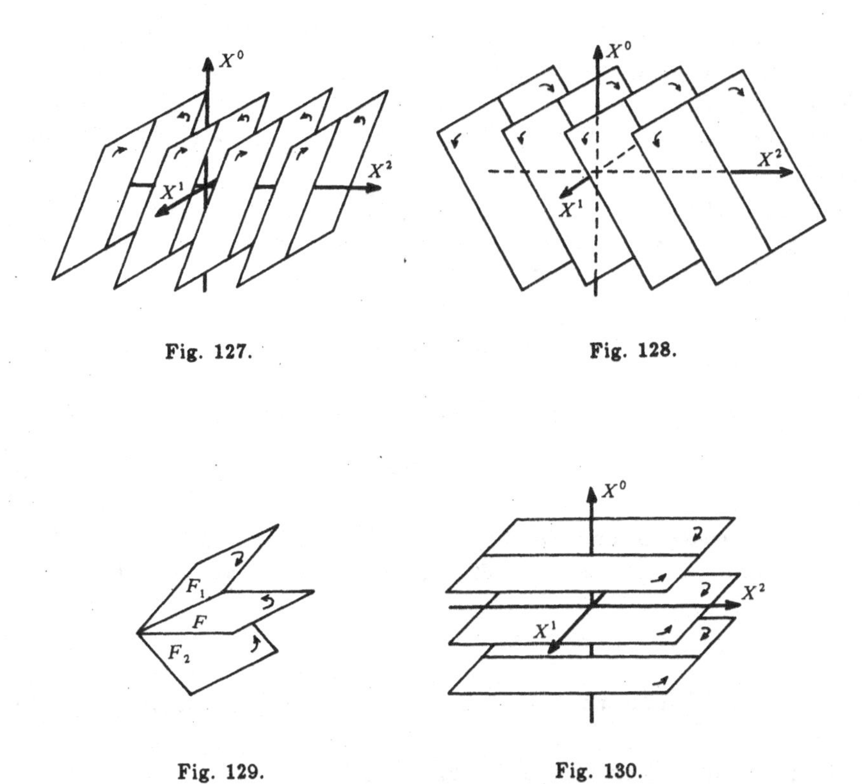

Fig. 127. Fig. 128.

Fig. 129. Fig. 130.

In this manner, we obtain a perceptible explanation of how the purely magnetic field arises from the electric field of separate charges. The use of force surfaces is helpful for this purpose.

2.4. The Plane Field

We are looking for solutions to the homogeneous Maxwell equation

$$\partial F = 0 \tag{30}$$

such that the bivector F depends on the space-time radius vector x only through the scalar $\xi = \kappa \cdot x$, where κ is a constant four-vector

$$F = H(\xi) = H(\kappa \cdot x) . \tag{31}$$

(Here H is a function of the scalar variable ξ.) This assumption means that the loci of points of constant field are three-dimensional hyperplanes orthogonal to κ, hence we may state that $F = H(\xi)$ satisfying (30) are *plane electromagnetic fields* in the space-time. We call κ the *propagation four-vector* for the plane field F.

For field (31), $\partial F = e^{\mu}\partial_{\mu}H(\xi) = e^{\mu}\kappa_{\mu}\frac{dH}{d\xi} = \kappa H'$, where prime signifies the derivative. Equation (30) implies $\kappa H' = \kappa \wedge H' + \kappa \cdot H' = 0$. The vector and trivector parts must vanish separately:

$$\kappa \wedge H' = 0, \qquad \kappa \cdot H' = 0 . \tag{32}$$

The first equation asserts that the bivector H' is parallel to κ,

$$H' = \kappa \wedge a(\xi) , \tag{33}$$

where a is a vectorial function of ξ. (Equation (33) asserts, moreover, that H' is a volutor.) The second Eq. (32) states that H' is orthogonal to κ:

$$a \cdot \kappa = 0, \qquad \kappa \cdot \kappa = 0 . \tag{34}$$

The latter condition means that κ is a light-like vector, hence H' is a light-like volutor.

We gather the information obtained in the formula

$$F(\xi) = \kappa g(\xi) , \tag{35}$$

where $\kappa^2 = 0$ and g is a vectorial function orthogonal to κ. The orthogonality condition is linear, so g' is also orthogonal to κ, hence we may identify $g' = a$ for a standing in (33). Each vector orthogonal to the light-like κ can be written as $g = \alpha\kappa + a_1 + a_2$, where α is real, a_1, a_2 are space-like vectors,

$a_1 \perp a_2$, and $a_1, a_2 \perp \kappa$. The term $\alpha\kappa$ disappears when multiplied by κ, so we may take $g = a_1 + a_2$ and write

$$F(\xi) = \kappa[g^1(\xi)e_1 + g^2(\xi)e_2] , \tag{36}$$

where $e_1^2 = e_2^2 = -1, e_1 \cdot e_2 = 0, e_1 \cdot \kappa = e_2 \cdot \kappa = 0$ and g^1, g^2 are scalar functions. This signifies that the plane field (35) is determined by two scalar functions of the scalar argument ξ and is expressed by the statement that the plane electromagnetic field *has two degrees of freedom.*

We learn from (36) that the plane electromagnetic field is a volutor tangent to the light cone along its propagation vector κ. Figure 114 is suitable for this situation — we repeat it here as Fig. 131 with the appropriate change of notation. The pair: vector κ and volutor $F = \kappa g$, characterizing the plane field is referred to as the *flag*.

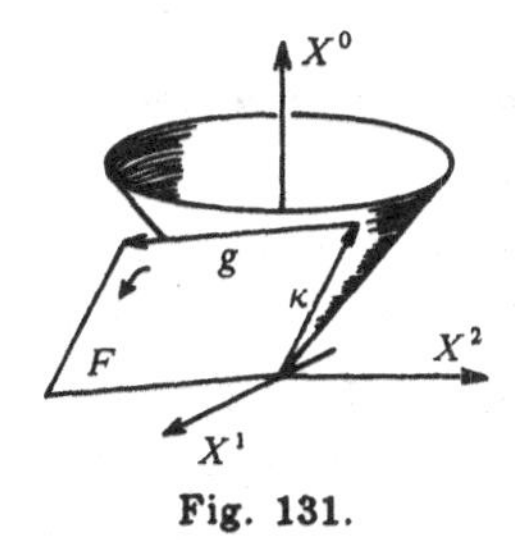

Fig. 131.

One can easily find that the field (35) has $A(\xi) = g(\xi)$ as its potential satisfying (12).

If we wish to compare solution (35) with the ones discussed in Chap. 4, we represent the propagation vector as $\kappa = (\kappa^0, \mathbf{k})$ and g as purely space vector $g = (0, \mathbf{g})$, where $\mathbf{g} \perp \mathbf{k}$. The condition $\kappa^2 = 0$ implies $\kappa^0 = \pm|\,\mathbf{k}\,|$, we choose $\kappa^0 = |\,\mathbf{k}\,|$. Then $\xi = \kappa \cdot x = \kappa^0 x^0 - \mathbf{k} \cdot \mathbf{r}$ and we should compare (35) only with those solutions from Chap. 4 which have scalar arguments. This denotes that (35) is a *plane travelling field.* The product κg is

$$\kappa g = (\kappa^0 e_0 + k)g = e_0|k|g + kg ,$$

where $k = (0, \mathbf{k})$. We get from the representation (10b)

$$E = -c|k|g , \qquad B = -kg .$$

Having introduced the unit vector $n = k/|k|$ we obtain

$$cB = nE , \tag{37}$$

which corresponds to the condition $\widehat{\mathbf{b}} = \mathbf{n}\,\mathbf{e}$ from Chap. 4 for the plane travelling fields. The equality $|E| = c|B|$ following from (37) agrees with the fact that F is light-like, i.e. $F \cdot F = \frac{1}{c^2}|E|^2 - |B|^2 = 0$.

If g is a periodic function, F can be called the *plane electromagnetic wave* and κ the *wave four-vector*. In particular, the plane harmonic wave with linear polarization can be written as

$$F = \kappa a \cos \kappa \cdot x = \kappa a \cos(|k|x^0 + k \cdot r) = \kappa a \cos(\omega t - \mathbf{k} \cdot \mathbf{r}) \; ,$$

where a is a constant pure space vector orthogonal to κ and $\omega = c|k|$. Then $\mathbf{k}\,\mathbf{a} = \mathbf{k} \wedge \mathbf{a}$ determines the plane of light oscillations.

For a circularly polarized wave, we can use the exponential function

$$F = e^{J(\kappa \cdot x)} \kappa a \; . \tag{38}$$

Indeed, if one introduces the unit trivector $I = e_1 e_2 e_3$ of the three-space orthogonal to e_0, one may write $J = e_0 I = -I e_0$, so

$$\begin{aligned} F &= (\cos\xi + J \sin\xi)(e_0 + n)|k|a \\ &= (e_0 + n)|k|a \cos\xi + (-I e_0^2 + e_0 I n)|k|a \sin\xi \; . \end{aligned}$$

We perform some manipulations

$$\begin{aligned} F &= |k|[(e_0 + n)a \cos\xi + (-1 + e_0 n) I a \sin\xi] \\ &= |k|[(e_0 + n)a \cos\xi + (e_0 + n) n I a \sin\xi] \\ &= |k|(e_0 + n)(\cos\xi + I n \sin\xi) a \; , \\ F &= (e_0 + n) e^{In\xi} |k| a \; . \end{aligned} \tag{39}$$

Thanks to (10), we can write

$$\mathbf{E} = -c e^{I\mathbf{n}\xi} |\mathbf{k}|\, \mathbf{a}, \qquad \widehat{\mathbf{B}} = -e^{I\mathbf{n}\xi} \mathbf{k}\, \mathbf{a} \; ,$$

which is equivalent to

$$f = \mathbf{e} + \widehat{\mathbf{b}} = -e^{I\mathbf{n}(\omega t - \mathbf{k} \cdot \mathbf{r})} (1 + \mathbf{n}) \frac{|\mathbf{k}|}{\sqrt{\mu_0}}\, \mathbf{a} \; .$$

We recognize in this the first term in (4.62), i.e. the plane travelling wave with right-handed circular polarization.

The wave (38) can also be written in the form (35). It suffices to inscribe (39) as

$$F = \kappa e^{In\xi} a \, ,$$

and notice that $g(\xi) = e^{In\xi}a$ is a space-like vector function.

§3. Charge in the Electromagnetic Field

3.1. Charge in the Uniform Field

Our aim is to solve the Lorentz equation (17) $m du/d\tau = qF \cdot u$ into which we substitute $\Omega = -(q/m)F$:

$$\frac{du}{d\tau} = -\Omega \cdot u = u \cdot \Omega \, . \tag{40}$$

For the uniform field, Ω is a constant bivector, hence (40) has the following solution

$$u(\tau) = e^{-\frac{1}{2}\Omega\tau} u_{(0)} e^{\frac{1}{2}\Omega\tau} \, . \tag{41}$$

Indeed,

$$\frac{du}{d\tau} = -\frac{1}{2}\Omega u + \frac{1}{2}u\Omega = \frac{1}{2}(u\Omega - \Omega u) = u \cdot \Omega \, .$$

When Ω is a volutor and $\Omega^2 \neq 0$, then, by Lemma 2 in §6.3, we decompose the initial celerity $u_{(0)}$ into components parallel and orthogonal to Ω : $u_{(0)} = u_{(0)\|} + u_{(0)\perp}$. The identities $u_{(0)\|}\Omega = -\Omega u_{(0)\|}$, $u_{(0)\perp}\Omega = \Omega u_{(0)\perp}$ hold, hence (41) may be rewritten as

$$u(\tau) = u_{(0)\|} e^{\Omega\tau} + u_{(0)\perp} \, . \tag{42}$$

Formulae (41) and (42) demand discussion in four possible cases.

(i) Ω is a time-like volutor. Substitute $\Omega^2 = \delta^2$ with real δ and choose a volutor D such that $\Omega = \delta D$. Then $D^2 = 1$ and the exponential takes the form

$$e^{\Omega\tau} = \cosh \delta\tau + D \sinh \delta\tau \, .$$

In this case, (42) yields

$$u(\tau) = u_{(0)\|} \cosh \delta\tau + u_{(0)\|} D \sinh \delta\tau + u_{(0)\perp} \, .$$

This expression can be integrated

$$x(\tau) = \frac{1}{\delta} u_{(0)\|} \sinh \delta\tau + \frac{1}{\delta} u_{(0)\|} D \cosh \delta\tau + u_{(0)\perp}\tau + x_{(0)} \, , \tag{43a}$$

$$x(\tau) = \frac{1}{\delta} u_{(0)\|} D e^{\delta \tau D} + u_{(0)\perp} \tau + x_{(0)} \ . \qquad (43b)$$

For $u_{(0)\perp} = 0$, we recognize in this the hyperbolic motion in the plane of D.

We know from considerations in §1 that for the time-like F a reference frame exists in which only the electric field remains. In such a frame, we choose the basis such that $D = e_0 e_1$ and $u_{(0)\|} = u^0_{(0)} e_0 + u^1_{(0)} e_1, u_{(0)\perp} = u^2_{(0)} e_2 + u^3_{(0)} e_3$. Without any loss of generality, we may assume $u^1_{(0)} = 0$, (an appropriate initial point on the time scale ensures this) and $u^3_{(0)} = 0$, then (43b) takes the form

$$x(\tau) = \frac{1}{\delta} u^0_{(0)} e_1 e^{\delta \tau e_0 e_1} + u^2_{(0)} e_2 \tau + x_{(0)} \ . \qquad (44)$$

We recognize motion (6.86) on the catenary curve. Now a Lorentz transformation changes the purely electric field into a time-like electromagnetic field and superposes this motion with the uniform motion.

Let us do this explicitly (for simplicity, we take $x_{(0)} = 0$). We apply the boost along X^2

$$\begin{aligned} x'^0 &= \gamma x^0 + \beta \gamma x^2 \,, \\ x'^1 &= x^1 \,, \\ x'^2 &= \beta \gamma x^0 + \gamma x^2 \ , \end{aligned}$$

to the coordinate of vector (44),

$$\begin{aligned} x'^0(\tau) &= \frac{\gamma}{\delta} u^0_{(0)} \sinh \delta \tau + \beta \gamma u^2_{(0)} \tau \ , \\ x'^1(\tau) &= \frac{1}{\delta} u^0_{(0)} \cosh \delta \tau \ , \\ x'^2(\tau) &= \frac{\beta \gamma}{\delta} u^0_{(0)} \sinh \delta \tau + \gamma u^2_{(0)} \tau \ . \end{aligned}$$

This is not the most general case, because the motion takes place in the plane of $\mathbf{e}_1 \mathbf{e}_2$, that is, in the common plane of the fields $\mathbf{E}' \| \mathbf{e}_1$ and $\mathbf{B}' \| \mathbf{e}_1 \mathbf{e}_2$, nevertheless it is still worth discussing.

We calculate the relevant coordinates of the velocity through the formula $v'^i = \frac{dx'^i}{d\tau} \frac{d\tau}{dt'} = c \frac{dx'^i}{d\tau} / \frac{dx'^0}{d\tau}$:

$$v'^1(\tau) = \frac{c}{\gamma} \frac{u^0_{(0)} \sinh \delta \tau}{u^0_{(0)} \cosh \delta \tau + \beta u^2_{(0)}} \ , \qquad v'^2(\tau) = c \frac{\beta u^0_{(0)} \cosh \delta \tau + u^2_{(0)}}{u^0_{(0)} \cosh \delta \tau + \beta u^2_{(0)}} \ . \qquad (45)$$

The initial velocity $\mathbf{v}'_{(0)} = c\frac{\beta u^0_{(0)}+u^2_{(0)}}{u^0_{(0)}+\beta u^2_{(0)}}\mathbf{e}_2$ is perpendicular to $\mathbf{E}'$. The limiting velocities for infinite time are

$$\mathbf{v}'_{\pm} = \lim_{\tau\to\pm\infty} \mathbf{v}'(\tau) = \pm\frac{c}{\gamma}\,\mathbf{e}_1 + \beta c\mathbf{e}_2 \; ,$$

so $|\mathbf{v}'_{\pm}| = c$.

Expressions (45) satisfy the equation of an ellipse:

$$(v'^1)^2 + \frac{(v'^2 - A)^2}{C^2} = \text{const} \; ,$$

where $A = \frac{\beta c^3}{(u^0_{(0)})^2-\beta(u^2_{(0)})^2}$, $C = \frac{u^2_{(0)}}{\gamma^2\sqrt{(u^0_{(0)})^2-\beta^2(u^2_{(0)})^2}}$. We show the two possible hodographs of this motion in Figs. 132a and b for $u^2_{(0)} < 0$ and $\beta > 0$. The case "a" corresponds to $v'^2(0) < 0$, the case "b" to $v'^2(0) > 0$. The fact that v'^2 changes sign in case "a" implies that the trajectory of the charged particle has a loop — we depict this in Fig. 133a. The other trajectory corresponding to case "b" is displayed in Fig. 133b.

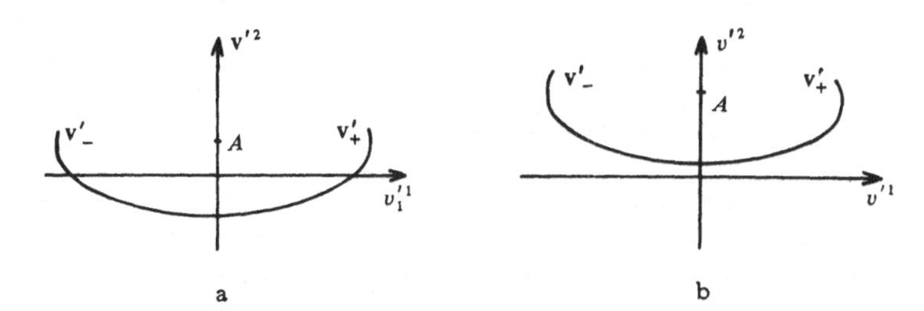

Fig. 132.

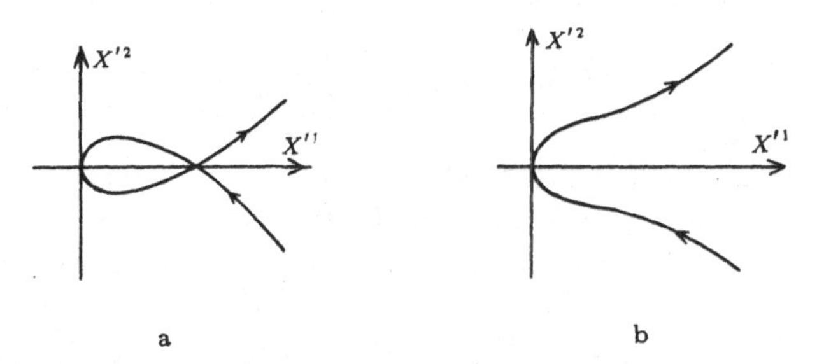

Fig. 133.

Summarizing, we may claim that the motion of a charge in a time-like electromagnetic field is the superposition of the motion (6.86) with the motion of velocity (22). In particular, when $\mathbf{u}_{(0)} = 0$, the motion is hyperbolic.

(ii) Ω is a space-like volutor. Denote $\Omega^2 = -\omega^2$ with real ω and substitute $\Omega = \omega D$, which implies $D^2 = -1$. Then the exponential takes the form $e^{\Omega\tau} = \cos\omega\tau + D\sin\omega\tau$ and (42) yields

$$u(\tau) = u_{(0)\parallel}\cos\omega\tau + u_{(0)\parallel}D\sin\omega\tau + u_{(0)\perp}\ .$$

The integral is

$$x(\tau) = \frac{1}{\omega}(u_{(0)\parallel}\sin\omega\tau - u_{(0)\parallel}D\cos\omega\tau) + u_{(0)\perp}\tau + x_{(0)}\ . \qquad (46)$$

Since F is space-like, a reference frame exists in which only the magnetic field is present. We choose a basis such that $D = e_1e_2$, then $u_{(0)\parallel} = u^1_{(0)}e_1 + u^2_{(0)}e_2, u_{(0)\perp} = u^0_{(0)}e_0 + u^3_{(0)}e_3$. Without loss of generality, we may assume $u^2_{(0)} = 0$, so (46) becomes

$$\begin{aligned} x(\tau) &= \frac{1}{\omega}\,u^1_{(0)}(e_1\sin\omega\tau + e_2\cos\omega\tau) + u_{(0)\perp}\tau + x_{(0)} \\ &= \frac{1}{\omega}\,u^1_{(0)}e_2e^{\omega\tau e_1e_2} + (u^0_{(0)}e_0 + u^3_{(0)}e_3)\tau + x_{(0)}\ . \end{aligned}$$

We decompose this into the time and space parts:

$$\begin{aligned} ct(\tau) &= u^0_{(0)}\tau\,, \\ r(\tau) &= \frac{u^1_{(0)}}{\omega}\,e_2e^{\omega\tau e_1e_2} + u^3_{(0)}e_3\tau + r_{(0)}\ . \end{aligned}$$

(We have chosen $x^0_{(0)} = 0$ in order that the time scales of t and τ have the same origin.) After denoting $u^0_{(0)} = \gamma$, we replace τ by t taken from the first formula

$$r(t) = \frac{u^1_{(0)}}{\omega}e_2e^{\frac{\omega t}{\gamma}e_1e_2} + \frac{u^3_{(0)}}{\gamma}e_3t + r_{(0)}\ .$$

We recognize here the formulae (3.5), i.e. the uniform motion on a helix with its axis perpendicular to the magnetic fields as the volutor. Therefore we ascertain that the motion of charge in a space-like uniform electromagnetic field is a superposition of the motion (3.5) with the uniform motion of velocity (21): $\mathbf{V} = \widehat{\mathbf{B}}\cdot\mathbf{E}/|\,\widehat{\mathbf{B}}\,|^2$ parallel to the magnetic field. Now it is not strange that this velocity coincides with the electric drift velocity $\mathbf{v}_d$ and that $\mathbf{v}_d$ does not depend on the charge. This velocity is characteristic of the electromagnetic field and arises only from the transformation properties of the field.

(iii) Ω is a light-like volutor. Now we cannot decompose u_0 into components parallel and orthogonal to Ω. We now develop the exponential function using $\Omega^2 = 0$

$$e^{\frac{1}{2}\Omega\tau} = 1 + \frac{1}{2}\Omega\tau ,$$

and insert this into (41):

$$\begin{aligned} u(\tau) &= (1 - \frac{1}{2}\tau\Omega)u_{(0)}(1 + \frac{1}{2}\tau\Omega) , \\ u(\tau) &= u_0 + \tau u_{(0)} \cdot \Omega - \frac{1}{4}\tau^2 \Omega u_0 \Omega . \end{aligned} \tag{47}$$

After substituting $\Omega = -\frac{q}{m}F = \frac{q}{m}(e_0\frac{E}{c} + B)$ and $u_{(0)} = c\gamma e_0 + \gamma v$, we obtain

$$\begin{aligned} u(\tau) = u_{(0)} &+ \frac{q\gamma\tau}{m}\Big[E - \frac{1}{c}e_0(v \cdot E) + v \cdot B\Big] \\ &- \frac{q^2\gamma\tau^2}{4m^2}\Big(2B \cdot E - 2ce_0|B|^2 + \frac{1}{c^2}EvE - \frac{1}{c}e_0 BvE + \frac{1}{c}e_0 EvB + BvB\Big) . \end{aligned}$$

This function can be easily integrated yielding $x(\tau)$ as a polynomial of third degree — this can be carried out by the reader. An interesting particular solution is for $v \perp B$ (by $E \| B$, this implies $v \perp E$). Then $vB = Bv$, $vE = -Ev$ and we obtain

$$u(\tau) = e_0\Big(1 + \frac{q^2\tau^2}{2m^2}|B|^2\Big) + \gamma v + \frac{q\gamma\tau}{m}E - \frac{q^2\tau^2\gamma}{2m^2}B \cdot E .$$

This formula indicates that for large times τ, the quadratic terms dominate, hence the velocity tends towards $c\,\widehat{\mathbf{B}} \cdot \mathbf{E}/|\,\widehat{\mathbf{B}} \cdot \mathbf{E}\,|$ as $\tau \to \pm\infty$, which is also called *drift velocity*.

(iv) Ω is not simple. As a result of the discussion in §6.3 we decompose it:$\Omega = \delta D_1 + \omega D_2$ with reals δ, ω and orthogonal volutors D_1, D_2 such that $D_1^2 = 1, D_2^2 = -1, D_1 D_2 = D_2 D_1$. The last property allows us to write the exponential as

$$e^{\frac{1}{2}\Omega\tau} = e^{\frac{1}{2}\delta\tau D_1} e^{\frac{1}{2}\omega\tau D_2}$$

and, consequently, rewrite (41) as

$$u(\tau) = e^{-\frac{1}{2}\omega\tau D_2}\Big(e^{-\frac{1}{2}\delta\tau D_1} u_{(0)} e^{\frac{1}{2}\delta\tau D_1}\Big) e^{\frac{1}{2}\omega\tau D_2} . \tag{48}$$

Space-time vector $u_{(0)}$ may be decomposed into two orthogonal components $u_{(0)} = u_{(1)} + u_{(2)}$ such that $u_{(i)} \| D_i, i \in \{1,2\}, u_{(1)}$ commutes with $D_2, u_{(2)}$ with D_1. In such a case, we get from (48),

$$u(\tau) = e^{-\frac{1}{2}\delta\tau D_1} u_{(1)} e^{\frac{1}{2}\delta\tau D_1} + e^{-\frac{1}{2}\omega\tau D_2} u_{(2)} e^{\frac{1}{2}\omega\tau D_2} ,$$

$$u(\tau) = u_{(1)} e^{\delta\tau D_1} + u_{(2)} e^{\omega\tau D_2} . \tag{49}$$

The first exponential contains the hyperbolic functions, the second one — trigonometric ones. Integration of (49) yields

$$x(\tau) = \frac{1}{\delta} u_{(1)} D_1 e^{\delta\tau D_1} - \frac{1}{\omega} u_{(2)} D_2 e^{\omega\tau D_2} + x_{(0)} .$$

This is a superposition of the hyperbolic motion in the plane of D_1 with the periodic motion in the plane of D_2.

3.2. Charge in the Plane Field

Motion of a charge in the plane electromagnetic field can only be found approximately. The first approximation concerns the space part of the argument $\xi = \kappa \cdot x$ of the field. We ought to substitute $\xi = \kappa \cdot x(\tau)$, where $\tau \to x(\tau)$ is the world line of the charged particle, hence $\xi = |\mathbf{k}|ct - \mathbf{k} \cdot \mathbf{r}(t)$. Notice that if $r_{(0)} = r(0)$,

$$\mathbf{k} \cdot \mathbf{r} = \mathbf{k} \cdot \mathbf{r}_{(0)} + \int_0^t \mathbf{k} \cdot \mathbf{v}(s) ds = \mathbf{k} \cdot \mathbf{r}_{(0)} + c|\mathbf{k}| \int_0^t \mathbf{n} \cdot \frac{\mathbf{v}(s)}{c} ds$$

$$= \mathbf{k} \cdot \mathbf{r}_{(0)} + \omega \int_0^t \mathbf{n} \cdot \boldsymbol{\beta}(s) ds .$$

Thus $\xi = \omega t - \mathbf{k} \cdot \mathbf{r}_{(0)} - \omega \int_0^t \mathbf{n} \cdot \boldsymbol{\beta}(s) ds$. If the velocity of the particle is much less than c, the last term can be neglected in comparison with the first one

$$\xi = \omega t - \mathbf{k} \cdot \mathbf{r}_{(0)}$$

and this is the first approximation. The second one concerns the relation between the coordinate time and the proper time τ of the particle. The condition

$$|\boldsymbol{\beta}| \ll 1 \tag{50}$$

implies $\gamma \cong 1$ and we accept $t \cong \tau$. In this manner, the vector-valued function g present in expression (35) describing the plane field depends only on the proper time: $g = g(\tau)$.

Now the Lorentz equation (40) is modified to the form

$$\frac{du}{d\tau} = u(\tau) \cdot \Omega(\tau) .$$

Its solution is similar to (41) with the only alteration that the definite integral $\Psi(\tau) = \int_0^\tau \Omega(s) ds$ should be in the exponents

$$u(\tau) = e^{-\frac{1}{2}\Psi(\tau)} u_{(0)} e^{\frac{1}{2}\Psi(\tau)} . \tag{51}$$

Indeed,

$$\frac{du}{d\tau} = -\frac{1}{2}\dot{\Psi} e^{-\frac{1}{2}\Psi} u_{(0)} e^{\frac{1}{2}\Psi} + \frac{1}{2} e^{-\frac{1}{2}\Psi} u_0 e^{\frac{1}{2}\Psi} \dot{\Psi} = \frac{1}{2}(u\Omega - \Omega u) = u \cdot \Omega .$$

After substituting $\Omega = -\frac{q}{m} F = -\frac{q}{m} \kappa g$, we get

$$\Psi(\tau) = -\frac{q}{m} \kappa G(\tau) , \tag{52}$$

where $G(\tau) = \int_0^\tau g(s) ds$, thus Ψ is a light-like volutor and we may refer to case (iii) discussed previously. The formula

$$u(\tau) = u_{(0)} + u_{(0)} \cdot \Psi(\tau) - \frac{1}{4} \Psi(\tau) u_0 \Psi(\tau) ,$$

corresponds to (47). We insert (52) in this:

$$u(\tau) = u_{(0)} - \frac{q}{m}(u_{(0)} \cdot \kappa) G(\tau) + \frac{q}{m}(u_{(0)} \cdot G)\kappa - \frac{q^2}{2m^2} G(\tau)^2 (u_{(0)} \cdot \kappa)\kappa . \tag{53}$$

For the electromagnetic wave with linear polarization, $g(\tau) = a \cos \omega\tau$, hence $G(\tau) = \frac{a}{\omega} \sin \omega\tau$ and (53) assumes the form

$$\begin{aligned} u(\tau) = u_{(0)} &- \frac{q}{m\omega}(u_{(0)} \cdot \kappa) a \sin \omega\tau + \frac{q}{m\omega}(u_{(0)} \cdot a)\kappa \sin \omega\tau \\ &- \frac{q^2 a^2}{2m^2\omega^2}(u_{(0)} \cdot \kappa)\kappa \sin^2 \omega\tau . \end{aligned} \tag{54}$$

The time coordinate of this four-velocity (we replace τ by t)

$$u^0(t) = c + \frac{q}{mc}(u_{(0)} \cdot a) \sin \omega t - \frac{q^2 a^2}{2m^2 c} \sin^2 \omega t$$

should satisfy our assumption $\gamma \cong 1$, i.e. $u^0 \cong c$, hence the additional terms should be small in comparison with c. This yields the condition

$$\frac{q|\mathbf{a}|}{mc} \ll 1\,, \tag{55}$$

which is a restriction on the intensity of the electromagnetic wave.

The space part of (54) is

$$\mathbf{v}(t) = \mathbf{v}_{(0)} - \frac{q}{m}\mathbf{a}\sin\omega t - \frac{q}{m}(\boldsymbol{\beta}_{(0)}\cdot\mathbf{a})\,\mathbf{n}\sin\omega t + \frac{q^2\mathbf{a}^2}{2m^2c}\mathbf{n}\sin^2\omega t\,. \tag{56}$$

The condition (55) ensures that this velocity is much less than c if $|\mathbf{v}_{(0)}| \ll c$. Notice that the additional terms to $\mathbf{v}_{(0)}$ are of two kinds: (i) parallel to the amplitude $\mathbf{a}$ of the wave, (ii) parallel to the direction $\mathbf{n}$ of the propagation. The terms of the latter group are much smaller than the terms of group (i), since the coefficient $q^2a^2/2m^2c^2$ is of the second order of approximation (55) and $|\frac{q}{mc}(\boldsymbol{\beta}_{(0)}\cdot\mathbf{a})| \leq \frac{q|\mathbf{a}|}{mc}|\boldsymbol{\beta}_{(0)}|$ is bounded by the product of conditions (50) and (55).

If we choose the mean velocity of (56): $\mathbf{v}_{(0)} + \mathbf{n}q^2\mathbf{a}^2/4m^2c$ to be equal to zero, the expression

$$\mathbf{v}(t) = -\frac{q}{m}\,\mathbf{a}\sin\omega t - \frac{q^2\mathbf{a}^2}{4m^2c}\mathbf{n}\cos 2\omega t$$

remains. Its integral is

$$\mathbf{r}(t) = \mathbf{d} + \frac{q}{m\omega}\mathbf{a}\cos\omega t - \frac{q^2\mathbf{a}^2}{8m^2c\omega}\mathbf{n}\sin 2\omega t\,,$$

where $\mathbf{d}$ is an integration constant. The charged particle performs oscillations in two perpendicular directions of $\mathbf{n}$ and $\mathbf{a}$ with the ratio of periods equal to 2. The ratio of amplitudes is $q|\mathbf{a}|/8mc$ so, by virtue of (55), the trajectory is very narrow in the direction of $\mathbf{n}$. We illustrate it in Fig. 134, where the initial position is marked by a dot upstairs.

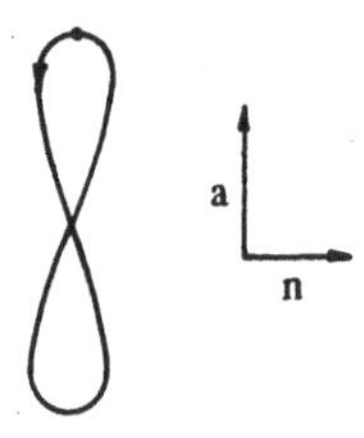

Fig. 134.

For the wave with circular polarization, $g(\tau) = e^{In\omega\tau}a$, hence $G(\tau) = (1 - e^{In\omega\tau})Ina/\omega$ and (53) takes the form

$$\begin{aligned} u(\tau) = u_{(0)} &- \frac{q}{m\omega}(u_{(0)} \cdot \kappa)(1 - e^{In\omega\tau})Ina \\ &+ \frac{q}{m\omega}\Big[v_{(0)} \cdot (Ina) - v_{(0)} \cdot (e^{In\omega\tau}Ina)\Big]\kappa \\ &- \frac{q^2}{m^2\omega^2}\Big[(Ina) \cdot (Ina) - (e^{In\omega\tau}Ina) \cdot (Ina)\Big](u_{(0)} \cdot \kappa)\kappa \,. \end{aligned}$$

Its space part is

$$\begin{aligned} \mathbf{v}(t) = \mathbf{v}_{(0)} &- \frac{q}{m}(1 - e^{-I\,\mathbf{n}\omega t})I\,\mathbf{n}\,\mathbf{a} + \frac{q}{m}[\boldsymbol{\beta}_{(0)} \cdot (\hat{\mathbf{n}}\,\mathbf{a})(1 - \cos\omega t) \\ &- (\boldsymbol{\beta}_{(0)} \cdot \mathbf{a})\sin\omega t]\,\mathbf{n} + \frac{q^2\mathbf{a}^2}{m^2c}(1 - \cos\omega t)\,\mathbf{n} \,. \end{aligned}$$

We neglect terms of the second order in the approximations, so

$$\mathbf{v}(t) = \mathbf{v}_{(0)} - \frac{q}{m}I\,\mathbf{n}\,\mathbf{a} + \frac{q}{m}e^{-I\,\mathbf{n}\omega t}I\,\mathbf{n}\,\mathbf{a} \,.$$

The integral of this expression is

$$\mathbf{r}(t) = \mathbf{d} + \left(\mathbf{v}_{(0)} - \frac{q}{m}I\,\mathbf{n}\,\mathbf{a}\right)t - \frac{q}{m\omega}e^{-I\,\mathbf{n}\omega t}\mathbf{a} \,,$$

where $\mathbf{d}$ is the integration constant. If the initial velocity is zero, the motion becomes

$$\mathbf{r}(t) = \mathbf{d} - \frac{q}{m}\hat{\mathbf{n}}\,\mathbf{a}\,t - \frac{q}{m\omega}e^{-I\,\mathbf{n}\omega t}\mathbf{a} \,.$$

We ascertain that this is a superposition of the circular motion described by the last term with the uniform motion perpendicular to the electric field at moment $t = 0$. This is a planar motion — it takes place in the plane perpendicular to the direction $\mathbf{n}$ of the wave propagation. Since the velocity assumes periodically the value zero, the trajectory is a cycloid similar to one shown in Fig. 76.

If the initial velocity is $\mathbf{v}_{(0)} = \frac{q}{m}I\,\mathbf{n}\,\mathbf{a}$, then

$$\mathbf{v}(t) = \frac{q}{m}e^{-I\,\mathbf{n}\omega t}I\,\mathbf{n}\,\mathbf{a} \,, \tag{57}$$

and the motion becomes

$$\mathbf{r}(t) = \mathbf{d} - \frac{q}{m\omega}e^{-I\,\mathbf{n}\omega t}\mathbf{a} \,,$$

so it is a purely circular motion around point $\mathbf{d}$ in the rhythm of the electric oscillations with the opposite phase. Velocity (57) is perpendicular to the magnetic induction $\widehat{\mathbf{B}}(t) = e^{-I\,\mathbf{n}\,\omega t}\mathbf{k}\,\mathbf{a}$, thus the magnetic term $q\mathbf{v}\cdot\widehat{\mathbf{B}}$ does not contribute to the Lorentz force, whereas the electric term $q\,\mathbf{E}$ plays the role of the centripetal force.

Problems

1. Find a covariant formula for the gauge transformation of the electromagnetic potential.
2. Check directly the invariance of the expressions $\frac{1}{c^2}|\mathbf{E}|^2 - |\mathbf{B}|^2$ and $\mathbf{E}\wedge\widehat{\mathbf{B}}$ under transformation (20).
3. In a given reference frame, let electric field $\mathbf{E}$ and magnetic field of induction $\widehat{\mathbf{B}}$ exist, giving the Poynting vector $\mathbf{S} = \frac{1}{\mu_0}\widehat{\mathbf{B}}\cdot\mathbf{E}$. Find velocity $\mathbf{V}$ of another frame in which $\mathbf{S}' = 0$. Answer: $2\,\mathbf{V}/(1+\beta^2) = \mathbf{S}/w$.
4. Verify that the outer derivative of potential (29) is the field (26). Does this potential satisfy the Lorentz condition?
5. In a given reference frame, let a current flow in an electrically neutral conductor. Using the Minkowski diagram, demonstrate that in another frame moving parallely to the current, the conductor cannot be neutral.
6. Let Δs be a volutor of some small space-like directed surface. Show that $F\cdot\Delta s$ is equal to the magnetic flux through Δs.
7. Find Hodge map of volutor (35).
8. The magnetic field of a rectilinear current is given by formula (1.39) in the frame where the conductor rests. Use Eqs. (20) to find the electromagnetic field in another frame where the conductor is moving.

Appendix I

BEHAVIOUR OF THE INTEGRAL (1.33)

We intend to show that the scalar part of the integral at the right-hand side of (1.33) is equal to zero for stationary charges and currents. Let

$$K = \int_V dv' \frac{(\mathbf{r} - \mathbf{r}') \cdot \mathbf{j}(\mathbf{r}')}{|\mathbf{r} - \mathbf{r}'|^3} .$$

On the basis of the formula $\nabla'(1/|\mathbf{r}-\mathbf{r}'|) = (\mathbf{r}-\mathbf{r}')/|\mathbf{r}-\mathbf{r}'|^3$ and the identity $\nabla \cdot (\mathbf{j}\,\psi) = \psi \nabla \cdot \mathbf{j} + \mathbf{j} \cdot \nabla \psi$, we may write

$$\frac{\mathbf{j}(\mathbf{r}') \cdot (\mathbf{r} - \mathbf{r}')}{|\mathbf{r} - \mathbf{r}'|^3} = \nabla' \cdot \frac{\mathbf{j}(\mathbf{r}')}{|\mathbf{r} - \mathbf{r}'|} - \frac{1}{|\mathbf{r} - \mathbf{r}'|} \nabla' \cdot \mathbf{j}(\mathbf{r}') .$$

The continuity equation (1.28) is reduced to $\nabla \cdot \mathbf{j} = 0$ due to the assumed stationary sources, thus

$$K = \int_V dv' \nabla' \cdot \frac{\mathbf{j}(\mathbf{r}')}{|\mathbf{r} - \mathbf{r}'|} = \int_{\partial V} d\mathbf{s}' \cdot \frac{\mathbf{j}(\mathbf{r}')}{|\mathbf{r} - \mathbf{r}'|} .$$

We used the Gauss theorem in the last transition.

When the currents flow in a bounded region, we choose the integration set V great enough for the currents to vanish on its boundary ∂V. In this case, $K = 0$.

When, on the other hand, the currents flow up to infinity, we consider the limiting procedure in which the surface ∂V tends to infinity in each direction. In such a case, we may reach $|\mathbf{r}-\mathbf{r}'| > a$ for some $a > 0$ on the whole surface ∂V. Then

$$|K| \leq \int_{\partial V} \frac{|d\mathbf{s}' \cdot \mathbf{j}(\mathbf{r}')|}{|\mathbf{r}-\mathbf{r}'|} < \frac{1}{a} \int_{\partial V} |d\mathbf{s}' \cdot \mathbf{j}(\mathbf{r}')| \, .$$

The last integral gives the sum of absolute values of the currents outgoing to infinity or incoming from infinity. When this sum is finite and is equal to J, we obtain $|K| < \frac{J}{a}$. In the limit $a \to \infty$, we obtain $\frac{J}{a} \to 0$, so $K = 0$.

Appendix II

THE EXISTENCE OF FORCE SURFACES

We consider conditions at which the force surfaces of the magnetic field exist. We use here the knowledge from Sec. II.5 of Sulanke and Wintgen[12] which we adopt to our situation, that is, to the volutor field.

Definition 1.

Let points x run over the region $V \subset R^n$. The mapping $M : x \to M_x$, where M_x is two-dimensional subspace of R^n, is called the *field of two-directions.*

Definition 2.

A manifold $Y \subset V$ is called the *integral manifold* of the field of two-directions M when for each point $y \in Y$, the tangent space $T_y(Y)$ (at the point y to Y) is contained in M_y.

Definition 3.

A field of two-directions is called *integrable* in the region $V \subset R^n$ if, for any point $x \in V$, a two-dimensional integral manifold of this field exists passing through x.

This two-dimensional integral manifold is the magnetic force surface we are looking for. The field of two-directions M may be represented by a set of

$n-2$ linearly independent 1-forms $\omega^l, l \in \{3,\dots,n\}$ such that in each point $x \in V$

$$\omega^l_x(t) = 0 \text{ for each } t \in M_x \text{ and } l \in \{3,\dots,n\}\ . \tag{II.1}$$

This set of 1-forms is called the *Pfaff system for M*.

The main tool for solving our problem is the *Frobenius theorem* which we give in the form suited for our needs.

Theorem 1.

If, for the field of two-directions M, such 1-forms $\theta^l_m, l, m \in \{3,\dots,m\}$ exist that the Pfaff system has the property $d\omega^l = \theta^l_m \wedge \omega^m$ in a region $V \subset R^n$, the field M is integrable in V.

For scalar product spaces, each differential 1-form ω may be replaced by a vector field $\boldsymbol{\omega}$ and the outer differential $d\omega$ by the outer derivative $\nabla \wedge \boldsymbol{\omega}$ giving a bivector field. The condition (II.1) has the form $\boldsymbol{\omega}^l \cdot \mathbf{t} = 0$ in this language, the dot meaning the scalar product of vectors. Therefore the Pfaff system is a set of $n-2$ linearly independent vector fields perpendicular to the field of two-directions.

In the three-dimensional space, i.e. for $n = 3$, the Pfaff system consists of only one vector field $\boldsymbol{\omega}$ perpendicular to the field of two-directions M. In our case, two-directions are planes of volutors $\widehat{\mathbf{B}}$ and, by means of the Hodge map, a vector field $\mathbf{B} = -I\widehat{\mathbf{B}}$ can be found which is perpendicular to $\widehat{\mathbf{B}}$. We use the physical fact that vectors $\mathbf{B}$ and $\mathbf{H}$ are parallel to each other in isotropic media, therefore we choose $\boldsymbol{\omega} = \mathbf{H}$ as the Pfaff system.

We now reformulate Theorem 1 with the use of the physical quantities introduced.

Theorem 2.

If, for the field of two-directions M (given by the magnetic induction $\widehat{\mathbf{B}}$ and represented by the magnetic field $\mathbf{H}$), a vector field $\boldsymbol{\theta}$ exists such that

$$\nabla \wedge \mathbf{H} = \boldsymbol{\theta} \wedge \mathbf{H} \tag{II.2}$$

in a region $V \subset R^3$, the field M is integrable in V.

The condition (II.2) may be rewritten with the help of the vector product as $\nabla \times \mathbf{H} = \boldsymbol{\theta} \times \mathbf{H}$ or, by virtue of (0.24) and (0.121), as

$$\nabla \cdot \widehat{\mathbf{H}} = \boldsymbol{\theta} \cdot \widehat{\mathbf{H}}\ . \tag{II.3}$$

After comparing this with the Maxwell equation (1.17c), we ascertain that the properties of the vector $\mathbf{K} = -\frac{\partial \mathbf{D}}{\partial t} - \mathbf{j}$ are essential here, since the Maxwell equation has the form

$$\nabla \cdot \widehat{\mathbf{H}} = \mathbf{K}\ . \tag{II.4}$$

Lemma.

A necessary and sufficient condition for equality (II.3) to be satisfied is that $\mathbf{K}$ is parallel to $\hat{\mathbf{H}}$.

Proof (i) The necessary condition. If the equality (II.3) is satisfied, it follows from the Maxwell equation (II.4) that $\boldsymbol{\theta} \cdot \hat{\mathbf{H}} = \mathbf{K}$ and the inner product of any vector with $\hat{\mathbf{H}}$ is always parallel to $\hat{\mathbf{H}}$.

(ii) The sufficient condition. If $\mathbf{K}$ is parallel to $\hat{\mathbf{H}}$, the equality $\mathbf{K} \cdot \hat{\mathbf{H}}^{-1} = \mathbf{K}\,\hat{\mathbf{H}}^{-1}$ is satisfied. Then we substitute $\boldsymbol{\theta} = \mathbf{K} \cdot \hat{\mathbf{H}}^{-1}$ and obtain

$$\boldsymbol{\theta} \cdot \hat{\mathbf{H}} = \boldsymbol{\theta}\hat{\mathbf{H}} = (\mathbf{K} \cdot \hat{\mathbf{H}}^{-1})\hat{\mathbf{H}} = \mathbf{K}\,\hat{\mathbf{H}}^{-1}\hat{\mathbf{H}} = \mathbf{K}\,.$$

Therefore, by virtue of (II.4), $\boldsymbol{\theta} \cdot \hat{\mathbf{H}} = \boldsymbol{\nabla} \cdot \hat{\mathbf{H}}$.

Comparing this Lemma with Theorem 2, we are convinced that we have proved

Theorem 3.

If the field $\frac{\partial \mathbf{D}}{\partial t} + \mathbf{j}$ is zero or is parallel to $\hat{\mathbf{H}}$ in some region V of the isotropic medium, the force surfaces for the magnetic field exist in V.

It is worth giving an example of a situation in which condition (II.3) is not satisfied. This may be the superposition of the magnetostatic field inside a cylindrical conductor with a uniform spatial current (example 6.6 of Chap. 1), with the uniform external field perpendicular (as the volutor) to the cylinder axis.

Notice also that when condition (II.3) is satisfied trivially, that is when curl $\mathbf{H} = 0$, then in analogy to electrostatics, one may introduce a pseudoscalar potential ψ such that $\mathbf{H} = \text{grad}\ \psi$ and then the force surfaces are merely equipotential surfaces, i.e. sets of points $\mathbf{r}$ satisfying $\psi(\mathbf{r}) = \text{constant}$.

Appendix III

SPECIAL EXAMPLES OF FORCE SURFACES

We shall find the approximate force surfaces for the magnetic field produced by the circular current J_1 along with the linear current J_2. We assume that the straight line of the current J_2 (we choose it as the Z-axis with the positive sense determined by J_2) coincides with the symmetry axis of the circle (we choose its plane as the X, Y-plane), see Fig. III.1).

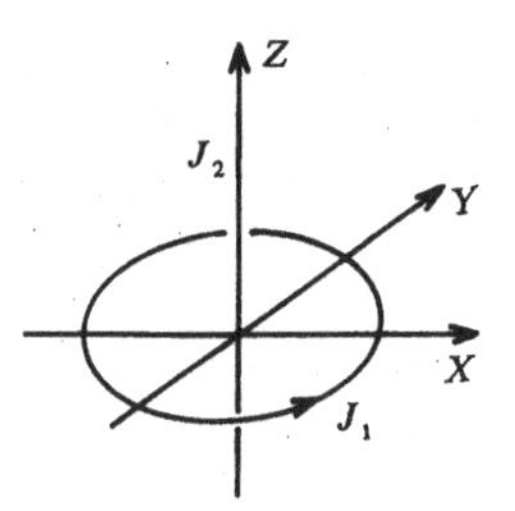

Fig. III.1.

Our approximation consists in the assumption that the force surfaces of the circular current are parts of spheres passing through the circuit. We start with the formula for the solid angle Ω_1, at which the circle is seen from the point z_0 on the Z-axis

$$\Omega_1 = 2\pi(1 - \cos\vartheta) ,$$

(see Fig. III.2). We express the cosine by z_0 and a, the radius of the circle

$$\Omega_1 = 2\pi\left(1 - \frac{z_0}{\sqrt{z_0^2 + a^2}}\right) . \tag{III.1}$$

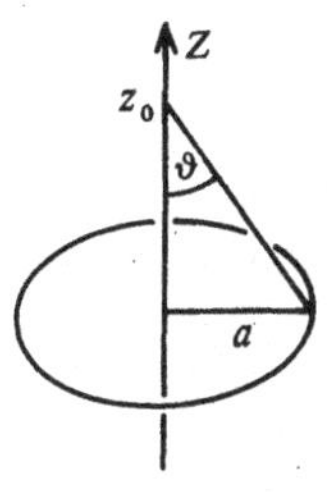

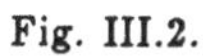
Fig. III.2.

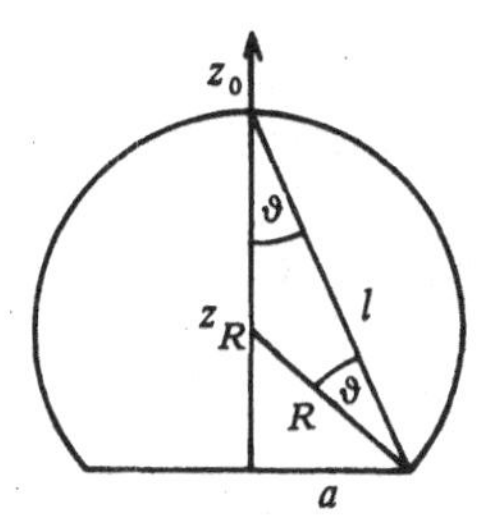

Fig. III.3.

This function ranges from 0 to 4π for z_0 changing from $+\infty$ to $-\infty$. We need an equation for the sphere passing through the point z_0 and the circle. Figure III.3 helps us to write this equation as

$$(z - z_R)^2 + x^2 + y^2 = R^2 , \tag{III.2}$$

where

$$R = \frac{l}{2\cos\vartheta} = \frac{z_0^2 + a^2}{2z_0} , \qquad z_R = z_0 - R = \frac{z_0^2 - a^2}{2z_0} .$$

We now express z_0 as the function of Ω_1 from (III.1)

$$z_0 = a\frac{2\pi - \Omega_1}{\sqrt{(2\pi)^2 - (2\pi - \Omega_1)^2}} . \tag{III.3}$$

Then R and z_R may also be calculated and inserted into (III.2). In this way, an equation for the locus of points of the constant visual angle Ω_1 can be obtained for the circular current. Its explicit form is not important, therefore we do not write it here.

We now superpose this magnetic field with the field of the straight-line current. We do this by considering sets of points determined by the equation

$$\widehat{\Psi}_1(\mathbf{r}) + \widehat{\Psi}_2(\mathbf{r}) = \text{const} , \tag{III.4}$$

where $\widehat{\Psi}_1 = -\Omega_1 J_1 I/4\pi$ and, by (2.52), $\widehat{\Psi}_2 = (\alpha - \pi)J_2 I/2\pi$. (Here α is the cylindrical coordinate: $\alpha = \arctan \frac{y}{x}$.) For a single specific force surface of the combined field, we choose to put $-\frac{1}{2}J_2 I$ on the right-hand side of (III.4)

$$-\frac{\Omega_1}{4\pi}J_1 I + \frac{\alpha - \pi}{2\pi}J_2 I = -\frac{1}{2}J_2 I\,,$$

whence $\Omega_1 = 2\alpha J_2/J_1$. We plug this into (III.3) and obtain

$$z_0 = a\frac{2\pi - 2k\alpha}{\sqrt{(2\pi)^2 - (2\pi - 2k\alpha)^2}} = a\frac{\pi - k\alpha}{\sqrt{\pi^2 - (\pi - k\alpha)^2}}\,,$$

where $k = J_2/J_1$. A new variable $\xi = \frac{\pi - k\alpha}{\pi}$ is worth introducing; then

$$z_0 = \frac{a\xi}{\sqrt{1-\xi^2}}\,.$$

R and z_R may also be expressed by this variable and inserted into (III.2), where we use another cylindrical coordinate $\rho = \sqrt{x^2 + y^2}$. Now, the equation reads

$$\left(z - \frac{a}{2\xi}\frac{2\xi^2 - 1}{\sqrt{1-\xi^2}}\right)^2 + \rho^2 = \frac{a^2}{4\xi^2(1-\xi^2)}\,. \tag{III.5}$$

This is the equation for which we were seeking to describe a force surface of the superposed magnetic field of the two conductors shown in Fig. III.1. The drawings may be obtained with the aid of a computer. The case with $k = 1$ was shown in Fig. 72. Figures III.4 and III.5 represent the cases $k = 2$ and $k = \frac{1}{2}$, respectively. The red lines in them designate intersections of the surface with horizontal planes and the green lines delineate intersections of the surface with vertical half-planes passing through the rectilinear conductor.

A general feature occurs that for k integer, exactly k half-planes exist which are "asymptotes" for the single force surface and the surface winds k times around the circular conductor when passing one turn along it. If $k = \frac{1}{n}$ for the n integer, the force surface has one asymptotic half-plane and, after n turns along the circle, winds one time around it — only then does it again become close to the asymptotic half-plane. On the other hand, for k irrational, the force surface does not close after passing a finite number of times along the circle — in such a case, it has infinitely many asymptotic half-planes and densely fills the whole space.

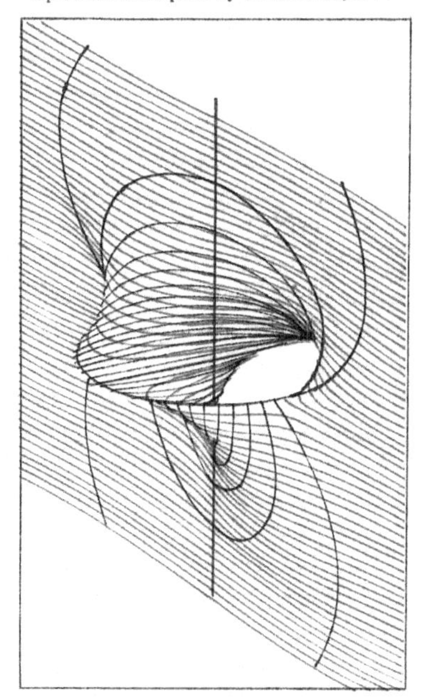

Fig. III.4.

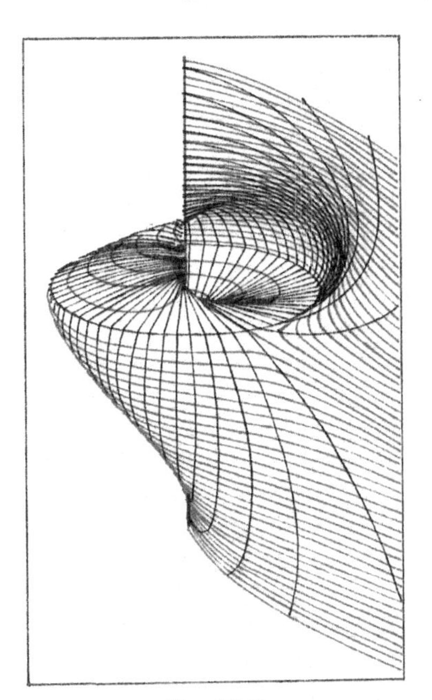

Fig. III.5.

REFERENCES

1. Charles W. Misner, Kip S. Thorne and John Archibald Wheeler, "Gravitation" (Freeman and Co., San Francisco, 1973).
2. Walter Thirring, "Course in Mathematical Physics", vol. 2: "Classical Field Theory" (Springer Verlag, Vienna, 1982).
3. Roman Ingarden and Andrzej Jamiołkowski, "Classical Electrodynamics" (Elsevier, Amsterdam, 1985).
4. Marcel Riesz, "Clifford Numbers and Spinors", Lecture Series No. 38, University of Maryland, College Park, 1958.
5. David Hestenes, "Space-Time Algebra" (Gordon and Breach, New York, 1966).
6. David Hestenes and Garret Sobczyk, "Clifford Algebra to Geometric Calculus" (D. Reidel, Dordrecht, 1984).
7. David Hestenes, "New Foundations for Classical Mechanics" (D. Reidel, Dordrecht, 1986).
8. Walter Rudin, "Functional Analysis" (McGraw Hill, New York, 1973). Chap. 10.
9. Krzysztof Maurin, "Analysis", part II (D. Reidel, Dordrecht, 1980).
10. Laurent Schwartz, "Théorie des distributions" (Hermann, Paris, 1966).
11. Vasily S. Vladimirov, "Generalized Functions in Mathematical Physics"(Nauka, Moscow, 1979 (in Russian)).
12. R. Sulanke and P. Wintgen, "Differentialgeometrie und Faserbündel"(Deutscher Verlag der Wiss., Berlin, 1972).
13. S. Teitler, *J. Math. Phys.* **7** (1966) 1730.
14. K. R. Greider, *Found. Phys.* **14** (1984) 467.
15. N. Salingaros, *Phys. Rev.* **D31** (1985) 3150.
16. D. Hestenes, *J. Math. Phys.* **15** (1984) 1768.

17. Hermann Weyl, "Symmetry" (Princeton Univ. Press, Princeton, 1952) pp. 19.
18. B. A. Robson, "The Theory of Polarization Phenomena" (Clarendon Press, Oxford, 1974).
19. E. Kamke, "Differentialgleichungen, Lösungsmethoden und Lösungen", vol. I, Leipzig 1959, item 2.14.
20. Nikolai M. Matveev, "Collection of Problems on Ordinary Differential Equations" (Vysheishaya Shkola, Minsk, 1970 (in Russian)), pp. 189.
21. Kuni Imaeda, "Quaternionic Formulation of Classical Electrodynamics and Theory of Functions of a Biquaternion Variable" (Okayama University of Science, Okayama, 1983).

Multivectors & Clifford Algebra in Electrodynamics

Bernard Jancewicz

INDEX

World Scientific

INDEX

ERRATA

Page,	line	instead of	should read
21	6↓	Pg. 77	Pg. 10
67	2↑	$(\widehat{\mathbf{A}}\widehat{\mathbf{B}})$	$(\widehat{\mathbf{A}} \cdot \widehat{\mathbf{B}})$
74	1↓	Pg. 71	Pg. 69
76	3↓	Ref. 2	Ref. 3
161	(Fig. 80)	$\mathbf{M}, \mathbf{B}, \mathbf{K}$	$\widehat{\mathbf{M}}, \widehat{\mathbf{B}}, \widehat{\mathbf{K}}$
168	5↑	$2p\mathbf{A}$	$2q\mathbf{A}$
168	12↑	$2\,\dfrac{\partial^2 \chi}{\partial x_i\,\partial x_j}$	$+\ 2\,\dfrac{\partial^2 \chi}{\partial x_i\,\partial x_j}$
168	3↓	$\dfrac{\partial^2 A_i}{\partial x_k\,\partial x_i} =$	$\dfrac{\partial^2 A_i}{\partial x_k\,\partial x_j} =$
194	7↑	running wave	travelling wave
202	7↓	$-\mathbf{b} \wedge \dfrac{\partial \mathbf{e}}{\partial t}$	$-\widehat{\mathbf{b}} \wedge \dfrac{\partial \mathbf{e}}{\partial t}$
202	8↓	$= \dfrac{1}{u}\,\dfrac{\partial \mathbf{e}}{\partial t}$	$= -\,\dfrac{1}{u}\,\dfrac{\partial \mathbf{e}}{\partial t}$
215	5↑	$\cos(s \log n)$	$\cos(-s \log n)$
215	6↑	$\sin(s \log n)$	$\sin(-s \log n)$
223	1↓	$\lvert \mathbf{n} \cdot \ddot{\widehat{\mathbf{M}}} \rvert$	$\lvert \mathbf{n} \cdot \ddot{\widehat{\mathbf{M}}} \rvert^2$
241	4↑	$= x\Lambda$	$= \Lambda x$
242	14↑	*contragradient*	*contragredient*
310	5↑	X,Y-plane), see	X,Y-plane, see